Lecture Notes in Mathematics 2031

Editors:
J.-M. Morel, Cachan
B. Teissier, Paris

T0202952

For further volumes:
http://www.springer.com/series/304

Jin Akiyama · Mikio Kano

Factors and Factorizations of Graphs

Proof Techniques in Factor Theory

Jin Akiyama
Tokai University
Research Institute
of Educational Development
Tomigaya 2-28-4
151-8677 Shibuya-ku
Japan
ja@jin-akiyama.com

Mikio Kano
Ibaraki University
Computer and Information Sciences
Nakanarusawa
Ibaraki-ken
316-8511 Hitachi
Japan
kano@mx.ibaraki.ac.jp

ISBN 978-3-642-21918-4   e-ISBN 978-3-642-21919-1
DOI 10.1007/978-3-642-21919-1
Springer Heidelberg Dordrecht London New York

Lecture Notes in Mathematics ISSN print edition: 0075-8434
                             ISSN electronic edition: 1617-9692

Library of Congress Control Number: 2011932316

Mathematics Subject Classification (2010): 05C, 05CXX, 05C75

Cover design: deblik, Berlin

Printed on acid-free paper

Springer is part of Springer Science+Business Media (www.springer.com)

to Frank Harary, my teacher
and
to my wife Yoko

Preface

A spanning subgraph of a graph is its subgraph whose vertex set is the same as the original graph. In this book, spanning subgraphs of graphs possessing some given properties are studied, and these spanning subgraphs are called *factors*. For example, for a positive integer k, a k-factor of a graph is its spanning subgraph each of whose vertices has a constant degree k. So a 1-factor is nothing but a perfect matching, where a matching in a graph G is a subgraph of G whose edges are pairwise disjoint and a perfect matching is a matching that covers all the vertices of G. Furthermore a 2-factor is a set of vertex-disjoint cycles which together cover the vertices of the graph. Similarly, a graph each of whose vertices has a constant degree k is called a k-regular graph, and if k is even, such a graph is said to be even regular. If the edges of a graph G can be decomposed into k-factors, then G is said to be k-factorable.

Petersen's results on graph factors and factorizations date back to the 19th century. Petersen's Theorem, which he proved in order to solve a problem on Diophantine equations, states: *Every even regular graph is 2-factorable*. Petersen then went on to prove another theorem: *Every 2-edge connected 3-regular graph has a 1-factor*. This theorem arose from a counterexample Petersen constructed to Tait's 'theorem': *Every 3-regular graph which has no bridge is 1-factorable*. This counterexample is the well-known Petersen graph. Petersen's theorems are byproducts of attempts to address problems outside of graph theory. Perhaps Petersen himself had no inkling that these theorems would open the door to a very promising area of graph theory.

Later came Hall's Marriage Theorem, a result obtained when Hall was studying the structure of subsets. *Let G be a bipartite graph with bipartition (A, B). Then G has a matching that saturates A (i.e., a matching covering all the vertices of A) if and only if $|N_G(S)| \geq |S|$ for all $S \subseteq A$.* König's Theorem followed: *Every regular bipartite graph is 1-factorable*. And then Tutte's 1-Factor Theorem: *A graph G has a 1-factor if and only if $odd(G - S) \leq |S|$ for all $S \subset V(G)$.* These five theorems form the foundation of the study of factors and factorizations.

König's theorem was the result of a conscious effort to answer a graph theory problem: *Does every bipartite regular graph have a 1-factor?* It seems that, at that time, König already had a good idea of how graph factors could be applied. In his paper "On graphs and their applications in determinant theory and set theory", he starts by saying "The following article deals with problems from analysis situs, the theory of determinants and set theory." He then goes on to assert that in fact, the notion of a graph and its usefulness as a method of representation actually links these three disparate areas.

A second stage of development is marked by Tutte's f-Factor Theorem: *Let G be a graph and $f : V(G) \to \mathbb{Z}^+ = \{0, 1, 2, 3, \ldots\}$. Then G has an f-factor (i.e., a spanning subgraph F satisfying $\deg_F(v) = f(v)$ for all vertices v of G) if and only if for all disjoint subsets S and T of $V(G)$,*

$$\sum_{x \in S} f(x) + \sum_{x \in T} (\deg_{G-S}(x) - f(x)) - q(S,T) \geq 0,$$

where $q(S,T)$ denotes the number of components C of $G - (S \cup T)$ such that $\sum_{x \in V(C)} f(x) + e_G(C,T) \equiv 1 \pmod 2$; and Lovász's (g, f)-Factor Theorem: *Let G be a graph and $g, f : V(G) \to \mathbb{Z}$ such that $g(x) \leq f(x)$ for all $x \in V(G)$. Then G has a (g, f)-factor (i.e., a spanning subgraph H satisfying $g(v) \leq \deg_H(v) \leq f(v)$ for all vertices v of G) if and only if for all disjoint subsets S and T of $V(G)$,*

$$\sum_{x \in S} f(x) + \sum_{x \in T} (\deg_{G-S}(x) - g(x)) - q^*(S,T) \geq 0,$$

where $q^*(S,T)$ denotes the number of components C of $G - (S \cup T)$ such that $g(x) = f(x)$ for all $x \in V(C)$ and $\sum_{x \in V(C)} f(x) + e_G(C,T) \equiv 1 \pmod 2$. Many subsequent theorems were proved based on these results.

A third stage of development began in the 1980's and is marked by the introduction of the notions of semi-regular factors and component factors. Since that time, many substantial results have been obtained and much progress has been made in this area. The results are detailed in this book.

As far as we know, there was no comprehensive text on factors and factorizations when we started to write this book. This is one compelling reason for writing this volume. Since we wrote our survey paper entitled "Factors and Factorizations of Graphs" published in *Journal of Graph Theory*, vol. 9 (1985), we collected and analyzed most of the results in the area. In fact, we started to write this book ten years ago. A first version of this text coincided with KyotoCGGT2007.

This book also chronicles the development of mathematical graph theory in Japan, a development which began with many important results in factors and factorizations of graphs.

One of the main themes of this text is the observation that many theorems can be proved using only a few standard proof techniques. Namely,

(i) We reduce a given factor problem of a graph G into a simpler factor problem of a related graph G^*, and apply a known theorem on a simpler factor to G^*.

(ii) We consider several standard cases; for example, when we consider regular factors, a standard case analysis is useful.

(iii) If a criterion for a graph to have a factor is given as "$f(X) \leq 0$ for all $X \subset V(G)$", then we consider the following two cases: (a) there exists a nonempty vertex set S such that $f(S) = 0$; and (b) $f(X) < 0$ for all nonempty vertex sets X.

(iv) If a criterion for a graph to have a factor is given as "$\rho(G - X) \leq f(X)$ for all $X \subset V(G)$", then we define a number

$$\beta = \min\{f(X) - \rho(G - X) : \emptyset \neq X \subset V(G), \ \rho(G - X) > 0\}.$$

Next, we consider a maximal vertex set S of G such that $\rho(G - S) > 0$ and $f(S) - \rho(G - S) = \beta$.

(v) We use alternating paths or trails. These arguments were often used in early stages of the developments of the theory of graph factors; they were eventually replaced by the above methods, since the proofs using alternating paths or trails are usually long and complicated. However, this method is still useful for algorithms for finding the desired factors.

Other key features of this book are the following:

1. It is comprehensive and covers most of the important results since 1980.
2. It is self-contained. One who wants to begin research in graph factors and factorizations can confine himself to this one book to follow the history and development of the area, and to find conjectures and open problems.
3. Many detailed illustrations are given to accompany the proofs.

As we mentioned above, the first version of this book was privately published on the occasion of KyotoCGGT2007 conference in honor of Jin Akiyama and Vašek Chvátal on their 60th birthdays. Many parts of the book have since been revised and two more chapters, "Component Factors" and "Spanning Trees", have been added.

Frank Harary predicted that graph theory would grow so much that each chapter of his book *Graph Theory* would eventually expand to become a book on its own. He was right. This book is an expansion of his *Chapter 9, Factorization*.

We also predict that the area of factors and factorizations will continue to grow because of many applications to BIBD, Steiner Designs, Matching Theory, OR, etc. that have been found.

The following references were very helpful to us during the preparation of this work:

- Bela Bollobás, *Extremal Graph Theory*, Academic Press, (1978)
- Mekkia Kouider and Preben D. Vestergaard, Connected factors in graphs — A survey, *Graphs Combinatorics* **21** (2005) 1–26.
- László Lovász, *Combinatorial Problems and Exercises*, North-Holland, (1979).
- László Lovász and Michael D. Plummer, *Matching Theory*, Annals of Discrete Mathematics **29**, North-Holland, (1986).
- Kenta Ozeki and Tomoki Yamashita, Spanning trees : A survey, *Graphs Combinatorics* **27** (2011) 1–26.
- Michael D. Plummer, Graph factors and factorization: 1985–2003: A survey, *Discrete Mathematics* **307** (2007) 791–821.
- Lutz Volkmann, Regular graphs, regular factors, and the impact of Petersen's Theorems, *Jber. d. Dt. Math.-Verein* **97** (1995) 19–42.
- Lutz Volkmann, *Graphen und Digraphen: Eine Einführung in die Graphentheorie* (Springer-Verlag) (1991).
- Qinglin Roger Yu and Guizhen Liu, *Graph Factors and Matching Extensions* (Springer) (2009).

We would like to thank Vašek Chvátal, Mark Goldsmith, Haruhide Matsuda, Michael D. Plummer, Mari-Jo Ruiz, Lutz Volkmann and our many friends who work in factor theory for their valuable help in making this book.

Tokyo and Hitachi,   *Jin Akiyama*
April, 2011   *Mikio Kano*

Contents

1

Basic Terminology

We introduce some graph theory definitions and notation which are needed to discuss factor theory in graphs. We begin with some notation on sets. If X is a subset of Y, then we write $X \subseteq Y$; if X is a **proper subset** of Y, we write $X \subset Y$. For two disjoint subsets A and B of Y, we denote $A \cup B$ by $A + B$. Moreover, if C is a subset of A, then we write $A - C$ for $A \setminus C$. The number of elements in a set X is denoted by $|X|$ or $\#X$.

A **graph** G consists of a **vertex set** $V(G)$ and an **edge set** $E(G)$. Each edge joins two vertices, which are not necessarily distinct. An edge joining a vertex x to a vertex y is denoted by xy or yx. A special edge joining a vertex to itself is called a **loop**. For two distinct vertices, there is typically at most one edge joining them. However, if there are two or more edges joining them, then the graph is said to have **multiple edges**.

We can define three types of graphs as follows: a graph that may have loops and multiple edges is called a **general graph**. A graph that has no loops but may have multiple edges is called a **multigraph**. A graph having neither loops nor multiple edges is called a **simple graph** (Fig. 1.1). Each of these types of graphs plays an important role in this book and we are careful to use these terms precisely. However, when the context is clear, we simply use the term graph.

The number of vertices of a graph G is called the **order** of G and denoted by $|G|$. The number of edges of G is called the **size** of G and denoted by $||G||$. Thus

$$|G| = |V(G)| = \text{the order of } G,$$
$$||G|| = |E(G)| = \text{the size of } G.$$

We consider only graphs of finite order, which are called **finite graphs**. When there exists an edge $e = xy$ joining a vertex x to a vertex y, we say that e is **incident** with x and y and vice versa, and x and y are **adjacent**. Moreover, x and y are called the **end-points** of e. If two edges have a common end-point, then they are said to be **adjacent**. For a vertex v of a graph G, the

J. Akiyama and M. Kano, *Factors and Factorizations of Graphs*,
Lecture Notes in Mathematics 2031, DOI 10.1007/978-3-642-21919-1_1,
© Springer-Verlag Berlin Heidelberg 2011

number of edges of G incident with v is called the **degree** of v in G and is denoted by $\deg_G(v)$. Each loop incident with v contributes two edges to $\deg_G(v)$ (Fig. 1.1 (a)). For a vertex v of G, the set of vertices adjacent to v is called the **neighborhood** of v and denoted by $N_G(v)$. Thus, if G is a simple graph, then $\deg_G(v) = |N_G(v)|$. The **closed neighborhood** $N_G[v]$ is defined to be $N_G(v) \cup \{v\}$. For a vertex set X of G, the neighborhood of X is defined as $N_G(X) = \cup_{x \in X} N_G(x)$.

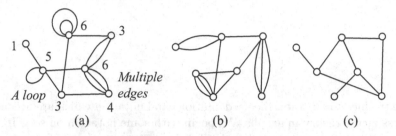

Fig. 1.1. (a) A general graph with order 7 and size 14, where numbers denote the degrees of the vertices. (b) A multigraph. (c) A simple graph.

The maximum degree among all vertices of G is called the **maximum degree** of G and denoted by $\Delta(G)$. Similarly, the **minimum degree** of G, denoted by $\delta(G)$, is the minimum degree among all vertices of G. If every vertex of a graph G has constant degree r, then G is called an r-**regular graph**. A 3-regular graph is often called a **cubic graph**.

A vertex of odd degree is called an **odd vertex**, and a vertex of even degree is called an **even vertex**. Moreover a vertex of degree one is called an **end-vertex** or **leaf**, and the edge incident with an end-vertex is called a **pendant edge**.

For two given graphs G and H, if there exists a bijection $f : V(G) \to V(H)$ such that $f(x)$ and $f(y)$ are adjacent in H if and only if x and y are adjacent in G, then we say that G and H are **isomorphic** and write $G \cong H$.

The name of the following theorem (The Handshaking Theorem) comes from imagining a graph in which each vertex represents a person, and an edge is drawn between two vertices if the corresponding two people shake hands. The sum of degrees is then the total number of hands shaken, which is of course equal to twice the number of handshakes.

Theorem 1.1 (Handshaking Theorem). *Let G be a general graph. The sum of degrees of vertices of G is equal to twice the number of edges of G. Namely,*

$$\sum_{x \in V(G)} \deg_G(x) = 2\|G\|.$$

Proof. Assume first that G is a multigraph. Let

$$\Omega = \{(v, e) : v \in V(G),\ e \in E(G),\ v \text{ is incident with } e\},$$

which is the set of pairs of vertices and incident edges. Since for every vertex v exactly $\deg_G(v)$ edges are incident with v, the number of pairs (v, e) in Ω with fixed v is $\deg_G(v)$. Furthermore, since for each edge e exactly two vertices are incident with e, the number of pairs (v, e) in Ω with fixed e is two. Therefore

$$|\Omega| = \sum_{x \in V(G)} \deg_G(x) = 2|E(G)| = 2\|G\|.$$

If G has some loops then we remove the loops one by one. In each step, the left side and the right side of the equation each decrease by two. In the final graph, which is a multigraph, the equality holds as shown above. Therefore the equality holds. □

Corollary 1.2. *Every general graph has an even number of odd vertices.*

Proof. Let G be a general graph, U the set of odd vertices and W the set of even vertices. Then by the handshaking theorem, we have

$$2\|G\| = \sum_{x \in U} \deg_G(x) + \sum_{x \in W} \deg_G(x)$$

$$0 \equiv |U| \pmod 2.$$

Hence $|U|$ is even. □

A graph H is called a **subgraph** of G if $V(H) \subseteq V(G)$ and $E(H) \subseteq E(G)$. For a vertex subset $X \subseteq V(G)$, the subgraph of G whose vertex set is X and whose edge set consists of the edges of G joining vertices of X is called the **subgraph of G induced by** X and denoted by $\langle X \rangle_G$. A subgraph K of G is called an **induced subgraph** of G if there exists a vertex subset Y such that $K = \langle Y \rangle_G$. We write $G - X$ for the subgraph of G induced by $V(G) - X$, which is obtained from G by removing all the vertices in X together with all the edges incident with vertices in X (Fig. 1.2).

For an edge set A of a graph G, the subgraph of G whose edge set is A and whose vertex set consists of the vertices of G incident with at least one edge of A is called the **subgraph of G induced by** A and denoted by $\langle A \rangle_G$. Moreover, $G - A$ denotes the subgraph of G obtained from G by removing all the edges in A.

A subgraph H of G is called a **spanning subgraph** of G if $V(H) = V(G)$. A spanning subgraph that possesses a certain given property is called a **factor**.

For a given graph H, if a graph G has no induced subgraph isomorphic to H, then we say that G is **H-free**. The star $K_{1,3}$ is often called a **claw**, and so a $K_{1,3}$-free graph G is often called **claw-free**.

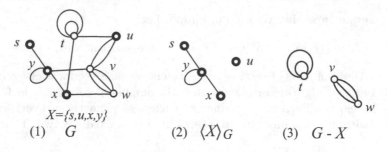

$$X=\{s,u,x,y\}$$
(1) G (2) $\langle X\rangle_G$ (3) G - X

Fig. 1.2. (1) A general graph G and its vertex subset $X = \{s,u,x,y\}$;   (2) The induced subgraph $\langle X\rangle_G$; (3) The subgraph $G - X$.

We now give some important classes of simple graphs. Let G be a simple graph. If any two distinct vertices of G are adjacent, then G is called a **complete graph**. The complete graph of order n is written K_n or $K(n)$. The graph K_1, which consists of one vertex and no edge, is called a **trivial graph**, and so a **non-trivial graph** has order at least two. If the vertex set of G is partitioned into m disjoint non-empty subsets $V(G) = X_1 \cup X_2 \cup \cdots \cup X_m$ such that any two vertices contained in distinct subsets are adjacent and no two vertices in the same subset are adjacent, then G is called a **complete m-partite graph**. If $|X_i| = n_i$, $1 \le i \le m$, then such a complete m-partite graph is denoted by $K(n_1, n_2, \ldots, n_m)$ or $K_{n_1,n_2,\ldots,n_m}$. In particular, $K(n_1, n_2) = K_{n_1,n_2}$ is called a **complete bipartite graph**. The complete bipartite graph $K(1, n)$ is called the **star** of order $n + 1$. The **path** P_n consists of n vertices $v_1, v_2, \ldots, v_n$ and $n - 1$ edges $v_i v_{i+1}$, $1 \le i \le n - 1$, where the two vertices v_1 and v_n of degree one are called the **end-vertices** of P_n. The **cycle** C_n is obtained from the path P_n by adding a new edge joining the two end-vertices of P_n.

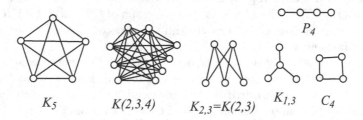

$$K_5 \qquad K(2,3,4) \qquad K_{2,3}=K(2,3) \qquad K_{1,3} \qquad C_4$$

Fig. 1.3. Some graphs.

For two graphs G and H, the **union** $G \cup H$ is the graph with vertex set $V(G) \cup V(H)$ and edge set $E(G) \cup E(H)$, where $V(G) \cap V(H) = \emptyset$. The union $G \cup G$ is often denoted by $2G$, and for any integer $n \ge 3$, we inductively define nG by $(n-1)G \cup G$. The **join** $G + H$ denotes the graph with vertex $V(G) \cup V(H)$ and edge set

$$E(G + H) = E(G) \cup E(H) \cup \{xy : x \in V(G) \text{ and } y \in V(H)\},$$

i.e., $G + H$ is obtained from $G \cup H$ by adding all the edges joining a vertex of G to a vertex of H (Fig. 1.4). For a simple graph G, the **complement** of G, denoted by $\overline{G}$, is the graph with vertex set $V(G)$ such that two vertices are adjacent in $\overline{G}$ if and only if they are not adjacent in G. In particular, $G \cup \overline{G} = K_{|G|}$.

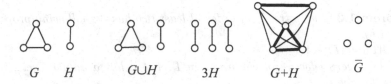

$G \quad H \qquad G \cup H \qquad 3H \qquad G+H \qquad \overline{G}$

Fig. 1.4. The union, the join, and the complement.

We introduce some notions concerning paths and cycles in a graph. A **walk** in a graph G is a sequence of vertices and edges

$$(v_0, e_1, v_1, e_2, v_2, \ldots, v_{n-1}, e_n, v_n),$$

such that every e_i, $1 \le i \le n$, is an edge joining vertices v_{i-1} and v_i. We say that this walk connects two vertices v_0 and v_n. Notice that a walk may pass through some edges and some vertices twice or more. The **length** of a walk is the number of edges appearing in it, and so the walk given above has length n. A walk that passes through every one of its edges exactly once is called a **trail**, and a walk that passes through every one of its vertices exactly once is called a **path**. It is easy to see that every path is a trail, but a trail is not always a path. We often denote a trail by a sequence $(e_1, e_2, \ldots, e_n)$ of edges and a path by a sequence $(v_0, v_1, \ldots, v_n)$ of vertices. If the initial vertex v_0 and the terminal vertex v_n of a walk are the same, then we say that such a walk is **closed**. A closed trail is called a **circuit**, and a closed path is called a **cycle**. A cycle of even length, which is of even order, is called an **even cycle**. An **odd cycle** can be defined similarly.

A graph G is said to be **connected** if any two vertices are connected by a path in G. If G is **disconnected**, that is, if G is not connected, then G is decomposed into **components** that are the maximal connected subgraphs of G. A connected simple graph having no cycle is called a **tree**, and a simple graph having no cycle is called a **forest**. So each component of a forest is a tree. A forest is called a **linear forest** if its all components are paths, and is called a **star forest** if all its components are stars.

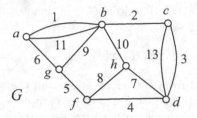

Fig. 1.5. A path (a, b, h, f, d) passing through 1; a cycle (b, h, f, d, c, b) passing through 3; and a circuit $(9, 6, 11, 10, 7, 3, 2)$.

Theorem 1.3. *Every tree T of order at least two has the following properties (Fig. 1.6).*
(i) $\|T\| = |T| - 1$.
(ii)   For every edge e not contained in T, $T + e$ has a unique cycle, which passes through e.
(iii) T has at lest two leaves.

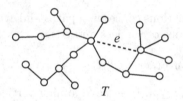

Fig. 1.6. A tree T with 9 leaves and an edge e not contained in T.

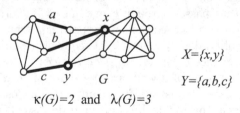

Fig. 1.7. A 2-connected and 3-edge connected graph G, for which $G - X$ and $G - Y$ are disconnected.

A connected graph G, which is not a complete graph, is n-**connected** if for every vertex set X containing at most $n - 1$ vertices, $G - X$ is connected. The **connectivity** of G is defined to be the maximum n for which G is n-connected, and is denoted by $\kappa(G)$. Hence if $\kappa(G) = n$, then for every

$X \subset V(G)$ with $|X| \leq n - 1$, $G - X$ is connected and there exists a subset $S \subseteq V(G)$ with $|S| = n$ such that $G - S$ is disconnected. The connectivity of the complete graph K_m is defined to be $m - 1$, that is, $\kappa(K_m) = m - 1$. We can analogously define edge connectivity as follows. A graph G is said to be n-**edge connected** if for every edge set X containing at most $n - 1$ edges, $G - X$ is connected. The **edge-connectivity** of G is defined to be the maximum n for which G is n-edge connected, and is denoted by $\lambda(G)$ (see Fig. 1.7). The edge-connectivity of the complete graph K_m is $m - 1$ by the definition. It is easy to see that a connected multigraph G satisfies the following inequality, which was found by Whitney [245].

$$\kappa(G) \leq \lambda(G) \leq \delta(G).$$

Equality holds in cubic graphs, that is, the next proposition follows.

Proposition 1.4. *For a cubic multigraph G, $\kappa(G) = \lambda(G)$.*

The proof of the above proposition is left to the reader (see Fig. 1.8).

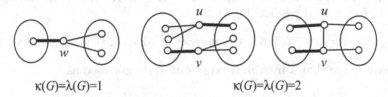

$$\kappa(G)=\lambda(G)=1 \qquad\qquad \kappa(G)=\lambda(G)=2$$

Fig. 1.8. Connected cubic multigraphs.

Let G be a connected graph and let n be an integer such that $n \geq 2$. For a vertex set S of G with $|S| \geq n$ and a vertex $v \in V(G) - S$, if there exist n paths $Q_1, Q_2, \ldots, Q_n$ such that (i) every Q_i connects v and S, (ii) $V(Q_i) \cap S = \{u_i\}$ and these u_i's are distinct, and (iii) $V(Q_i) \cap V(Q_j) = \{v\}$ for all $i \neq j$, then the set $\{Q_1, \ldots, Q_n\}$ of these n paths is called a (v, S)-**fan** of size n (Fig. 1.9). When paths satisfy the condition (iii), we say these paths are **internally disjoint**.

The following two theorems have important consequences. In particular, Theorem 1.6 will often be used in Chapter 8.

Theorem 1.5 (Menger's Theorem [189] (1927)**).** *Let $n \geq 2$ be an integer, and G be an n-connected simple graph. Then for $2n$ distinct vertices $u_1, u_2, \ldots, u_n, v_1, v_2, \ldots, v_n$ of G, G has n vertex disjoint paths $P_1, P_2, \ldots, P_n$ such that every P_i connects u_i and v_i for all $1 \leq i \leq n$ (Fig. 1.9 (i)).*

Theorem 1.6 (Dirac (1960) [61] (1960)**).** *Let $n \geq 2$ be an integer, and let G be an n-connected simple graph. Then for every vertex set S with $|S| \geq n$ and every vertex $v \in V(G) - S$, there exists a (u, S)-fan of size n.*

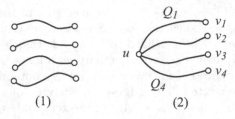

Fig. 1.9. (i) Four vertex disjoint paths; (2) a $(u, \{v_1, v_2, v_3, v_4\})$-fan with size four.

Let G be a connected graph. Then a vertex v is called a **cut vertex** if $G - v$ is not connected, and an edge e of G is called a **bridge** or **cut edge** if $G - e$ is not connected. For a disconnected graph, we can similarly define a cut vertex and a bridge. More generally, a vertex subset $S \subset V(G)$ of a connected graph G is called a **vertex cut** if $G - S$ is disconnected, and an edge subset $T \subseteq E(G)$ is called an **edge cut** if $G - T$ is disconnected.

For two disjoint subsets X and Y of $V(G)$, we denote the set of edges of G joining a vertex of X to a vertex of Y by $E_G(X, Y)$, and the number of edges in $E_G(X, Y)$ by $e_G(X, Y)$, i.e.,

$$e_G(X, Y) = |E_G(X, Y)| = \text{the number of edges of } G \text{ joining } X \text{ to } Y.$$

It is easy to see that a **minimal edge cut** T is expressed as

$$T = E_G(X, Y), \quad \text{where } V(G) = X \cup Y, \ X \cap Y = \emptyset.$$

A graph G is called a **bipartite graph** if $V(G)$ is partitioned into two disjoint sets $A \cup B$ so that every edge of G joins a vertex of A to a vertex of B. The two sets A and B are called **partite sets** and (A, B) is called a **bipartition** of G. Thus a bipartite graph with bipartition (A, B) can be regarded as a subgraph of $K(|A|, |B|)$. Bipartite graphs play an important role throughout this book, and the following proposition is useful.

Proposition 1.7 (König [155]). A multigraph G is a bipartite multigraph if and only if every cycle of G has even length.

Proof. Suppose that G is a bipartite multigraph with bipartition (A, B). Then every cycle of G passes alternately through A and B, and thus its length must be even.

Conversely, suppose that every cycle of G is of even length. We may assume that G is connected since if G is disconnected, then its components can be considered separately. Assume that G has two distinct vertices such that there are two paths connecting them whose lengths are odd and even, respectively. Choose such a pair (x, y) of vertices so that the sum of lengths of two such paths connecting them is minimum. Let $P_1(x, y)$ and $P_2(x, y)$

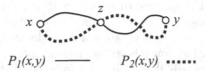

$P_1(x,y)$ ——— $P_2(x,y)$ ••••••

Fig. 1.10. A path $P_1(x,y)$ of odd length and another path $P_2(x,y)$ of even length.

be paths connecting x and y, whose length are odd and even respectively. If $P_1(x,y)$ and $P_2(x,y)$ meet a vertex z on the way, then either $P_1(x,z)$ and $P_2(x,z)$ or $P_1(z,y)$ and $P_2(z,y)$ have distinct parities of lengths, which contradicts the choice of (x,y) since the sum of their lengths is shorter than that of $P_1(x,y)$ and $P_2(x,y)$ (see Fig. 1.10). Thus $P_1(x,y)$ and $P_2(x,y)$ do not meet on the way, which implies that $P_1(x,y) + P_2(x,y)$ forms a cycle of odd length. This contradicts the assumption. Therefore, for any two distinct vertices of G, the length of every path connecting them has the same parity.

Choose one vertex v of G, and define two vertex subsets of G as follows:
$A = \{x \in V(G) : \exists$ a path of even length connecting v to $x \}$,
$B = \{x \in V(G) : \exists$ a path of odd length connecting v to $x \}$.
Then by the above argument, $A \cap B = \emptyset$, and it is easy to see that every edge of G joins A to B, and thus G is a bipartite graph with bipartition (A, B).
□

If a simple graph G can be drawn in the plane so that no two edges intersect except at their end-points, then G is called a **planar graph**, and the graph drawn in the plane is called a **plane graph** (Fig. 1.11). When we consider a planar graph G, we always assume that G is a simple graph. Note that a planar graph can be drawn in different ways as a plane graph, and a plane graph G means one of such plane graphs. For example the two plane graphs G_1 and G_2 in Fig. 1.11 are isomorphic but different plane graphs.

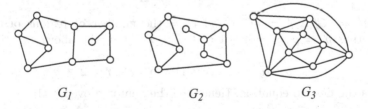

G_1 G_2 G_3

Fig. 1.11. A plane graph G_1 with five regions, whose degrees are 3,3,5,6 and 7; another plane graph representation G_2 of the same graph; and a maximal planar graph G_3.

Given a plane graph G in the plane, a **region** of G is a maximal portion of the plane in which any two points can be joined by a curve that does not cross any edge of G. The **degree of a region** R in G, denoted by degree(R), is defined as the number of edges of G in the boundary of R, where bridges are counted twice. If a region of G is not a triangle, then we can add an edge to G so that the resulting graph is still a plane graph. A plane graph is called a **maximal plane graph** if we cannot add a new edge without violating the planarity (Fig. 1.11). Hence every region of a maximal plane graph is a triangle and, inversely, a plane graph whose regions are all triangles is a maximal plane graph.

By a similar argument as in the proof of the handshaking theorem, we obtain the following.

Lemma 1.8. *For a connected plane simple graph G, let $\Omega(G)$ be the set of regions of G. Then*

$$\sum_{R \in \Omega(G)} degree(R) = 2||G||.$$

The following theorem is very important for plane graphs.

Theorem 1.9 (Euler's Formula). *For a connected plane simple graph G, let $p = |G|$, $q = ||G||$ and r be the number of regions of G. Then*

$$p - q + r = 2.$$

Proof. We prove the theorem by induction on $||G||$ for all connected plane graphs G of fixed order $n \geq 2$. Since G is connected, a smallest such plane graph is a tree. If G is a tree, then $r = 1$ and $q = p - 1$ by Theorem 1.3, and so

$$p - q + r = p - (p - 1) + 1 = 2.$$

Thus we may assume that G is not a tree. Then G has a cycle C. Let e be an edge of C. Then $G' = G - e$ is a connected plane graph, and so by the inductive hypotheses, we have

$$p' - q' - r' = 2,$$

where $p' = |G'|$, $q' = ||G'||$ and $r' = $ the number of regions of G'. By substituting $p' = p$, $q' = q - 1$ and $r' = r - 1$ in the above equation, we have

$$p - (q - 1) + (r - 1) = p - q + r = 2,$$

which is the desired equation. Hence the theorem is proved. □

Using the above theorem, we can easily prove the following theorem.

Theorem 1.10. *Let G be a connected plane simple graph, and let $p = |G|$ and $q = ||G||$. Then*
(i) $q \leq 3p - 6$,   where the equality holds if G is a maximal plane graph.
(ii) If G has no triangles, in particular, if G is a bipartite plane graph, then $q \leq 2p - 4$.

Proof. Let Ω be the set of regions of G. By Lemma 1.8, we obtain

$$3r \leq \sum_{R \in \Omega} \text{degree}(R) = 2||G|| = 2q.$$

Hence $r \leq (2q)/3$. By substituting this into Euler's formula, we have

$$p - q + \frac{2q}{3} \geq 2.$$

Hence $3p - 6 \geq q$. If G has no triangles, then every region has degree at least four, and so

$$4r \leq \sum_{R \in \Omega} \text{degree}(R) = 2||G|| = 2q.$$

This inequality together with Euler's formula implies $2p - 4 \geq q$. □

A **vertex coloring** of a graph G is an assignment of positive integers $1, 2, \ldots, n$ to the vertices of G such that no two adjacent vertices have the same number. A vertex coloring is sometimes called a **proper vertex coloring**. If there exists a vertex coloring with n colors, then we say that G is n**-colorable**. The minimum number n for which G is n-colorable is called the **chromatic number** of G and is denoted by $\chi(G)$. Analogous definitions for edges can be introduced as follows: an **edge coloring** of a graph G is an assignment of positive integers $1, 2, \ldots, n$ to the edges of G such that no two adjacent edges have the same number. An edge coloring is also called a **proper edge coloring**. We mainly consider **edge colorings** in this book. If there exists an edge coloring with n colors, then we say G is n**-edge colorable**. The minimum number n for which G is n-edge colorable is called the **chromatic index** of G and denoted by $\chi'(G)$. For edge coloring, the following famous theorem holds.

Theorem 1.11 (Vizing [232] (1964)). *A simple graph G can be edge colored with $\Delta(G) + 1$ colors. In particular, $\chi'(G) = \Delta(G)$ or $\Delta(G) + 1$.*

We introduced connectivity and edge-connectivity to evaluate the strength of a graph. There are some other concepts which might evaluate the strength of a graph from a different point of view. We explain two of them below.

The **toughness** of a connected non-complete graph G is defined by the following and denoted by $tough(G)$.

$$tough(G) = \min\left\{ \frac{|X|}{\omega(G - X)} : \emptyset \neq X \subseteq V(G), \ \omega(G - X) \geq 2 \right\},$$

where $\omega(G - X)$ denotes the number of components of $G - X$. A graph G is said to be t**-tough** if $tough(G) \geq t$. Hence, if G is t-tough, then for a subset $S \subset V(G)$ with $\omega(G - S) \geq 2$, it follows that

$$t \le tough(G) \le \frac{|S|}{\omega(G-S)}, \qquad \omega(G-S) \le \frac{|S|}{t}.$$

The **binding number** $bind(G)$ of a graph G, which was introduced by Anderson [17], is defined by

$$bind(G) = \min\left\{ \frac{|N_G(X)|}{|X|} : \emptyset \ne X \subseteq V(G), \ N_G(X) \ne V(G) \right\}.$$

Thus for every subset $\emptyset \ne S \subset V(G)$ with $N_G(S) \ne V(G)$, it follows that

$$bind(G) \le \frac{|N_G(S)|}{|S|}, \qquad bind(G)|S| \le |N_G(S)|.$$

We conclude this introduction with two results on circuits and cycles of graphs. For a graph G, a circuit that passes through every edge of G precisely once is called an **Euler circuit** of G. A graph having an Euler circuit is called an **Eulerian graph**. On the other hand, a cycle that passes through every vertex exactly once is called a **Hamiltonian cycle**. A graph having a Hamiltonian cycle is called a **Hamiltonian graph**. Analogously a path in G is called a **Hamiltonian path** if it passes through all the vertices of G. It is easy to see that if a graph G is Hamiltonian, then $tough(G) \ge 1$. So the non-existence of a Hamiltonian cycle in G_4 in Fig. 1.12 is shown by $3 = \omega(G_4 - X) > |X| = 2$ for a set X, which implies $tough(G_4) < 1$.

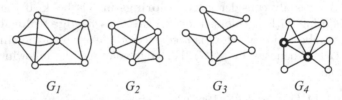

$$G_1 \qquad\qquad G_2 \qquad\qquad G_3 \qquad\qquad G_4$$

Fig. 1.12. An Eulerian multigraph G_1, a non-Eulerian graph G_2, a Hamiltonian graph G_3, and a non-Hamiltonian graph G_4 with a set X consisting of two bold vertices.

Theorem 1.12. *A connected general graph G is Eulerian if and only if every vertex of G has even degree.*

Proof. Let H be the graph obtained from G by removing all its loops. Then G is Eulerian if and only if H is Eulerian, and every vertex of G has even degree if and only if every vertex of H has even degree. Hence it suffices to prove the theorem for a multigraph.

Let G be a connected multigraph. Assume that G has an Eulerian circuit C. We move along C and remove an edge when we pass through it. Then for every vertex v except an initial vertex, two edges incident with v are removed when we pass through v. In the final stage, all the edges incident with v are removed, and thus the degree of v is even. For the initial vertex, by a similar argument, we can show that it has even degree.

Conversely, assume that every vertex of G has even degree. Then we prove that G has an Eulerian circuit by induction on $\|G\|$. If $\|G\| = 2$, then G consists of two vertices and two multiple edges joining them, and so G has an Eulerian circuit. Assume that $\|G\| \geq 3$. Since every vertex has degree at least two, G has a cycle C. Every vertex of each non-trivial component D of $G - E(C)$ has even degrees, where non-trivial component means a component of order at least two. Hence by induction, D has an Eulerian circuit $Cir(D)$. We arrange all the non-trivial components of $G - E(C)$ $D_1, D_2, \ldots, D_m$ in the order that C passes through at least one of its vertices. Then we can obtain an Eulerian circuit of G as follows: start at a vertex w of C and proceed along C to reach a vertex v_1 of D_1; pass through $Cir(D_1)$ and come back to v_1; proceed along C to reach a vertex v_2 of D_2; passes through $Cir(D_2)$ and come back to v_2. Continue this procedure until we return to w (see Fig. 1.13 (1)). Finally, we can obtain a circuit that passes through all the edges of G.   □

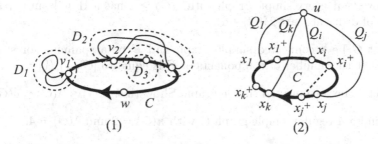

Fig. 1.13. (1) A connected graph with each vertex of even degree; (2) a connected graph with $\alpha(G) \leq \kappa(G)$.

We now give a well-known theorem about Hamiltonian cycles. The cardinality of a maximum independent set of a graph G is called the **independence number** of G and denoted by $\alpha(G)$.

Theorem 1.13 (Chvátal and Erdős [55] (1972)). *Every connected simple graph G with $\alpha(G) \leq \kappa(G)$ has a Hamiltonian cycle unless $G = K_2$.*

Proof. Let $k = \kappa(G)$. Then we may assume $k \geq \alpha(G) \geq 2$ since $\alpha(G) = 1$ implies $G = K_n$ with $n \geq 3$. Let C be a longest cycle of G. We may assume that C is not a Hamiltonian cycle. Since $\delta(G) \geq \kappa(G) = k$, the length of C is at least $k+1$ (see Problem 1.2). Let u be a vertex of $V(G) - V(C)$. By Theorem 1.6 there exists a $(u, V(C))$-fan $\{Q_1, \ldots, Q_k\}$ of size k. Let $x_i = V(C) \cap V(Q_i)$ be a vertex of C (see Fig. 1.13 (2)).

Without loss generality, we may assume that $x_1, x_2, \ldots, x_k$ each lie on C in this order, and we regard C as a directed cycle with this orientation. Let x_i^+ be the successor of x_i, which immediately follows x_i in C. Since C is a longest cycle, $x_i^+ \neq x_{i+1}$ and u and x_i^+ are not adjacent in G for all i. If two vertices x_i^+ and x_j^+, $i \neq j$, are adjacent in G, then we can find a cycle longer than C, which passes through $(u, Q_j, x_j, x_i^+, x_j^+, x_i, Q_i, u)$ (Fig. 1.13 (2)). This contradicts the maximality of C. Hence x_i^+ and x_j^+ are not adjacent in G. Therefore, $\{u, x_1^+, x_2^+, \ldots, x_k^+\}$ is an independent set of G with cardinality $k + 1$, which contradicts $\alpha(G) \leq \kappa(G) = k$. Consequently G has a Hamiltonian cycle. $\square$

Problems

1.1. Prove that if G is a connected cubic multigraph, then $\kappa(G) = \lambda(G)$. Namely, prove Proposition 1.4.

1.2. (a) Show that every simple graph with $\delta(G) \geq 2$ has a cycle.
(b) Prove that every simple graph with $\delta(G) \geq 2$ has a Hamiltonian cycle or a cycle of order at least $\delta(G) + 1$.
(c) Prove that every simple graph with $\delta(G) \geq 2$ has a Hamiltonian path or a path of order at least $2\delta(G) + 1$.

1.3. Let G be a connected simple graph and F be a spanning forest of G. Show that the number of components of F is equal to $|G| - \|F\|$.

1.4. Let G be a connected simple graph. Show that $\kappa(G) \leq \lambda(G) \leq \delta(G)$.

1.5. Find a 4-regular simple graph G with $\kappa(G) = 2$ and $\lambda(G) = 4$.

1.6. Let $r \geq 2$ be an integer. Prove that every connected r-regular bipartite multigraph is 2-edge connected.

1.7. Prove that every connected graph G satisfies $\kappa(G) \geq 2 \cdot tough(G)$.

1.8. Let G be a connected simple graph. Show that if $bind(G) > 0$, then $|N_G(X)| \geq |G| - (|G| - |X|)/bind(G)$ for all $\emptyset \neq X \subseteq V(G)$.

1.9. Prove Theorem 1.6 by using Menger's Theorem 1.5.

Matchings and 1-Factors

In this chapter we consider matchings and 1-factors of graphs, and these results will frequently be used in latter chapters. Of course, they are interesting on their own, and have many applications in various areas of mathematics and computer science. The reader is referred to two books [182] by Lovász and Plummer and [254] by Yu and Liu for a more detailed treatment of matchings and 1-factors.

2.1 Matchings in bipartite graphs

In this section we investigate matchings in bipartite graphs, and the results shown in this section will play an important role throughout this book.

Two edges of a general graph are said to be **independent** if they have no common end-point and none of them is a loop. A **matching** in a general graph G is a set of pairwise independent edges of G (Fig. 2.1). If M is a matching in a general graph G, then the **subgraph of G induced by** M, denoted by $\langle M \rangle_G$ or $\langle M \rangle$, is the subgraph of G whose edge set is M and whose vertex set consists of the vertices incident with some edge in M. Then every vertex of $\langle M \rangle$ has degree one. Thus it is possible to define a matching as a subgraph whose vertices all have degree one, and we often regard a matching M as its induced subgraph $\langle M \rangle$.

Let M be a matching in a graph G. Then a vertex of G is said to be **saturated** or **covered** by M if it is incident with an edge of M; otherwise, it is said to be **unsaturated** or **not covered** by M. If every vertex of a vertex subset U of G is saturated by M, then we say that U is **saturated** by M. A matching with maximum cardinality is called a **maximum matching**. A matching that saturates all the vertices of G is called a **perfect matching** or a **1-factor** of G (Fig. 2.1). It is easy to see that a **maximal matching**, which is a maximal set of independent edges, is not a maximum matching, and a maximum matching is not a perfect matching. On the other hand, if

J. Akiyama and M. Kano, *Factors and Factorizations of Graphs*,
Lecture Notes in Mathematics 2031, DOI 10.1007/978-3-642-21919-1_2,
© Springer-Verlag Berlin Heidelberg 2011

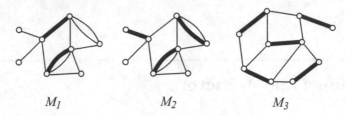

$$M_1 \qquad\qquad M_2 \qquad\qquad M_3$$

Fig. 2.1. A maximal matching M_1;   a maximum matching M_2;   and a perfect matching M_3.

a matching in a bipartite graph saturates one of its partite sets, then it is clearly a maximum matching.

For a matching M, we write

$\|M\| =$ the size of $\langle M\rangle =$ the number of edges in M,

$|M| =$ the order of $\langle M\rangle =$ the number of vertices saturated by M.

We first give a criterion for a bipartite graph to have a matching that saturates one of its partite sets. Then we apply this criterion to some problems on matchings in bipartite graphs.

The criterion mentioned above is given in the following theorem, which was found by Hall [98] and is called the marriage theorem. This theorem appears throughout this book, and the proof given here is due to Halmos and Vaughan [97]. An algorithm for finding a maximum matching in a bipartite graph will be given in Algorithm 2.25.

Theorem 2.1 (The Marriage Theorem, Hall [98] (1935)**).** *Let G be a bipartite multigraph with bipartition (A, B). Then G has a matching that saturates A if and only if*

$$|N_G(S)| \ge |S| \qquad for\ all \quad S \subseteq A. \tag{2.1}$$

Proof. We first construct a bipartite simple graph H from the given bipartite multigraph G by replacing all the multiple edges of G by single edges. Then it is obvious that G has the desired matching if and only if H has such a matching, and that G satisfies (2.1) if and only if H satisfies it. Therefore, we may assume that G itself has no multiple edges by considering H as the given bipartite graph.

Suppose that G has a matching M that saturates A (Fig. 2.2). Then for every subset $S \subseteq A$, we have

$$|N_G(S)| \ge |N_M(S)| = |S|.$$

We next prove sufficiency by induction on $|G|$. It is clear that we may assume $|A| \ge 2$. We consider the following two cases.

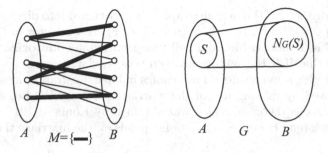

Fig. 2.2. A matching M that saturates A; S and $N_G(S)$.

Case 1. *There exists $\emptyset \neq S \subset A$ such that $|N_G(S)| = |S|$.*

Let $H = \langle S \cup N_G(S) \rangle_G$ and $K = \langle (A - S) \cup (B - N_G(S)) \rangle_G$ be induced subgraphs of G (Fig. 2.3). It is clear that H satisfies condition (2.1), and so H has a matching M_H that saturates S by induction. For every subset $X \subseteq A - S$, we have

$$|N_K(X)| = |N_G(X \cup S)| - |N_G(S)| \geq |X \cup S| - |S| = |X|.$$

Hence, by induction, K also has a matching M_K that saturates $A - S$. Therefore $M_H \cup M_K$ is the desired matching in G which saturates A.

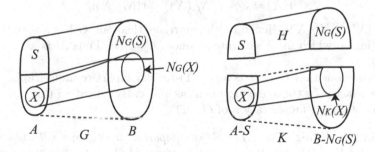

Fig. 2.3. The induced subgraphs H and K.

Case 2. $|N_G(S)| > |S|$ *for all $\emptyset \neq S \subset A$.*

Let $e = ab$ $(a \in A, b \in B)$ be an edge of G, and let $H = G - \{a, b\}$. Then for every subset $\emptyset \neq X \subseteq A - \{a\}$, by the assumption of this case, we have

$$|N_H(X)| \geq |N_G(X) \setminus \{b\}| > |X| - 1, \qquad (2.2)$$

which implies $|N_H(X)| \geq |X|$. Therefore, H has a matching M' that saturates $A - \{a\}$ by induction. Then $M' + e$ is the desired matching in G. Consequently the theorem is proved. $\square$

If the edge set $E(G)$ of a multigraph G is partitioned into disjoint 1-factors $E(G) = F_1 \cup F_2 \cup \cdots \cup F_r$, where each $F_i = \langle F_i \rangle$ is a 1-factor of G, then we say that G is **1-factorable**, and call this partition a **1-factorization** of G. It is trivial that if G is 1-factorable, then G is regular.

We now give some results on matchings in bipartite graphs, most of which can be proved by making use of the marriage theorem. We begin with the following famous theorem, which was obtained by König in 1916 before the marriage theorem. However, our proof depends on the marriage theorem.

Theorem 2.2 (König [155]). *Every regular bipartite multigraph is 1-factorable, in particular, it has a 1-factor (Fig. 2.4).*

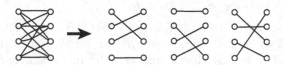

Fig. 2.4. A 3-regular bipartite multigraph and its 1-factorization.

Proof. Let G be an r-regular bipartite multigraph with bipartition (A, B). Then $|A| = |B|$ since $r|A| = e_G(A, B) = r|B|$. For every subset $X \subseteq A$, we have

$$r|X| = e_G(X, N_G(X)) \leq r|N_G(X)|,$$

and so $|X| \leq |N_G(X)|$. Hence by the marriage theorem, G has a matching M saturating A, which must saturate B since $|A| = |B|$. Thus M is a 1-factor of G.

It is obvious that $G - M$ is a $(r - 1)$-regular bipartite multigraph, and so it has a 1-factor by the same argument as above. By repeating this procedure, we can obtain a 1-factorization of G.   □

Lemma 2.3 (König [155]). *Let G be a bipartite multigraph with bipartition (A, B). If $|A| \geq |B|$ and the maximum degree of G is Δ, then there exists a Δ-regular bipartite multigraph which contains G as a subgraph and A as one of its partite sets.*

Proof. By adding $|A| - |B|$ new vertices to B if $|A| > |B|$, we obtain a bipartite graph with bipartition (A, B') such that $|A| = |B'|$. Add new edges joining vertices in A to vertices in B' whose degrees are less than Δ, one at a time until no new edge can be added. We show that by this procedure, we get the desired Δ-regular bipartite multigraph.

Suppose that the bipartite multigraph H obtained in this way has a vertex $a \in A$ such that $\deg_H(a) < \Delta$. Then there exists a vertex $b \in B'$ such that $\deg_H(b) < \Delta$ because

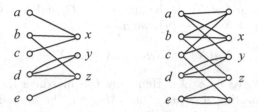

Fig. 2.5. A bipartite multigraph and a regular bipartite graph containing it.

$$\sum_{x \in B'} \deg_H(x) = e_H(B', A) = \sum_{x \in A} \deg_H(x) < \Delta |A| = \Delta |B'|.$$

Hence we can add a new edge ab to H, which is a contradiction. Therefore every vertex of A has degree Δ in H, which implies that every vertex of B' has degree Δ since $|A| = |B'|$. Consequently, H is the desired Δ-regular bipartite multigraph. $\square$

The next theorem says that the chromatic index of a bipartite multigraph G is equal to the maximum degree of G.

Theorem 2.4 (König [155]). *Let G be a bipartite multigraph with maximum degree Δ. Then $E(G)$ can be partitioned into $E(G) = E_1 \cup E_2 \cup \cdots \cup E_\Delta$ so that each E_i ($1 \le i \le \Delta$) is a matching in G.*

Proof. By Lemma 2.3, there exists a Δ-regular bipartite multigraph H which contains G as a subgraph. Then by Theorem 2.2, $E(H)$ can be partitioned into 1-factors $F_1 \cup F_2 \cup \cdots \cup F_\Delta$. It is obvious that $F_i \cap E(G)$ ($1 \le i \le \Delta$) is a matching in G, and their union is equal to $E(G)$. Therefore the theorem follows. $\square$

We now state various properties of matchings in bipartite graphs, some of which are generalizations of the marriage theorem, and some of which are properties that are specific to bipartite graphs.

Theorem 2.5. *A bipartite multigraph G has a matching that saturates all the vertices of degree $\Delta(G)$.*

Proof. Let $\Delta = \Delta(G)$. By Lemma 2.3, there exists a Δ-regular bipartite multigraph H that contains G as a subgraph. Then by Theorem 2.2, H has a 1-factor F. It is easy to see that $F \cap E(G)$ is a matching of G that saturates all the vertices v of G with degree Δ since every edge of H incident with v is an edge of G. $\square$

Theorem 2.6. *Let G be a bipartite multigraph with bipartition (A, B). If $|N_G(S)| > |S|$ for all $\emptyset \ne S \subset A$, then for each edge e of G, G has a matching that saturates A and contains e.*

Proof. Let $e = ab$ $(a \in A, b \in B)$ be any edge of G, and $H = G - \{a, b\}$. Then for every subset $\emptyset \neq X \subseteq A - \{a\}$, we have

$$|N_H(X)| \geq |N_G(X) \setminus \{b\}| > |X| - 1,$$

which implies $|N_H(X)| \geq |X|$. Hence by the marriage theorem, H has a matching M saturating $A - \{a\}$. Thus, $M + e$ is the desired matching in G that saturates A and contains e. $\quad\square$

Theorem 2.7 (Hetyei, Theorem 4.1.1 of [182]). *Let G be a bipartite multigraph with bipartition (A, B) such that $|A| = |B|$. Then the following three statements are equivalent.*
(i) G is connected, and for each edge e, G has a 1-factor containing e.
(ii) For every subset $\emptyset \neq X \subset A$, $|N_G(X)| > |X|$.
(iii) For every two vertices $a \in A$ and $b \in B$, $G - \{a, b\}$ has a 1-factor.

Proof. (i)$\Rightarrow$(ii) Suppose that $|N_G(Y)| \leq |Y|$ for some subset $\emptyset \neq Y \subset A$. Since G has a 1-factor F, we have $|N_G(Y)| = |Y|$ as $|N_G(Y)| \geq |N_F(Y)| = |Y|$. Since G is connected, G has an edge e joining a vertex in $A - Y$ to a vertex in $N_G(Y)$. However there exists no 1-factor in G containing e since $|N_G(Y)| - 1 < |Y|$. This contradicts (i). Thus (ii) holds.

(ii)$\Rightarrow$(iii) Let $H = G - \{a, b\}$ and $X \subseteq A - \{a\}$. Then

$$|N_H(X)| = |N_G(X) \setminus \{b\}| > |X| - 1,$$

which implies $|N_H(X)| \geq |X|$. Hence H has a matching saturating $A - \{a\}$, which is obviously a 1-factor of H as $|A - \{a\}| = |B - \{b\}|$.

(iii)$\Rightarrow$(i) Let $e = ab$ $(a \in A, b \in B)$ be an edge of G. Since $G - \{a, b\}$ has a 1-factor F, G has a 1-factor $F + e$, which contains e. The proof of connectivity is left to the reader. $\quad\square$

Theorem 2.8 (Exercise 3.1.32 of [244]). *Suppose that a bipartite multigraph G with bipartition (A, B) has a matching saturating A. Then there exists a vertex $v \in A$ possessing the property that every edge incident with v is contained in a matching in G saturating A.*

Proof. We prove the theorem by induction on $|A|$. It is clear that we may assume $|A| \geq 2$.

If $|N_G(X)| > |X|$ for all $\emptyset \neq X \subset A$, then each vertex in A has the required property by Theorem 2.6. Hence we may assume that $|N_G(S)| \leq |S|$ for some $\emptyset \neq S \subset A$. Since G has a matching saturating A, it follows from the marriage theorem that $|N_G(S)| = |S|$.

By the inductive hypothesis, the induced subgraph $H = \langle S \cup N_G(S) \rangle_G$ has a vertex $v \in S$ such that every edge of H incident with v is contained in a matching in H saturating S (Fig. 2.6). Note that every edge of G incident with v is contained in H.

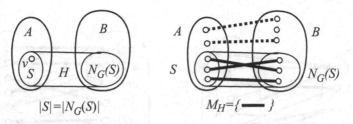

$$|S|=|N_G(S)|$$ $$M_H=\{\;\rule[0.5ex]{1.5em}{0.5pt}\;\}$$

Fig. 2.6. A bipartite multigraph G and its subgraph H; A matching M in G saturating A.

Let M be a matching in G saturating A, and M_H be a matching in H saturating S. Then it is easy to see that $(M \setminus E(H)) \cup M_H$ is a matching in G that saturates A (see Fig. 2.6). Therefore the vertex v in S has the desired property.   □

Theorem 2.9. *Let G be a bipartite multigraph with bipartition (A, B) such that $|N_G(X)| \geq |X|$ for every $X \subseteq A$. If two subsets S, $T \subseteq A$ satisfy $|N_G(S)| = |S|$ and $|N_G(T)| = |T|$, then*

$$|N_G(S \cup T)| = |S \cup T| \quad and \quad |N_G(S \cap T)| = |S \cap T|.$$

In particular, if such subsets exist, there exists a unique maximum subset $A_0 \subseteq A$ such that $|N_G(A_0)| = |A_0|$

Proof. Since $N_G(S \cup T) = N_G(S) \cup N_G(T)$ and $N_G(S) \cap N_G(T) \supseteq N_G(S \cap T)$, we have

$$\begin{aligned}
|N_G(S \cup T)| &= |N_G(S)| + |N_G(T)| - |N_G(S) \cap N_G(T)| \\
&\leq |S| + |T| - |N_G(S \cap T)| \\
&\leq |S| + |T| - |S \cap T| = |S \cup T|.
\end{aligned}$$

On the other hand, $|N_G(S \cup T)| \geq |S \cup T|$ by the assumption. Hence $|N_G(S \cup T)| = |S \cup T|$, and also $|N_G(S \cap T)| = |S \cap T|$ by the above inequality.   □

The following theorem is a generalization of the marriage theorem since the subgraph H with $f(x) = 1$ given in the following theorem is a matching.

Theorem 2.10 (Generalized Marriage Theorem). *Let G be a bipartite multigraph with bipartition (A, B), and let $f : A \to \mathbb{N}$ be a function. Then G has a subgraph H such that*

$$\deg_H(x) = f(x) \qquad for\ all \quad x \in A, \quad and \tag{2.3}$$
$$\deg_H(y) \leq 1 \qquad for\ all \quad y \in B \tag{2.4}$$

if and only if

$$|N_G(S)| \geq \sum_{x \in S} f(x) \qquad for\ all \quad S \subseteq A. \tag{2.5}$$

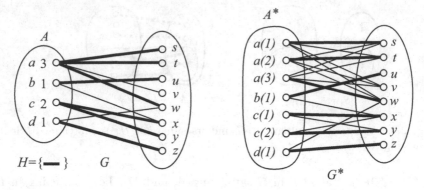

Fig. 2.7. A bipartite graph G and its subgraph H, where numbers denote $f(v)$; and G^*.

Proof. We first assume that G has a subgraph H that satisfies (2.3) and (2.4) (Fig. 2.7). Then for every $S \subseteq A$, we have

$$|N_G(S)| \geq |N_H(S)| = \sum_{x \in S} \deg_H(x) = \sum_{x \in S} f(x).$$

Hence (2.5) holds.

In order to prove the sufficiency, we construct a new bipartite simple graph G^* with bipartition $A^* \cup B$ as follows. For every vertex $v \in A$, define $f(v)$ vertices $v(1), v(2), \ldots, v(f(v))$ of A^*, and connect them to all the vertices of $N_G(v)$ by edges (Fig. 2.7). Then

$$|A^*| = \sum_{x \in A} f(x) \quad \text{and} \quad N_{G^*}(v(i)) = N_G(v) \quad \text{for all} \quad v(i) \in A^*.$$

For every subset $\emptyset \neq S^* \subseteq A^*$, let $S = \{x \in A : x(i) \in S^* \text{ for some } i\}$. Then we have by (2.5) that

$$|N_{G^*}(S^*)| = |N_G(S)| \geq \sum_{x \in S} f(x) \geq |S^*|.$$

Hence by the marriage theorem, G^* has a matching M^* that saturates A^*. It is clear that the subgraph H of G induced by M^* satisfies

$$\deg_H(x) = f(x) \quad \text{for all} \quad x \in A, \quad \text{and} \quad \deg_H(y) \leq 1 \quad \text{for all} \quad y \in B.$$

Hence the theorem is proved. □

The next theorem gives a formula for the size of a maximum matching in a bipartite graph. Moreover, this formula includes the marriage theorem since $|N_G(S)| \geq |S|$ for all $S \subseteq A$ implies $||M|| = |A|$ as $|\emptyset| - |N_G(\emptyset)| = 0$.

Theorem 2.11 (Ore [204]). *Let G be a bipartite multigraph with bipartition (A, B), and M a maximum matching in G. Then the size of M is given by*

$$||M|| = |A| - \max_{X \subseteq A}\{|X| - |N_G(X)|\}.$$

Proof. Let M be a maximum matching in G, $d = \max_{X \subseteq A}\{|X| - |N_G(X)|\}$, and $S \subseteq A$ such that $|S| - |N_G(S)| = d$. Then

$$
\begin{aligned}
||M|| &= |N_M(A - S)| + |N_M(S)| \\
&\leq |A - S| + |N_G(S)| = |A| - (|S| - |N_G(S)|) \\
&= |A| - d.
\end{aligned}
$$

In order to prove the inverse inequality, we construct a new bipartite multigraph H with bipartition $(A, B \cup D)$ from G by adding a new vertex set D of d vertices and by joining every vertex of D to all the vertices of A (Fig. 2.8). Then for every $\emptyset \neq X \subseteq A$, since $d \geq |X| - |N_G(X)|$, we have

$$|N_H(X)| = |N_G(X)| + |D| = |N_G(X)| + d \geq |X|.$$

Hence, by the marriage theorem, H has a matching M_H saturating A. Then $M_H \cap E(G)$ is a matching in G that contains at least $|A| - d$ edges. Therefore $||M|| \geq |A| - d$, and the theorem is proved. $\square$

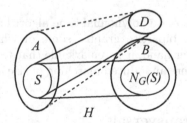

Fig. 2.8. A new bipartite multigraph H.

An **induced matching** in a general graph G is a matching that is an induced subgraph of G, that is, an induced matching can be expressed as $\langle U \rangle_G$ for some vertex set $U \subseteq V(G)$.

Theorem 2.12 (Liu and Zhou, [173]). *Let G be a connected bipartite simple graph with bipartition (A, B). Then the size of a maximum induced matching M in G is given by*

$$||M|| = \max\{|X| : X \subseteq A \text{ such that } N_G(Y) \neq N_G(X) \text{ for all } Y \subset X\}.$$

Proof. We say that a subset $X \subset A$ has the property $\mathcal{P}$ if $N_G(Y) \neq N_G(X)$ for all $Y \subset X$. Let

$$k = \max\{|X| : X \subseteq A, \ X \text{ has the property } \mathcal{P}\}.$$

Let M be a maximum induced matching in G, and let A_M and B_M be the sets of vertices of A and B, respectively, which are saturated by M. Since M is an induced matching, we have $E_G(A_M, B_M) = M$. Then for every $Y \subset A_M$, we have $N_G(Y) \neq N_G(A_M)$ since

$$|N_G(Y) \cap B_M| = |Y| < |A_M| = |N_G(A_M) \cap B_M|.$$

Hence A_M has the property $\mathcal{P}$, and thus $||M|| = |A_M| \leq k$.

We now show that $||M|| \geq k$. We may assume $k \geq 2$. Let $S = \{a_1, a_2, \ldots, a_k\}$ be a maximum subset of A that has the property $\mathcal{P}$. Then $N_G(Y) \neq N_G(S)$ for all $Y \subset S$. For every $1 \leq i \leq k$, we have

$$N_G(a_i) \not\subseteq \bigcup_{j \neq i} N_G(a_j) \qquad \text{by} \qquad N_G(S - \{a_i\}) \neq N_G(S),$$

and so we can choose

$$b_i \in N_G(a_i) \setminus \bigcup_{j \neq i} N_G(a_j).$$

Then $\{a_i b_i \mid 1 \leq i \leq k\}$ is an induced matching with k edges, which implies $||M|| \geq k$. Consequently the theorem is proved. $\square$

It is known that there are polynomial time algorithms for many problems involving matchings in bipartite graphs, some of which will be shown in Section 2.3. On the other hand, it is remarkable that the following problem is NP-complete [41]: Is there an induced matching of size k in a bipartite graph?

2.2 Covers and transversals

In this section we discuss covers of bipartite graphs and transversals of family of subsets. Some of the results are called **min-max theorems** because they say that the minimum value of some invariant is equal to the maximum value of another invariant.

When a vertex v is incident with an edge e, we say that v and e **cover** each other. A **vertex cover** of a graph G is a set of vertices that cover all the edges of G. A vertex cover of minimum cardinality is called a **minimum vertex cover**, and its cardinality is denoted by $\beta(G)$ (Fig. 2.9). Similarly, an **edge cover** of a graph G is defined to be a set of edges that cover all the vertices of G. An edge cover with minimum cardinality is called a **minimum edge cover**, and its cardinality is denoted by $\beta'(G)$.

Recall that a set of vertices of a graph G is said to be independent if no two of its vertices are adjacent, and a set of edges of G is said to be independent if no two of its edges have an end-point in common, that is, a set of edges is independent if and only if it forms a matching. We analogously define a **maximum independent vertex set** and a **maximum independent edge set** of G, and denote their cardinalities by $\alpha(G)$ and $\alpha'(G)$, respectively.

$\beta(G)$ = the minimum cardinality of a vertex cover
$\beta'(G)$ = the minimum cardinality of an edge cover
$\alpha(G)$ = the maximum cardinality of an independent set of vertices
$\alpha'(G)$ = the maximum cardinality of an independent set of edges

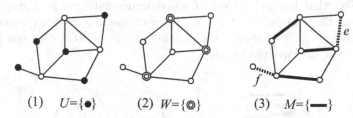

(1) $U=\{\bullet\}$ (2) $W=\{\circledcirc\}$ (3) $M=\{\rule[0.4ex]{1em}{0.4pt}\}$

Fig. 2.9. (1) A maximum independent set of vertices U ; (2) A minimum vertex cover $W = V(G) - U$; (3) A maximum matching M and a minimum edge cover $M + \{e, f\}$.

Lemma 2.13 (Gallai [91]). *A vertex set S of a simple graph G is independent if and only if $\overline{S} = V(G) - S$ is a vertex cover. Moreover,*

$$\alpha(G) + \beta(G) = |G|.$$

Proof. Suppose that S is independent. Then no edge of G joins two vertices of S, that is, for each edge e of G, at least one end-point of e is contained in $\overline{S}$. Hence $\overline{S}$ is a vertex cover.

Conversely, if $\overline{S}$ is a vertex cover, then every edge is incident with a vertex in $\overline{S}$, which implies that no edge joins two vertices in $S = V(G) - \overline{S}$, and thus S is independent.

Moreover, from the above arguments, it follows that S is a maximum independent set of vertices if and only if $\overline{S}$ is a minimum vertex cover. Thus $\alpha(G) + \beta(G) = |S| + |\overline{S}| = |G|$. $\square$

The next theorem shows that a similar equality also holds for edges.

Theorem 2.14 (Gallai [91]). *If a simple graph G has no isolated vertices, then $\alpha'(G) + \beta'(G) = |G|$.*

Proof. Let M be a maximum matching of G. For every vertex v of G unsaturated by M, by adding one edge incident with v to M, we can obtain an edge cover which contains the following number of edges:

$$||M|| + |G| - 2||M|| = |G| - ||M|| = |G| - \alpha'(G).$$

Hence $\beta'(G) \leq |G| - \alpha'(G)$, which implies $\alpha'(G) + \beta'(G) \leq |G|$.

Conversely, if L is a minimum edge cover of G, then $\langle L \rangle$ is a spanning subgraph of G and each component of $\langle L \rangle$ is a tree. Thus the number of components of $\langle L \rangle$ is $|G| - ||L||$ (see Problem 1.3). By choosing one edge from each component of $\langle L \rangle$, we can obtain a matching that contains $|G| - ||L||$ edges. Hence $\alpha'(G) \geq |G| - \beta'(G)$, and thus $\alpha'(G) + \beta'(G) \geq |G|$. Consequently, we have $\alpha'(G) + \beta'(G) = |G|$.   □

It follows that for each edge e of a maximum matching in a graph G, at least one of the end-points of e must be contained in a vertex cover of G. Hence $\alpha'(G) \leq \beta(G)$. However, the equality $\alpha'(G) = \beta(G)$ does not hold in general except for bipartite graphs.

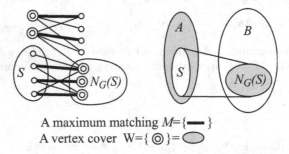

A maximum matching M={▬}
A vertex cover W={◎}=⬭

Fig. 2.10. A maximum matching and a minimum vertex cover in a bipartite graph.

Theorem 2.15 (König [156]). *In a bipartite simple graph G without isolated vertices, $\alpha'(G) = \beta(G)$. Moreover this implies $\alpha(G) = \beta'(G)$.*

Proof. Since $\alpha'(G) \leq \beta(G)$ as shown above, it suffices to prove $\alpha'(G) \geq \beta(G)$.

Let (A, B) be the bipartition of G, and let M be a maximum matching in G. Let S be a subset of A such that

$$|S| - |N_G(S)| = \max_{X \subseteq A}\{|X| - |N_G(X)|\},$$

and $W = (A - S) \cup N_G(S)$. Then W is a vertex cover of G since G has no edge joining S to $B - N_G(S)$ (see Fig. 2.10). By Theorem 2.11, we have

$$\beta(G) \leq |W| = |A| - |S| + |N_G(S)| = ||M|| = \alpha'(G).$$

Therefore the theorem is proved.   □

$$M = \begin{pmatrix} 1 & 1 & 0 & ① & 1 \\ 0 & ① & 0 & 0 & 0 \\ 0 & 1 & 0 & 0 & 0 \\ 1 & 0 & 0 & 1 & ① \end{pmatrix}$$

$|\{\bigcirc\}| = 3$
{1st row, 4th row, 2nd column}
contains all the 1-entries.

Fig. 2.11. A $(0,1)$-matrix M.

A $(0,1)$-matrix is a matrix whose entries are all 0 or 1. The preceding theorem leads to the following interesting property of $(0,1)$-matrices.

Theorem 2.16 (König-Egerváry [155], [74]). *Let $M = (m_{ij})$ be a $(0,1)$-matrix. Then the maximum number of 1-entries of M, such that no two of them lie on the same row or column, is equal to the minimum number of rows and columns that contain all the 1-entries of M (Fig. 2.11).*

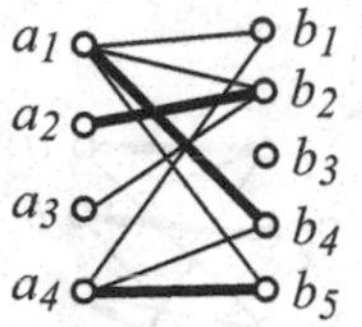

Fig. 2.12. The bipartite graph corresponding to the $(0,1)$-matrix in Fig. 2.11, where $\alpha'(G) = \#\{a_1b_4, a_2b_2, a_4b_5\} = 3$ and $\beta(G) = \#\{a_1, b_2, a_4\} = 3$.

Proof. From the given $n \times m$ $(0,1)$-matrix $M = (m_{ij})$, we construct a bipartite graph G with bipartite sets $A = \{a_1, a_2, \ldots, a_n\}$ and $B = \{b_1, b_2, \ldots, b_m\}$ as follows:

$$a_i b_j \in E(G) \quad \text{if and only if} \quad m_{ij} = 1. \quad \text{(see Fig. 2.12)}$$

Then a set of 1-entries, no two of which lie on the same row or column, corresponds to a matching of G, and a set of rows and columns that contains all the 1-entries corresponds to a vertex cover of G. Therefore the theorem follows from Theorem 2.15. $\square$

Let X be a finite set and $\mathcal{F} = \{S_1, S_2, \ldots, S_n\}$ a family of subsets of X, where the S_i's are not necessarily distinct. We say that $\mathcal{F}$ has a **transversal** if there exists a set of n distinct elements of X, one from each S_i. For example, a family $\{\{a, b, e\}, \{b\}, \{a, c, d\}, \{b, c\}\}$ of subsets of $\{a, b, c, d, e\}$ has a transversal $\{a, b, d, c\}$. On the other hand, a family $\{\{a, b, e\}, \{b, c\}, \{c\}, \{b, c\}\}$ has no transversal.

Theorem 2.17 (Hall [98]). *A family* $\mathcal{F} = \{S_1, S_2, \ldots, S_n\}$ *of subsets of* X *has a transversal if and only if*

$$\left| \bigcup_{i \in I} S_i \right| \geq |I| \qquad \text{for all} \quad I \subseteq \{1, 2, \ldots, n\}. \tag{2.6}$$

Proof. We prove only the sufficiency since the necessity is immediate. We construct a bipartite graph G with partite sets $\{S_1, S_2, \ldots, S_n\}$ and X as follows: a vertex S_i is adjacent to a vertex $x \in X$ in G if and only if $x \in S_i$ (see Fig. 2.13). Then for every subset $\{S_i \mid i \in I\}$ of $\{S_1, S_2, \ldots, S_n\}$, we have by (2.6) that

$$|N_G(\{S_i \mid i \in I\})| = \left| \bigcup_{i \in I} S_i \right| \geq |I|.$$

Hence by the marriage theorem, G has a matching M saturating $\{S_1, S_2, \ldots, S_n\}$. Then we can obtain a transversal from M by taking the set of vertices of X saturated by M. $\square$

Fig. 2.13. The bipartite graph corresponding to $\{S_1 = \{a, b, e\}, S_2 = \{b\}, S_3 = \{a, c, d\}, S_4 = \{b, c\}\}$, its matching saturating $\{S_1, S_2, S_3, S_4\}$, and a transversal $\{a, b, c, d\}$ obtained from it.

Let X be a finite set and $\mathcal{F} = \{S_1, S_2, \ldots, S_n\}$ and $\mathcal{H} = \{T_1, T_2, \ldots, T_n\}$ be two families of subsets of X. Then we say that $\mathcal{F}$ and $\mathcal{H}$ have a **common transversal** if there exists a set of n distinct elements of X that is a transversal of both $\mathcal{F}$ and $\mathcal{H}$, i.e., there exists a set $\{x_1, x_2, \ldots, x_n\}$ of n distinct elements of X such that $x_i \in S_i \cap T_{(i)}$ for all $1 \leq i \leq n$, where $\{T_{(1)}, T_{(2)}, \ldots, T_{(n)}\}$ is a rearrangement of $\mathcal{H} = \{T_1, T_2, \ldots, T_n\}$.

For example, $\{S_1' = S_2' = \{a, b\}, S_3' = \{a, c, e\}\}$ and $\{T_1' = \{b, c\}, T_2' = \{e\}, T_3' = \{a, b, e\}\}$ have a common transversal $\{b, a, e\}$, where $b \in S_1' \cap T_1'$, $a \in S_2' \cap T_3'$ and $e \in S_3' \cap T_2'$. On the other hand, $\{S_1 = S_2 = \{a, b\}, S_3 = \{a, c, e\}\}$ and $\{T_1 = \{b, c\}, T_2 = \{b, d\}, T_3 = \{c, d, e\}\}$ have transversals but do not have a common transversal. Let $n = 3$, $I = \{1, 2\}$ and $J = \{1, 2, 3\}$, and substitute these into the left and right sides of (2.7) given in the following theorem. Then

$$|(S_1 \cup S_2) \cap (T_1 \cup T_2 \cup T_3)| = |\{b\}| = 1 < |I| + |J| - n = 2 + 3 - 3 = 2.$$

Hence inequality (2.7) does not hold.

Theorem 2.18. *Let $\mathcal{F}$ and $\mathcal{H}$ be two families of subsets of a set X. Then $\mathcal{F}$ and $\mathcal{H}$ have a common transversal if and only if for all subsets I and J of $\{1, 2, \ldots, n\}$, it follows that*

$$\left|\left(\bigcup_{i\in I} S_i\right) \cap \left(\bigcup_{j\in J} T_j\right)\right| \geq |I| + |J| - n. \tag{2.7}$$

Proof. Let $\mathcal{F} = \{S_1, S_2, \ldots, S_n\}$ and $\mathcal{H} = \{T_1, T_2, \ldots, T_n\}$. Assume that $\mathcal{F}$ and $\mathcal{H}$ have a common transversal $\{x_1, x_2, \cdots, x_n\}$. Let I and J be subsets of $\{1, 2, \ldots, n\}$. We may assume that $|I| + |J| > n$ since otherwise (2.7) trivially holds. Then two sets $\{x_r \mid x_r \in S_i, \ i \in I\}$ and $\{x_r \mid x_r \in T_j, \ j \in J\}$ must have at least $|I| + |J| - n$ elements in common, and these elements are contained in

$$\left(\bigcup_{i\in I} S_i\right) \cap \left(\bigcup_{j\in J} T_j\right).$$

Hence (2.7) holds.

In order to prove sufficiency, we construct a bipartite graph G with bipartite sets

$$A = \{S_1, S_2, \ldots, S_n\} \cup X \qquad \text{and} \qquad B = \{T_1, T_2, \ldots, T_n\} \cup X'$$

as follows, where $X' = \{x' \mid x \in X\}$. Two vertices $S_i \in A$ and $x' \in X'$ are joined by an edge if $x \in S_i$. Similarly two vertices $x \in X$ and $T_j \in B$ are joined by an edge if $x \in T_j$. Moreover, $x \in X$ and $x' \in X'$ are joined by an edge, and G has no more edges.

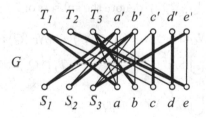

Fig. 2.14. The bipartite graph corresponding to two families of subsets.

For example, the bipartite graph G given in Fig. 2.14 corresponds to $X = \{a, b, c, d, e\}$, $\mathcal{F} = \{S_1 = S_2 = \{a, b\}, S_3 = \{b, c, e\}\}$ and $\mathcal{H} = \{T_1 = \{a, c\}, T_2 = \{d, e\}, T_3 = \{a, b, e\}\}$. G has a perfect matching, and these two families have a common transversal $\{a, b, e\}$ such that $a \in S_1 \cap T_1$, $b \in S_2 \cap T_3$ and $e \in S_3 \cap T_2$.

We first show that if the bipartite graph G has a perfect matching M, then $\mathcal{F}$ and $\mathcal{H}$ have a common transversal. If M contains an edge $S_i a_k'$ $(a_k' \in X')$, then M must contain an edge $a_k T_j$ $(a_k \in X)$ for some T_j, and thus $a_k \in S_i \cap T_j$. Hence it is easy to see that $\{a_k \mid S_i a'_k \in M\}$ forms a common transversal of $\mathcal{F}$ and $\mathcal{H}$. Note that if M contains an edge joining $a_t \in X$ to $a'_t \in X'$, then this fact implies that a_t is not chosen to be an element of the common transversal.

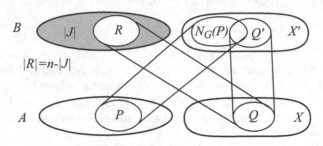

Fig. 2.15. Illustration of Theorem 2.18; $P = \{X_i \mid i \in I\}$, $J = \{j \mid T_j \cap Q = \emptyset\}$, and $R = \{T_j \mid T_j \cap Q \neq \emptyset\}$.

Let
$$P \subseteq \{S_1, S_2, \ldots, S_n\} \quad \text{and} \quad Q \subseteq X.$$

Then $P \cup X \subseteq A$. Put $I = \{i : S_i \in P\}$, $J = \{j : T_j \cap Q = \emptyset\}$ and $Q' = \{x' : x \in Q\} \subseteq X'$. Hence $Q' \subset B$ and

$$\left(\bigcup_{i \in I} S_i\right)' \setminus Q' = \left(\bigcup_{i \in I} S_i\right)' \cap (X' - Q') \supseteq \left(\bigcup_{i \in I} S_i\right)' \cap \left(\bigcup_{j \in J} T_j\right)',$$

where $(\bigcup_{i \in I} S_i)'$ and $(\bigcup_{j \in J} T_j)'$ denote the subsets of X' (Fig. 2.15). Therefore

$$
\begin{aligned}
|N_G(P \cup Q)| &= |(N_G(P) \cup N_G(Q)) \cap X'| + |N_G(Q) \cap \{T_1, T_2, \ldots, T_n\}| \\
&= \left|\left(\bigcup_{i \in I} S_i\right)' \cup Q'\right| + \#\{T_j : T_j \cap Q \neq \emptyset\} \\
&\geq \left|\left(\bigcup_{i \in I} S_i\right)' \setminus Q'\right| + |Q'| + n - |J| \\
&\geq \left|\left(\bigcup_{i \in I} S_i\right)' \cap \left(\bigcup_{j \in J} T_j\right)'\right| + |Q'| + n - |J| \\
&\geq |I| + |J| - n + |Q'| + n - |J| \qquad \text{(by (2.7))} \\
&= |I| + |Q'| = |P| + |Q| = |P \cup Q|.
\end{aligned}
$$

Consequently, G has a matching M saturating A by the marriage theorem. Since $|A| = |B|$, M is a perfect matching, and thus the proof is complete. $\square$

2.3 Augmenting paths and algorithms

In this section we consider matchings in graphs by using alternating paths instead of neighborhoods, which is a new approach to matchings and useful for algorithms. For two sets X and Y of edges of a graph G, we define

$$X \bigtriangleup Y = (X \cup Y) - (X \cap Y).$$

Lemma 2.19. *Let M_1 and M_2 be matchings in a simple graph. Then each component of $\langle M_1 \bigtriangleup M_2 \rangle$ is either (i) a path whose edges are alternately in M_1 and M_2, or (ii) an even cycle whose edges are alternately in M_1 and M_2 (Fig. 2.16).*

Proof. Let $H = \langle M_1 \bigtriangleup M_2 \rangle$ and v a vertex of H. Then $1 \le \deg_H(v) \le 2$, and $\deg_H(v) = 2$ implies that exactly one edge of M_1 and one edge of M_2 are incident with v. Hence each component C of H is a path or cycle, whose edges are alternately in M_1 and M_2. In particular, if C is a cycle, it must be an even cycle (see Fig. 2.16). $\square$

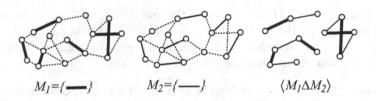

$$M_1 = \{ \text{———} \} \qquad M_2 = \{ \text{———} \} \qquad \langle M_1 \bigtriangleup M_2 \rangle$$

Fig. 2.16. Matchings M_1 and M_2; and $\langle M_1 \bigtriangleup M_2 \rangle$.

Let M be a matching in a simple graph G. Then a path of G is called an M-**alternating path** if its edges are alternately in M and not in M. If both end-vertices of an M-alternating path are unsaturated by M, then such an M-alternating path is called an M-**augmenting path**.

For example, in the graph G shown in Fig. 2.17, $M = \{a, b, c\}$ is a matching, (d, b, f, a) and (g, c, e) are M-alternating paths, and $P = (d, b, e, c, g)$ is an M-augmenting path, whose end-vertices u and v are not saturated by M. Furthermore, it is immediate that $M \bigtriangleup E(P) = \{a, d, e, g\}$ is a matching that contains $\|M\| + 1$ edges and is larger than M. The next theorem states a characterization of maximum matchings by using augmenting path.

Theorem 2.20 (Berge [25]). *A matching M in a simple graph G is maximum if and only if there exists no M-augmenting path in G.*

Proof. The contraposition of the statement, which we shall prove, is the following. A matching M is not maximum if and only if there exists an M-augmenting path in G.

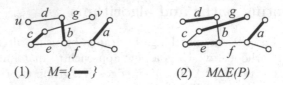

$$(1) \quad M=\{\text{━}\} \qquad\qquad (2) \quad M\triangle E(P)$$

Fig. 2.17. (1) A matching $M = \{a,b,c\}$ and an M-augmenting path $P = (d,b,$ $e,c,g)$; (2) $M \triangle E(P)$.

Suppose that there is an M-augmenting path P in G. Then $M \triangle E(P)$ is a matching which contains $\|M\| + 1$ edges, and so M is not a maximum matching.

We next assume that M is not a maximum matching in G. Let M' be a maximum matching in G. Then by Lemma 2.19, $\langle M \triangle M' \rangle$ contains a path P in which the number of edges in M' is greater than the number of edges in M. Hence the two pendant edges of P are contained in M', which implies that P is an M-augmenting path of G. Therefore the contraposition is proved. $\quad\square$

By making use of M-alternating paths and M-augmenting paths, we can obtain some properties of matchings in a graph and also an algorithm for finding a maximum matching in a bipartite graph. We begin with one basic theorem and one result on a game played on a graph.

Theorem 2.21. *For any matching M in a simple graph G, G has a maximum matching that saturates all vertices saturated by M.*

Proof. Suppose M is not a maximum matching. By Theorem 2.20, G has an M-augmenting path P. Then $M \triangle E(P)$ is a matching which contains $\|M\|+1$ edges and saturates all the vertices saturated by M. Since a maximum matching can be obtained by repeating this procedure, we can find a maximum matching that saturates all the vertices saturated by M. $\quad\square$

Proposition 2.22 (Exercise 5.1.4 of [34]). *Two players play a game on a connected simple graph G by alternately selecting distinct vertices $v_1, v_2, v_3, \ldots$ so that $(v_1 v_2 v_3 \cdots)$ forms a path. The player who cannot select a vertex loses, that is, the player who selects the last vertex wins (Fig. 2.18). The second player has a winning strategy if G has a perfect matching; otherwise the first player has a winning strategy.*

Proof. Assume that G has a perfect matching M. If the first player selects a vertex x in his turn, then the second player selects a vertex that is joined to x by an edge of M. Since M is a perfect matching, the second player can always select such a vertex. Therefore, he wins since the game is over in a finite number of moves.

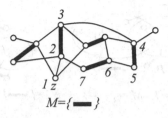

Fig. 2.18. A graph for which the first player can win, and its maximum matching M.

Next suppose that G has no perfect matching. Let M be a maximum matching and z be a vertex unsaturated by M. The first player selects z in the first turn, and if the second player selects a vertex y in his turn, then the first player selects a vertex that is joined to y by an edge of M. Then the vertices selected by the two players form an M-alternating path P starting at z. Since G has no M-augmenting path by Theorem 2.20, the path P does not pass through any other vertices unsaturated by M, which implies that the first player can always select a vertex. Therefore the first player wins. □

If a matching M in a graph G saturates no vertex in a subset $X \subset V(G)$, then we say that M **avoids** X.

Theorem 2.23 (Edmonds and Fulkerson [63]). *Let A and B be vertex subsets of a simple graph G such that $|A| < |B|$. Then*
(i) If there exist two matchings, one saturating A and the other saturating B, then there exists a matching that saturates A and at least one vertex in $B \setminus A$.
(ii) If there exist two maximum matchings, one avoiding A and the other avoiding B, then there exists a maximum matching that avoids A and at least one vertex in $B \setminus A$.

Proof. We first prove (i). Let M_A and M_B be matchings in G which saturate A and B respectively. If M_A saturates one vertex in $B \setminus A$, then M_A itself is the desired matching. Hence we may assume that M_A avoids $B \setminus A$.

Consider $\langle M_A \triangle M_B \rangle$. For every vertex b in $B \setminus A$, there exists a path in $\langle M_A \triangle M_B \rangle$ starting at b, which may end at a vertex in $A \setminus B$. Since $|A| < |B|$ and no path of $\langle M_A \triangle M_B \rangle$ ends at a vertex in $A \cap B$, there exists a path P in $\langle M_A \triangle M_B \rangle$ that starts at $b_1 \in B \setminus A$ and ends at $x \notin A$ (Fig. 2.19). Since P is an M_A-alternating path connecting b_1 and x, $M_A \triangle E(P)$ is the desired matching in G, which saturates A and $b_1 \in B \setminus A$.

We next prove (ii). Let N_A and N_B be maximum matchings in G which avoid A and B respectively. We may assume that N_A saturates $B \setminus A$ since otherwise N_A itself is the desired matching. Then by the same argument as above, there exists an N_A-alternating path P in $\langle N_A \triangle N_B \rangle$ starting at a vertex $b \in B \setminus A$ and ending at a vertex not contained in A. This implies that P does not pass through A. If the both pendant edges of P are contained in N_A, then

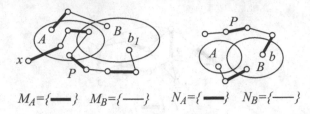

$$M_A=\{\text{—}\} \quad M_B=\{\text{—}\} \qquad N_A=\{\text{—}\} \quad N_B=\{\text{—}\}$$

Fig. 2.19. M_A-alternating path P and N_A-alternating path P.

$N_B \triangle E(P)$ is a matching with $\|N_B\| + 1$ edges, contrary to the maximality of N_B. Thus one pendant edge of P belongs to N_B, and thus $N_A \triangle E(P)$ is the desired maximum matching in G, which avoids A and $b \in B \setminus A$. □

Theorem 2.24 (Mendelsohn and Dulmage [187]). *Let G be a bipartite simple graph with bipartition (A, B), and $X \subseteq A$ and $Y \subseteq B$. If X and Y are saturated by matchings in G, respectively, then G has a matching that saturates $X \cup Y$.*

Proof. Let M_X and M_Y be matchings in G saturating X and Y, respectively. We may assume that M_X does not saturate Y since otherwise M_X is the desired matching.

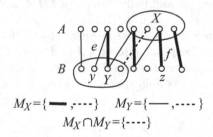

$$M_X=\{\text{—},\text{----}\} \quad M_Y=\{\text{—},\text{----}\}$$
$$M_X \cap M_Y=\{\text{----}\}$$

Fig. 2.20. A path $P = (y, e, \ldots, f, z)$ in $H = \langle M_X \triangle M_Y \rangle$.

Let $H = \langle M_X \triangle M_Y \rangle$, and choose a vertex $y \in Y$ unsaturated by M_X. Then there exists a path P in H starting at y, which can be expressed as

$$P = (y, e, \ldots, f, z), \quad y, \ldots, z \in V(H), \quad e, \cdots, f \in E(H),$$

where z is the other end-vertex of P and $e \in M_Y$. It is obvious that the vertices of P are alternately in B and A, and the edges of P are alternately in M_Y and M_X. If $z \in X$, then the pendant edge f of P must belong to M_Y, which contradicts the fact that X is saturated by M_X (Fig. 2.20). Hence

$z \notin X$. Similarly, if $z \in Y$, then $f \in M_X$, which contradicts the fact that Y is saturated by M_Y. Therefore $z \notin X \cup Y$.

Thus $M_X \bigtriangleup E(P)$ is a matching that saturates $V(\langle M_X \rangle)$ and $y \in Y \setminus V(M_X)$. By repeating this procedure, we can obtain the desired matching in G, which saturates $X \cup Y$. □

We now give an algorithm for finding a maximum matching in a bipartite graph, which is often called the Hungarian Method.

Algorithm 2.25 (Hungarian Method) *Let G be a bipartite simple graph with bipartition (A, B). Then a maximum matching in G can be obtained by the following procedure. Let M be any matching in G, and $A_0 = A \setminus V(\langle M \rangle)$ be the set of vertices in A unsaturated by M. Let $B_1 = N_G(A_0) \subseteq B$ and define*

$$A_i = N_M(B_i), \quad B_{i+1} = N_G(A_i) \setminus (B_1 \cup B_2 \cup \cdots \cup B_i)$$

for every $1 \le i \le k$ until B_{k+1} contains an M-unsaturated vertex or $B_{k+1} = \emptyset$. If B_{k+1} contains an M-unsaturated vertex, say b_{k+1}, then we find an M-augmenting path P joining $b_{k+1} \in B_{k+1}$ to $a \in A_0$, and obtain a larger matching $M \bigtriangleup E(P)$ containing $\|M\| + 1$ edges. We apply the above procedure to $M \bigtriangleup E(P)$. If $B_{k+1} = \emptyset$, then M is the desired maximum matching.

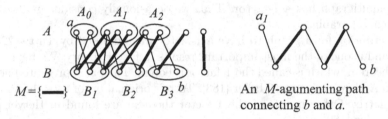

$M = \{ \rule{0.6cm}{0.4pt} \}$ B_1 B_2 B_3 b An M-agumenting path connecting b and a.

Fig. 2.21. An Algorithm for finding a maximum matching in a bipartite graph; and an improved algorithm.

Proof. It is easy to see that if B_{k+1} contains an M-unsaturated vertex, then we can find an M-augmenting path P, resulting in a larger matching $M \bigtriangleup E(P)$ (Fig. 2.21). Hence it suffices to show that if $B_{k+1} = \emptyset$, then M is a maximum matching.

Suppose $B_{k+1} = \emptyset$. Let $S = A_0 \cup A_1 \cup \cdots \cup A_k$ and $T = B_1 \cup B_2 \cup \cdots \cup B_k$. Then $N_G(S) = T$ and $N_M(T) = S - A_0$, and so $|T| = |N_M(T)| = |S - A_0| = |S| - |A_0|$. Hence

$$\|M\| = |A| - |A_0| = |A| - (|S| - |T|)$$
$$= |A| - (|S| - |N_G(S)|)$$
$$\ge |A| - \max_{X \subseteq A} \{|X| - |N_G(X)|\}.$$

By Theorem 2.11, this implies that M is a maximum matching in G. $\square$

Of course, we can apply the procedure given in Algorithm 2.25 for each vertex $a \in A_0$ individually, that is, put $B_1(a) = N_G(a) \subseteq B$ and obtain

$$A_i(a) = N_M(B_i(a)), \quad B_{i+1}(a) = N_G(A_i(a)) \setminus (B_1(a) \cup \cdots \cup B_i(a))$$

for every $1 \leq i \leq k$ until $B_{k+1}(a)$ contains an M-unsaturated vertex or $B_{k+1}(a) = \emptyset$. If $B_{k+1}(a)$ contains an M-unsaturated vertex, then we can get a larger matching than M. If $B_{k+1}(a) = \emptyset$, then we remove $S = \{a\} \cup A_1(a) \cup \cdots \cup A_k(a)$ and $T = B_1(a) \cup \cdots \cup B_k(a)$ from G, obtain $G - (S \cup T)$, and then apply the same procedure in $G - (S \cup T)$ for another vertex $a' \in A - \{a\}$.

2.4 1-factor theorems

In this section we investigate matchings and 1-factors of graphs. Since a 1-factor contains neither loops nor multiple edges, we can restrict ourselves to simple graphs when we consider 1-factors. So in this section, we mainly consider simple graphs. However, some results hold for multigraphs or general graphs, and these generalization are useful and interesting. For example, every 2-connected cubic simple graph has a 1-factor, but also every 2-connected cubic multigraph has a 1-factor. Thus we occasionally consider multigraphs and general graphs.

A criterion for a graph to have a 1-factor was obtained by Tutte [225] in 1947 and is one of the most important results in factor theory. We begin with this theorem, which is called the 1-factor theorem. The proof presented here is due to Anderson [16] and Mader [183]. Tutte's original proof uses the Pfaffian of a matrix. Other proofs of the 1-factor theorem are found in Hetyei [102] and Lovász [180].

After the 1-factor theorem, we discuss some other criteria for graphs to have 1-factors. For example, a criterion for a graph to have a 1-factor containing any given edge and a criterion for a tree to have a 1-factor, which is much simpler than the criterion of the 1-factor theorem, are shown.

For a vertex subset X of G and a component C of $G - X$, we simplify notation by denoting

$$E_G(C, X) = E_G(V(C), X) \quad \text{and} \quad e_G(C, X) = e_G(V(C), X).$$

A component of a graph is said to be **odd** or **even** according to whether its order is odd or even. For a graph G, $Odd(G)$ denotes the set of odd components of G, and $odd(G)$ denotes the number of odd components of G, that is,

$$odd(G) = |Odd(G)| = \text{the number of odd components of } G.$$

Lemma 2.26. *Let G be a general graph and $S \subseteq V(G)$. Then*

$$odd(G - S) + |S| \equiv odd(G - S) - |S| \equiv |G| \quad (\text{mod } 2). \qquad (2.8)$$

In particular, if G is of even order, then

$$odd(G - S) \equiv |S| \quad (\text{mod } 2), \qquad (2.9)$$

and $odd(G - v) \geq 1$ for every vertex v.

Proof. Let $C_1, C_2, \ldots, C_m$ be the odd components of $G - S$, and D_1, D_2, $\ldots, D_r$ the even components of $G - S$, where $m = odd(G - S)$. Then

$$|G| = |S| + |C_1| + \cdots + |C_m| + |D_1| + \cdots + |D_r| \equiv |S| + m \quad (\text{mod } 2).$$

Hence $odd(G - S) + |S| \equiv |G| \pmod 2$. It is obvious that $odd(G - S) + |S| \equiv odd(G - S) - |S| \pmod 2$. Therefore (2.8) holds. (2.9) is an immediate consequence of (2.8). $\square$

Before giving the 1-factor theorem, we make the following remark. A matching in a general graph contains no loops, and so does not contain a 1-factor. Then a 1-factor of a general graph is a spanning subgraph with all vertices degree one. Thus it is obvious that a general graph G has a 1-factor if and only if its underlying graph has a 1-factor, where the **underlying graph** of G is a simple graph obtained from G by removing all the loops and by replacing the multiple edges joining two vertices by a single edge joining them. Therefore, essential part of the following 1-factor theorem is that the theorem holds for simple graphs.

Theorem 2.27 (The 1-Factor Theorem, Tutte [225]**).** *A general graph G has a 1-factor if and only if*

$$odd(G - S) \leq |S| \qquad \text{for all} \quad S \subset V(G). \qquad (2.10)$$

Proof. As we remarked above, we may assume that the given general graph G is a simple graph. Assume that G has a 1-factor F. Let $\emptyset \neq S \subseteq V(G)$, and $C_1, C_2, \ldots, C_m$ be the odd components of $G - S$, where $m = odd(G - S)$. Then for every odd component C_i of $G - S$, there exists at least one edge in F that joins C_i to S (Fig. 2.22). Hence

$$odd(G - S) = m \leq e_F(C_1 \cup C_2 \cup \cdots \cup C_m, S) \leq |S|.$$

We now prove the sufficiency by induction on $|G|$. By setting $S = \emptyset$, we have $odd(G) = 0$, which implies that each component of G is even. If G is not connected, then every component of G satisfies (2.10), and so it has a 1-factor by the inductive hypothesis. Hence G itself has a 1-factor. Therefore, we may assume that G is connected and has even order.

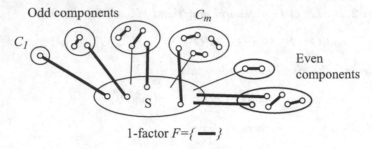

Fig. 2.22. For every odd component C_i of $G - S$, at least one edge of F joins C_i to S.

It follows from Lemma 2.26 and (2.10) that $odd(G - \{v\}) = |\{v\}| = 1$. Let S be a maximal subset of $V(G)$ with the property that $odd(G - S) = |S|$. Then $\emptyset \neq S \subset V(G)$ and

$$odd(G - X) < |X| \qquad \text{for all} \quad S \subset X \subseteq V(G). \qquad (2.11)$$

Claim 1. *Every component of $G - S$ is of odd order.*

Suppose that $G - S$ has an even component D. Let v be any vertex of D. Then by Lemma 2.26, $D - v$ has at least one odd component, and so

$$odd(G - (S \cup \{v\})) = odd(G - S) + odd(D - v) \geq |S| + 1,$$

which implies $odd(G - (S \cup \{v\})) = |S \cup \{v\}|$ by (2.10). This contradicts the maximality of S. Hence Claim 1 holds.

Claim 2. *For any vertex v of each odd component C of $G - S$, $C - v$ has a 1-factor.*

Let C be an odd component of $G - S$, and v any vertex of C. Then for every subset $X \subseteq V(C - v)$, we have by (2.11) that

$$
\begin{aligned}
|S| + 1 + |X| &> odd(G - (S \cup \{v\} \cup X)) \\
&= odd(G - S) - 1 + odd(C - (\{v\} \cup X)) \\
&= |S| - 1 + odd((C - v) - X).
\end{aligned}
$$

Thus $odd((C - v) - X) < |X| + 2$, which implies $odd((C - v) - X) \leq |X|$ by (2.9). Hence $C - v$ has a 1-factor by the induction hypothesis, and thus Claim 2 is proved.

Let $C_1, C_2, \ldots, C_m$ be the odd components of $G - S$, where $m = odd(G - S) = |S|$. We construct a bipartite graph B with partite sets S and $\{C_1, C_2, \ldots, C_m\}$ as follows: a vertex x of S and C_i are joined by an edge of B if and only if x and C_i are joined by an edge of G (Fig. 2.23).

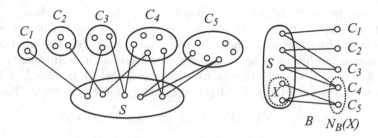

Fig. 2.23. $G - S$ and the bipartite graph B.

Claim 3. *The bipartite graph B has a 1-factor.*

It follows that $|N_B(S)| = |\{C_1, C_2, \ldots, C_m\}| = |S|$ since G is connected. Assume that $|N_B(X)| < |X|$ for some $\emptyset \neq X \subset S$. Then every vertex $C_i \in \{C_1, C_2, \ldots, C_m\} - N_B(X)$ is an isolated vertex of $B - (S - X)$, which implies C_i is an odd component of $G - (S - X)$, and thus we obtain

$$odd(G - (S - X)) \geq |\{C_1, C_2, \ldots, C_m\} - N_B(X)| > m - |X| = |S - X|.$$

This contradicts (2.10). Therefore $|N_B(X)| \geq |X|$ for all $X \subseteq S$, and so by the marriage theorem, B has a matching K saturating S. Since $m = |S|$, K must saturate $\{C_1, C_2, \ldots, C_m\}$, and thus K is a 1-factor of B.

For every edge $x_i C_i$ of K, choose a vertex $v_i \in V(C_i)$ that is adjacent to x_i in G, and take a 1-factor $F(C_i)$ of $C_i - v_i$, whose existence is guaranteed by Claim 2. Therefore, we obtain the following desired 1-factor of G:

$$\Big(F(C_1) \cup \cdots \cup F(C_m) \Big)$$
$$\cup \{x_i v_i : x_i C_i \in K,\ x_i \in S,\ v_i \in V(C_i),\ 1 \leq i \leq m\}.$$

Consequently the theorem is proved. □

A graph G is said to be **factor-critical** if $G - v$ has a 1-factor for every vertex v of G (Fig. 2.24). It is easy to see that if G is factor-critical, then G is of odd order, connected and not a bipartite graph (Problem 2.8).

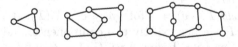

Fig. 2.24. Factor-critical graphs.

Theorem 2.28 (Edmonds [62]). *A simple graph G of even order has a 1-factor if and only if for every subset S of $V(G)$, the number of factor-critical components of $G - S$ is less than or equal to $|S|$.*

Proof. The necessity follows immediately from the 1-factor theorem since every factor-critical component of $G - S$ is an odd component of $G - S$.

We now prove the sufficiency. Suppose that G satisfies the condition in the theorem, but has no 1-factor. Then by the 1-factor theorem, there exists a subset $\emptyset \neq X \subseteq V(G)$ such that $odd(G - X) > |X|$, which implies $X \neq V(G)$. Take a maximal subset $\emptyset \neq S \subset V(G)$ such that $odd(G - S) > |S|$. Then

$$odd(G - Y) \leq |Y| \qquad \text{for all} \quad S \subset Y \subseteq V(G). \tag{2.12}$$

We first show that $G - S$ has no even component, since otherwise for a vertex u of an even component of $G - S$, we have by Lemma 2.26 that

$$odd(G - (S \cup \{u\})) \geq odd(G - S) + 1 > |S| + 1 = |S \cup \{u\}|,$$

contrary to (2.12). We shall next show that every odd component of $G - S$ is factor-critical. Let C be an odd component of $G - S$, and v any vertex of C. Then for every $X \subseteq V(C - v)$, (2.12) implies

$$\begin{aligned}
|S| + 1 + |X| &\geq odd(G - (S \cup \{v\} \cup X)) \\
&= odd(G - S) - 1 + odd(C - (\{v\} \cup X)) \\
&> |S| - 1 + odd((C - v) - X).
\end{aligned}$$

Hence $|X| + 2 > odd((C - v) - X)$, which implies $odd((C - v) - X) \leq |X|$ by (2.9). Therefore $C - v$ has a 1-factor by the 1-factor theorem, and so C is factor-critical. Consequently,

the number of factor-critical components of $G - S = odd(G - S) > |S|$.

This contradicts the assumption, and thus the theorem is proved.   □

The next theorem gives a necessary and sufficient condition for a tree to have a 1-factor, and the proof presented here is due to Amahashi [14].

Theorem 2.29 (Chungphaisan). *A tree T of even order has a 1-factor if and only if $odd(T - v) = 1$ for every vertex v of T.*

Proof. Suppose that T has a 1-factor F. Then for every vertex v of T, let w be the vertex of T joined to v by an edge of F. It follows that the component of $T - v$ containing w is odd, and all the other components of $T - v$ are even (Fig. 2.25). Hence $odd(T - v) = 1$.

Suppose that $odd(T - v) = 1$ for every $v \in V(T)$. It is obvious that for each edge e of T, $T - e$ has exactly two components, and both of them are simultaneously odd or even. Define a set F of edges of T as follows:

$$F = \{e \in E(T) : odd(T - e) = 2\}.$$

For every vertex v of T, there exists exactly one edge e that is incident with v and satisfies $odd(T - e) = 2$ since $T - v$ has exactly one odd component, where e is the edge joining v to this odd component (Fig. 2.25). Therefore e is an edge of F, and thus F is a 1-factor of G.   □

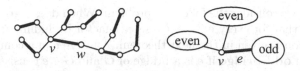

Fig. 2.25. A tree having a 1-factor, and a tree T satisfying $odd(T - v) = 1$.

We now give a variety of requirements for a graphs to have a 1-factor. A graph G is said to be **1-extendable** if for every edge e, G has a 1-factor containing e. More generally, for an integer $n \geq 1$, a graph G is said to be **n-extendable** if every matching of size n in G can be extended to a 1-factor of G.

Theorem 2.30 (Little, Grant and Holton, [168]). *A simple graph G is 1-extendable if and only if*

$$odd(G - S) \leq |S| - 2 \qquad for\ all \quad S \subset V(G) \qquad (2.13)$$

such that $\langle S \rangle_G$ contains an edge.

Proof. We first prove the necessity. Let $S \subseteq V(G)$ such that $\langle S \rangle_G$ contains an edge, say $e = xy$ $(x, y \in S)$. Since G has a 1-factor F containing e, for each odd component C of $G - S$, there exists an edge in F joining C to a vertex in $S - \{x, y\}$, and hence

$$odd(G - S) \leq e_F(G - S, S - \{x, y\}) \leq |S| - 2.$$

We next prove the sufficiency. Let $e = xy$ $(x, y \in V(G))$ be any edge of G. We shall show that $G - \{x, y\}$ has a 1-factor, which obviously implies that G has a 1-factor containing e.

Let $S \subseteq V(G) - \{x, y\}$. Since $\langle S \cup \{x, y\} \rangle_G$ contains an edge e, we have

$$odd((G - \{x, y\}) - S) - odd(G - (S \cup \{x, y\})) \leq |S \cup \{x, y\}| - 2 = |S|.$$

Therefore $G - \{x, y\}$ has a 1-factor by the 1-factor theorem. $\square$

A criterion for a graph to be n-extendable is given in the following theorem. Since this theorem can be proved in a similar fashion to the proof of the above theorem, we omit the proof (Problem 2.10).

Theorem 2.31 (Chen [47]). *Let $n \geq 1$ be an integer, and G be a simple graph. Then G is n-extendable if and only if*

$$odd(G - S) \leq |S| - 2n \qquad for\ all \quad S \subset V(G)\ with\ \alpha'(\langle S \rangle_G) \geq n, \qquad (2.14)$$

where $\alpha'(\langle S \rangle_G)$ denotes the size of a maximum matching in $\langle S \rangle_G$.

Similarly, we consider the following problem. When does a graph G possess the property that for every edge e, G has a 1-factor excluding e? The answer to this question is given in the next theorem. An edge e of a connected graph G is called an **odd-bridge** if e is a bridge of G and $G - e$ consists of two odd components. In particular, such a graph G has even order. The next theorem was obtained by C. Chen.

Theorem 2.32 (Chen [45],[47]). *Let G be a connected simple graph. Then for every edge e of G, G has a 1-factor excluding e if and only if*

$$odd(G - S) \le |S| - \epsilon_2 \qquad \text{for all } \ S \subset V(G), \qquad (2.15)$$

where $\epsilon_2 = 2$ if $G - S$ has a component containing an odd-bridge; otherwise $\epsilon_2 = 0$.

Proof. We first prove the necessity. Let $S \subseteq V(G)$. Then $odd(G - S) \le |S|$ by the 1-factor theorem. Assume that $G - S$ has a component D containing an odd-bridge e. Consider a 1-factor F of G excluding e. Then for each odd component C of $G - S$, F contains at least one edge joining C to S. Furthermore, for each odd component C' of $D - e$, at least one edge of F joins C' to S. Hence

$$odd(G - S) + 2 \le e_F(G - S, S) \le |S|.$$

Consequently, (2.15) holds.

Conversely, assume that G satisfies (2.15). We shall show that for any edge e of G, $G - e$ has a 1-factor, which is of course a 1-factor of G excluding e. Let $S \subseteq V(G - e) = V(G)$. If e is an odd-bridge of an even component of $G - S$, then $odd(G - e - S) = odd(G - S) + 2$; otherwise $odd(G - e - S) = odd(G - S)$. For example, if e is a bridge of an odd component C of $G - S$, then $C - e$ has exactly one odd component, and so $odd(G - S - e) = odd(G - S)$. Therefore

$$odd((G - e) - S) = odd(G - S - e) = odd(G - S) + \epsilon_2 \le |S|.$$

Hence $G - e$ has a 1-factor. $\square$

Theorem 2.33 (Corollary 1.6 of [32]). *Let G be a simple graph and W a vertex set of G. Then G has a matching that saturates W if and only if*

$$odd(G - S|W) \le |S| \qquad \text{for all } \ S \subseteq V(G), \qquad (2.16)$$

where $odd(G - S|W)$ denotes the number of those odd components of $G - S$ whose vertices are all contained in W (Fig. 2.26).

Proof. Assume that G has a matching M which saturates W. Then for every odd component C of $G - S$ such that $V(C) \subseteq W$, at least one edge of M joins C to S. Thus $odd(G - S|W) \le e_M(V(G) - S, S) \le |S|$.

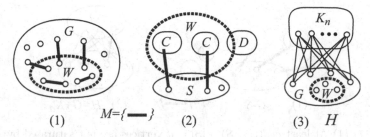

Fig. 2.26. (1) A matching M saturating W; (2) Each odd component C of $G - S$ is counted in $odd(G - S|W)$ but is not an odd component D of $G - S$; and (3) the graph H.

We next prove the sufficiency. By the 1-factor theorem, we may assume that W is a proper subset of $V(G)$, and so $V(G) - W \neq \emptyset$. Let $n = |G|$. We construct a new graph H from G by adding the complete graph K_n and by joining every vertex in $V(G) - W$ to every vertex of K_n (Fig. 2.26). Then H has even order, and it is easy to see that G has a matching saturating W if and only if H has a 1-factor.

Let $X \subseteq V(H)$. If $V(K_n) \subseteq X$, then $odd(H - X) \leq |G| \leq |X|$. If $V(K_n) \not\subseteq X$, then since $V(H) - (X \cup W)$ is contained in a component of $H - X$, we have by (2.16) that

$$odd(H - X) \leq odd(G - V(G) \cap X|W) + 1 \leq |V(G) \cap X| + 1 \leq |X| + 1,$$

which implies $odd(H - X) \leq |X|$ by (2.9). Therefore H has a 1-factor, and thus G has the desired matching saturating W. □

For a graph G,

$$\operatorname{def}(G) = \max_{X \subset V(G)} \{odd(G - X) - |X|\}$$

is called the **deficiency** of G. Note that the deficiency is non-negative as $odd(G - \emptyset) - |\emptyset| \geq 0$. This concept is introduced in the next theorem.

Theorem 2.34 (Berge [26]). *Let M be a maximum matching in a simple graph G. Then the number $|M|$ of vertices saturated by M is given by*

$$|M| = |G| - \max_{X \subset V(G)} \{odd(G - X) - |X|\}. \tag{2.17}$$

Proof. The proof given here is due to Bollobás [32]. Let d be the deficiency of G and S a subset of $V(G)$ such that $odd(G - S) - |S| = d$. Then $odd(G - X) - |X| \leq d$ for every $X \subset V(G)$.

We first show $|M| \leq |G| - d$. Let M be a maximum matching of G. Then for any odd component C of $G - S$, if $V(C)$ is saturated by M, then at least

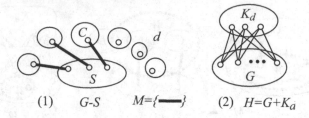

Fig. 2.27. (1) At least $odd(G - S) - |S| = d$ vertices are not saturated by M; (2) The graph $H = G + K_d$.

one edge of M joins C to S (Fig. 2.27). Thus at least $odd(G-S)-|S| = d$ odd components of $G - S$ are not saturated by M, which implies $|M| \leq |G| - d$.

In order to prove the inverse inequality, we construct the join $H = G + K_d$, where K_d is the complete graph of order d. Then for every $\emptyset \neq Y \subseteq V(H)$, if $V(K_d) \not\subseteq Y$, then $odd(H - Y) \leq 1 \leq |Y|$; and if $V(K_d) \subseteq Y$, then

$$odd(H - Y) = odd(G - V(G) \cap Y) \leq |V(G) \cap Y| + d = |Y|.$$

Hence H has a 1-factor F by the 1-factor theorem. Then $F \cap E(G) = F - V(K_d)$ is a matching in G and saturates at least $|G| - d$ vertices. This implies that $|M| \geq |F \cap E(G)| \geq |G| - d$. Consequently the theorem is proved. □

2.5 Graphs having 1-factors

We shall show some classes of graphs that have 1-factors, and give some results on the sizes of maximum matchings. Among these results, the following is well-known: every $(r - 1)$-edge connected r-regular graph of even order has a 1-factor that contains any given edge. The next lemma is useful when we prove the existence of 1-factors in regular graphs.

Lemma 2.35. *Let $r \geq 2$ be an integer. Let G be an r-regular general graph, and S a vertex subset of G. Then for every odd component C of $G - S$,*

$$e_G(C, S) \equiv r \pmod 2, \tag{2.18}$$

that is, $e_G(C, S)$ and r have the same parity. In particular, if G is an $(r - 1)$-edge connected r-regular multigraph, then $e_G(C, S) \geq r$.

Proof. Since $|C|$ is odd, congruence (2.18) follows from

$$r \equiv r|C| = \sum_{x \in V(C)} \deg_G(x) = e_G(C, S) + 2\|C\| \equiv e_G(C, S) \pmod 2,$$

where $\|C\|$ denotes the size of C. If G is an $(r - 1)$-edge connected r-regular multigraph, then $e_G(C, S) \geq r - 1$, and so by combining this inequality and (2.18), we have $e_G(C, S) \geq r$. □

The next theorem was first proved by Petersen in 1891 in a slightly weaker form. The stronger version, which we present here, is due to Errera and others (see Chapter 10 of [31]),

Theorem 2.36 (Petersen [209]). *Let G be a connected 3-regular general graph such that all the bridges of G, if any, are contained in a path of G. Then G has a 1-factor (Fig. 2.28). In particular, every 2-edge connected 3-regular multigraph has a 1-factor (Fig. 2.28).*

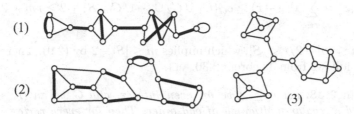

Fig. 2.28. (1) A 3-regular general graph having a 1-factor; (2) A 2-edge connected 3-regular multigraph having a 1-factor; and (3) A 3-regular simple graph having no 1-factor.

Proof. We prove only the first statement since the second statement follows immediately by noting that a 2-edge connected graph has no bridges. Let $\emptyset \neq S \subset V(G)$, and $C_1, C_2, \cdots, C_m$, $m = odd(G - S)$, be the odd components of $G - S$ such that

$$e_G(C_i, S) = 1 \qquad \text{for all} \quad 1 \leq i \leq t, \quad \text{and}$$
$$e_G(C_j, S) \geq 2 \qquad \text{for all} \quad t+1 \leq j \leq m.$$

An edge joining C_i $(1 \leq i \leq t)$ to S is a bridge of G and is contained in a path of G, and thus $t \leq 2$. By Lemma 2.35,

$$e_G(C_j, S) \geq 3 \qquad \text{for all} \quad t+1 \leq j \leq m.$$

Therefore

$$3|S| \geq e_G(C_1 \cup C_2 \cup \cdots \cup C_m, S) \geq t + 3(m - t) = 3m - 2t \geq 3m - 4.$$

Hence $m \leq |S| + 4/3 < |S| + 2$, which implies $m \leq |S|$ by (2.9). Therefore G has a 1-factor by the 1-factor theorem. □

The above theorem can be extended to r-regular graphs as follows. Note that an $(r-1)$-edge connected r-regular general graph contains no loops, and so it must be a multigraph.

Theorem 2.37 (Bäbler [22]). *Let $r \geq 2$ be an integer, and G be an $(r-1)$-edge connected r-regular multigraph of even order. Then for every edge e of G, G has a 1-factor containing e. In particular, G has a 1-factor.*

Proof. We use Theorem 2.30, which gives a necessary and sufficient condition for a multigraph to have a 1-factor containing any given edge. Let $\emptyset \neq S \subset V(G)$ such that $\langle S \rangle_G$ contains an edge. Let $C_1, C_2, \cdots, C_m$ be the odd components of $G - S$, where $m = odd(G - S)$. Then by Lemma 2.35, we have $e_G(C_i, S) \geq r$. Hence

$$r|S| = \sum_{x \in S} \deg_G(x) \geq e_G(C_1 \cup C_2 \cup \cdots \cup C_m, S) + 2 \geq rm + 2.$$

Hence $m \leq |S| - 2/r < |S|$, which implies $m \leq |S| - 2$ by (2.9). Therefore the theorem follows from Theorem 2.30. $\square$

Theorem 2.38. *Let $r \geq 2$ be an even integer, and G be an $(r-1)$-edge connected r-regular multigraph of odd order. Then for every vertex v, $G - v$ has a 1-factor, that is, G is factor-critical.*

Proof. Let $\emptyset \neq S \subset V(G - v) = V(G) - v$. Let $C_1, C_2, \cdots, C_m$ be the odd components of $(G - v) - S = G - (S \cup \{v\})$, where $m = odd((G - v) - S)$. Then by Lemma 2.35, we have $e_G(C_i, S \cup \{v\}) \geq r$. Thus

$$r(|S| + 1) = \sum_{x \in S \cup \{v\}} \deg_G(x) \geq e_G(C_1 \cup C_2 \cup \cdots \cup C_m, S \cup \{v\}) \geq rm.$$

Hence $m \leq |S| + 1$, which implies $m \leq |S|$ by (2.9). Consequently $G - v$ has a 1-factor by the 1-factor theorem. $\square$

Theorem 2.39 (Plesník [211]). *Let $r \geq 2$ be an integer, and G an $(r-1)$-edge connected r-regular multigraph of even order. Then for any $r - 1$ edges $e_1, e_2, \ldots, e_{r-1}$ of G, G has a 1-factor excluding $\{e_1, e_2, \ldots, e_{r-1}\}$.*

Proof. Let $H = G - \{e_1, \ldots, e_{r-1}\}$. It suffices to show that H has a 1-factor. Let $\emptyset \neq S \subset V(H) = V(G)$, and $C_1, C_2, \cdots, C_m$ be the odd components of $H - S$. Then by the same argument as in the proof of Lemma 2.35, we have

$$e_G(C_i, V(G) - V(C_i)) \geq r - 1 \quad \text{and} \quad e_G(C_i, V(G) - V(C_i)) \equiv r \pmod 2.$$

Hence

$$r \leq e_G(C_i, V(G) - V(C_i)) = e_G(C_i, S) + e_G(C_i, V(G) - S - V(C_i)). \quad (2.19)$$

Since C_i is a component of $H - S$, it follows that $E_G(C_i, V(G) - S - V(C_i)) \subseteq \{e_1, \ldots, e_{r-1}\}$, and it is clear that each e_j is contained in at most two such edge subsets, and thus

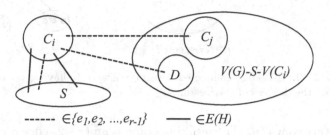

Fig. 2.29. $e_G(C_i, S)$ and $e_G(C_i, V(G) - S - V(C_i))$, where C_i and D denote an odd and even component of $H - S$, respectively.

$$\sum_{i=1}^{m} e_G(C_i, V(G) - S - V(C_i)) \leq 2\#\{e_1, e_2, \ldots, e_{r-1}\} = 2(r-1).$$

Therefore by (2.19), we have

$$rm < \sum_{i=1}^{m}(e_G(C_i, S) + e_G(C_i, V(G) - S - V(C_i)))$$

$$\leq \sum_{x \in S} \deg_G(x) + \sum_{i=1}^{m} e_G(C_i, V(G) - S - V(C_i))$$

$$\leq r|S| + 2(r-1).$$

Hence $m \leq |S| + 2(1 - 1/r) < |S| + 2$, which implies $m \leq |S|$ by (2.9). Consequently H has a 1-factor, and the theorem follows. □

By Theorem 2.37, every $(r-1)$-edge connected r-regular multigraph of even order has a 1-factor. We can say that this result is the best in the sense that there exist infinitely many $(r-2)$-edge connected r-regular multigraphs of even order that have no 1-factor. An example is given below.

Example Let $r \geq 3$ be an odd integer. Let $\overline{K_{r-2}} = (r-2)K_1$ be the totally disconnected graph of order $r - 2$ and R an $(r-2)$-edge connected simple graph of odd order such that $r - 2$ vertices of R have degree $r - 1$ and all the other vertices have degree r. Such an R can be obtained from the complete graph K_{r+2} by deleting one cycle with $r - 2$ edges and two independent edges. It is easy to see that there exist infinitely many such graphs R for any given r.

We construct an $(r-2)$-edge connected r-regular simple graph G from $\overline{K_{r-2}}$ and r copies of R. Join each vertex of $\overline{K_{r-2}}$ to one vertex of degree $r-1$

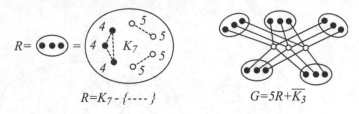

$$R= \; \bullet\bullet\bullet \; = \begin{pmatrix} 4 \\ 4 \\ 4 \end{pmatrix} K_7$$

$$R=K_7-\{\text{----}\} \qquad\qquad G=5R+\overline{K_3}$$

Fig. 2.30. A 3-edge connected 5-regular graph G having no 1-factors; numbers denote the degrees of vertices in R.

in every copy of R. The resulting graph is an $(r-2)$-edge connected r-regular simple graph G (Fig. 2.30). Then G has no 1-factor since

$$odd(G-V(\overline{K_{r-2}})) = \text{the number of copies of } R$$
$$= r > |V(\overline{K_{r-2}})| = r-2.$$

We can similarly construct such graphs for even integers r.

The next theorem shows that an n-edge connected r-regular simple graph with small order has a 1-factor even if it is not $(r-1)$-edge connected. This result with $n = 1$ was obtained by Wallis [240], and then generalized and extended by Zhao [255], Klinkenberg and Volkmann [154] and Volkmann [237].

Theorem 2.40 (Wallis [240]). *Let $r \geq 2$ be an integer, and G be an n-edge connected r-regular simple graph of even order. Define an integer $n' \in \{n, n+1\}$ so that $n' \equiv r \pmod 2$. Then the following two statements hold.*
(i) If r is odd and

$$\left\lfloor \frac{|G|-1}{r+2} \right\rfloor < \frac{2r}{r-n'},$$

then G has a 1-factor. In particular, every connected r-regular simple graph with order at most $3(r+2)$ has a 1-factor.
(ii) If r is even and

$$\left\lfloor \frac{|G|-1}{r+1} \right\rfloor < \frac{2r}{r-n'},$$

then G has a 1-factor. In particular, every connected r-regular simple graph with order at most $3(r+1)$ has a 1-factor. Furthermore, every connected 4-regular simple graph with order at most 20 has a 1-factor

Proof. We shall prove only (i) since (ii) can be proved in a similar way. Let $r \geq 3$ be an odd integer. Let $\emptyset \neq S \subset V(G)$, and $C_1, C_2, \ldots, C_m$ be the odd components of $G-S$, where $m = odd(G-S)$. If $|C_i| \leq r$, then

$$r|C_i| = \sum_{x \in V(C_i)} \deg_G(x) = e_G(C_i, S) + 2||C_i||$$
$$\leq e_G(C_i, S) + |C_i|(|C_i|-1) \leq e_G(C_i, S) + r(|C_i|-1).$$

Hence $e_G(S, C_i) \geq r$.

Since $r + 1$ is an even integer, no C_i has order $r + 1$. Let us define

m_1 = the number of C_i with $|C_i| \leq r$; and

m_2 = the number of C_i with $|C_i| \geq r + 2$.

Then $m = m_1 + m_2$, and $e_G(S, C_i) \geq r$ if $|C_i| \leq r$, and by Lemma 2.35, $e_G(S, C_i) \geq n'$ if $|C_i| \geq r + 2$. Hence we have

$$r|S| \geq e_G(S, C_1 \cup C_2 \cup \cdots \cup C_m) \geq rm_1 + n'm_2 = rm + (n' - r)m_2,$$

which implies $m \leq |S| + (r - n')m_2/r$. Therefore the inequality $m < |S| + 2$ holds if

$$m_2 < \frac{2r}{r - n'}. \tag{2.20}$$

On the other hand, it follows that

$$|G| \geq |S| + |C_1| + \cdots + |C_m| \geq |S| + (r + 2)m_2,$$

which implies

$$m_2 \leq \left\lfloor \frac{|G| - |S|}{r + 2} \right\rfloor \leq \left\lfloor \frac{|G| - 1}{r + 2} \right\rfloor.$$

Consequently, if the following inequality holds, then inequality (2.20) holds, which implies $m < |S| + 2$ and G has a 1-factor by (2.9).

$$\left\lfloor \frac{|G| - 1}{r + 2} \right\rfloor < \frac{2r}{r - n'}.$$

Therefore the first part of (i) is proved.

We next prove the second part of (i). If $|G| \leq 3(r + 2)$ and $n' = 1$, then

$$\left\lfloor \frac{|G| - 1}{r + 2} \right\rfloor = 2 < \frac{2r}{r - 1} = \frac{2r}{r - n'}.$$

Hence G has a 1-factor by the first part of (i).

Notice that when r is even, every r-regular graph is 2-edge connected. Thus we can apply the result with $n' = 2$ and obtain the latter part of (ii). $\square$

We next identify some non-regular graphs which have 1-factors. The next theorem uses the binding number of a graph to give a sufficient condition for a graph to have a 1-factor.

Theorem 2.41 (Woodall [249]). *Let G be a connected simple graph of even order. If for every $\emptyset \neq S \subseteq V(G)$,*

$$N_G(S) = V(G) \qquad or \qquad |N_G(S)| > \frac{4}{3}|S| - 1, \tag{2.21}$$

then G has a 1-factor. In particular, a connected simple graph H with $\mathrm{bind}(H) \geq 4/3$ has a 1-factor.

Proof. The proof is by contradiction. Assume that G has no 1-factor. Then by the 1-factor theorem and by (2.9), there exists a subset $\emptyset \neq S \subset V(G)$ such that $odd(G - S) \geq |S| + 2$.

Let X be the set of isolated vertices of $G - S$, and $C_1, C_2, \ldots, C_k$ the odd components of $G - S$ with order at least three. Let $V = V(G)$ and $Y = V(C_1) \cup V(C_2) \cup \cdots \cup V(C_k)$ (Fig. 2.31). Then

$$odd(G - S) = |X| + k \geq |S| + 2, \qquad |Y| \geq 3k \quad \text{and}$$
$$|V| \geq |S| + |X| + |Y|. \tag{2.22}$$

We consider two cases:

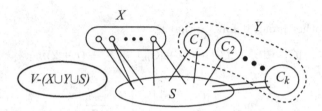

Fig. 2.31. The graph G with a subset S.

Case 1. $X \neq \emptyset$.

Since $N_G(V - S) \subseteq V - X \neq V$, it follows from (2.21) that

$$|N_G(V - S)| > \frac{4}{3}|V - S| - 1 = \frac{4|V|}{3} - \frac{4}{3}|S| - 1$$

and

$$|N_G(V - S)| \leq |V - X| = |V| - |X|.$$

By the previous two inequalities, we obtain

$$|V| < 4|S| + 3 - 3|X|. \tag{2.23}$$

On the other hand, it follows from (2.22) that

$$|V| \geq |S| + |X| + |Y| \geq |S| + |X| + 3k$$
$$\geq |S| + |X| + 3(|S| + 2 - |X|)$$
$$= 4|S| - 2|X| + 6.$$

Hence, by this inequality and (2.23), we have

$$4|S| - 2|X| + 6 \leq |V| < 4|S| + 3 - 3|X|,$$

which is a contradiction.

Case 2. $X = \emptyset$.

In this case $odd(G - S) = k$. Let $Z = V(C_2) \cup \cdots \cup V(C_k)$. Since $N_G(Z) \subseteq V(G) - V(C_1) \neq V(G)$, it follows from (2.21) that

$$\frac{4}{3}|Z| - 1 < |N_G(Z)| \leq |Z| + |S|.$$

Thus $|Z| < 3|S| + 3$. On the other hand, by (2.22) we have that

$$|Z| \geq 3(k - 1) = 3(odd(G - S) - 1) \geq 3(|S| + 2 - 1).$$

Therefore

$$3(|S| + 1) \leq |Z| < 3|S| + 3.$$

This is again a contradiction. Consequently the theorem is proved. $\square$

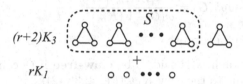

Fig. 2.32. A graph $G = (r + 2)K_3 + rK_1$, which has no 1-factor and whose binding number is $(4/3) - (1/3r)$.

Consider a graph $G = (r+2)K_3+rK_1$, where $r \geq 1$ is an integer (Fig. 2.32). Then we can easily show that G has no 1-factor, and setting $S = V((r+1)K_3)$, we have

$$bind(G) = \frac{|N_G(S)|}{|S|} = \frac{|V(G) - V(K_3)|}{|S|} = \frac{4(r + 1) - 1}{3(r + 1)} = \frac{4}{3} - \frac{1}{3(r + 1)}.$$

Hence the condition of Theorem 2.41 is sharp.

Recall that if G is t-tough, then $tough(G) \geq t$ and for a subset $S \subset V(G)$ with $\omega(G - S) \geq 2$, it follows that

$$\omega(G - S) \leq \frac{|S|}{tough(G)} \leq \frac{|S|}{t}.$$

The next theorem uses the idea of toughness to give a sufficient condition for a graph to have a 1-factor.

Theorem 2.42 (Chvátal [54], Exercise 3.4.11 of [182]). *Every 1-tough connected simple graph of even order has a 1-factor. Moreover, for every real number $\epsilon > 0$, there exist simple graphs G of even order that have no 1-factors and satisfy $tough(G) > 1 - \epsilon$.*

Proof. Let G be a 1-tough connected simple graph of even order, and $\emptyset \neq S \subset V(G)$. If $\omega(G - S) \geq 2$, then

$$odd(G - S) \leq \omega(G - S) \leq \frac{|S|}{tough(G)} \leq |S|.$$

If $\omega(G-S) = 1$, then $odd(G-S) \leq \omega(G-S) = 1 \leq |S|$. Hence $odd(G-S) \leq |S|$ always holds, and thus G has a 1-factor by the 1-factor theorem.

Let $G = K_m + tK_n$, which is a join of the complete graph K_m and the t copies of the complete graph K_n, where n is an odd integer (Fig. 2.33 (1)). Moreover, we can choose m, n and t so that

$$t > m, \quad \frac{m}{t} > 1 - \epsilon \quad \text{and} \quad m + nt \equiv 0 \pmod 2.$$

Then G has no 1-factor since $odd(G - V(K_m)) = t > |V(K_m)| = m$, and

$$tough(G) = \frac{|V(K_m)|}{\omega(G - V(K_m))} = \frac{m}{t} > 1 - \epsilon.$$

Therefore the theorem is proved. $\square$

Recall that a graph G is said to be **claw-free** if G contains no induced subgraph isomorphic to the claw $K_{1,3}$ (Fig. 2.33 (2)).

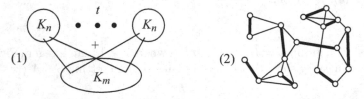

Fig. 2.33. (1) $K_m + tK_n$; (2) A claw-free graph and its 1-factor.

Theorem 2.43 (Sumner [218], Las Vergnas [164]). *Every connected claw-free simple graph of even order has a 1-factor.*

Proof. We prove the theorem by induction on $|G|$. We may assume $|G| \geq 3$. Let $P = (v_1, v_2, \ldots, v_k)$, $(v_i \in V(G))$ be a longest path of G. Then $k \geq 3$. It is immediate that $N_G(v_1) \subseteq V(P) - \{v_1\}$.

We shall show that $G - \{v_1, v_2\}$ is connected. If $\deg_G(v_2) = 2$, then $G - \{v_1, v_2\}$ is connected since $N_G(v_1) \subseteq V(P) - \{v_1\}$. Thus we may assume that $\deg_G(v_2) \geq 3$. For every $x \in N_G(v_2) - V(P)$, xv_3 must be an edge of G, since otherwise $\langle \{v_2, v_1, v_3, x\} \rangle_G \neq K_{1,3}$ implies that v_1x or v_1v_3 is an edge of G. Hence G contains a path longer than P, a contradiction. Therefore $G - \{v_1, v_2\}$ is a connected graph of even order, and is of course claw-free. By the induction hypothesis, $G - \{v_1, v_2\}$ has a 1-factor, and so does G. $\square$

It is known that every 4-connected planar graph has a Hamiltonian cycle [229], which implies that it has a 1-factor if it is of even order, but, there are infinitely many 3-connected planar graphs of even order that have no 1-factors. The next theorem gives a lower bound for the order of a maximum matching of a planar graph. The proof given here seems to be different from that of [203].

Theorem 2.44 (Nishizeki and Baybars [203]). *Let G be a connected planar simple graph with $\delta(G) \geq 3$. Then the number $|M|$ of vertices saturated by a maximum matching M in G is*

$$|M| \geq \frac{2|G| + 4}{3}. \tag{2.24}$$

If G is 2-connected, then

$$|M| \geq \frac{2|G| + 8}{3}. \tag{2.25}$$

Proof. We first prove (2.25). Let G be a 2-connected planar simple graph. We may assume that G is drawn in the plane as a plane graph. Let $\emptyset \neq S \subset V(G)$. Let us denote by X the set of isolated vertices of $G - S$, and by $C_1, C_2, \ldots, C_m$ the odd components of $G - S$ of order at least three. Then

$$odd(G - S) = |X| + m, \quad \text{and} \quad |G| \geq |S| + |X| + 3m. \tag{2.26}$$

If $|S| = 1$, then $odd(G - S) \leq 1 = |S|$ since G is 2-connected. So $odd(G - S) - |S| \leq 0$. Assume $|S| \geq 2$. We construct a planar bipartite graph B with bipartite sets S and $X \cup \{v_1, v_2, \ldots, v_m\}$ from G by contracting $C_1, C_2, \ldots, C_m$ into single vertices $v_1, v_2, \ldots, v_m$ and by replacing all multiple edges in the resulting graph by single edges, that is, $s \in S$ and v_i are joined by an edge of B if and only if s and C_i are joined by an edge of G (Fig. 2.34).

Since G is a 2-connected graph with $\delta(G) \geq 3$, it follows that

$$\deg_B(x) \geq 3 \quad \text{for all} \quad x \in X, \quad \text{and}$$
$$\deg_B(v_i) \geq 2 \quad \text{for all} \quad 1 \leq i \leq m. \tag{2.27}$$

Since B is a planar bipartite graph, it follows from Theorem 1.10 that

$$\|B\| \leq 2|B| - 4.$$

By combining this inequality and (2.27), we obtain

$$3|X| + 2m \leq \|B\| \leq 2|B| - 4 \doteq 2(|S| + |X| + m) - 4.$$

Hence $|X| - 2|S| \leq -4$DBy this inequality and (2.26), we get

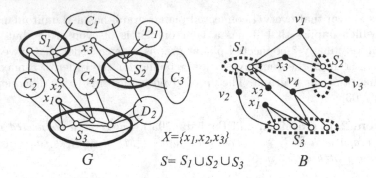

Fig. 2.34. G is a 2-connected plane graph, x_1, x_2, x_3 are the isolated vertices of $G - S$, $C_1, \ldots, C_4$ are the odd components of $G - S$ with order at least 3, and D_1, D_2 are the even components of $G - S$. B is the corresponding planar bipartite graph.

$$odd(G - S) - |S| = |X| + m - |S|$$

$$\leq |X| + \frac{|G| - |S| - |X|}{3} - |S| = \frac{|G| + 2(|X| - 2|S|)}{3}$$

$$\leq \frac{|G| - 8}{3}.$$

Consequently by (2.17) in Theorem 2.34, we have

$$|M| = |G| - \max_S(odd(G - S) - |S|) \geq |G| - \frac{|G| - 8}{3} = \frac{2|G| + 8}{3},$$

which implies the desired inequality (2.25).

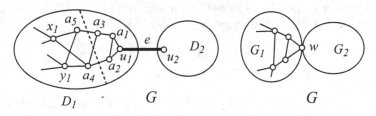

Fig. 2.35. A connected plane graph G with bridge e; and a connected plane graph G with cut vertex w.

We next prove (2.24) by induction on $|G|$, where G is a connected plane graph with $\delta(G) \geq 3$. Since (2.24) holds for a small graph G, we may assume that the order of G is not small (for example, $|G| \geq 10$). Suppose first that G has a bridge $e = u_1 u_2$ (Fig. 2.35). Let D_1 and D_2 be the two components of $G - e$ containing u_1 and u_2, respectively. We shall later show that each D_i has a matching that covers at least $(2|D_i| + 2)/3$ vertices of D_i. If the above statement holds, then G has a matching that covers at least

$$\frac{2|D_1| + 2}{3} + \frac{2|D_2| + 2}{3} = \frac{2|G| + 4}{3} \quad \text{vertices.}$$

Hence (2.24) holds.

If G has no bridges but has a cut vertex w, then by letting $G = G_1 \cup G_2$, $V(G_1) \cap V(G_2) = \{w\}$ (see Fig. 2.35), we shall show that each G_i has a matching that covers at least $(2|G_i| + 4)/3$ vertices of G_i. Then G has a matching that covers at least the following number of vertices:

$$\frac{2|G_1| + 4}{3} + \frac{2|G_2| + 4}{3} - 2 = \frac{2(|G_1| + |G_2| - 1) + 4}{3} = \frac{2|G| + 4}{3}.$$

Hence (2.24) holds.

Let D_1 and D_2 be the two components of $G - e$, where e is a bridge of G. We shall show that each D_i has a matching that covers at least $(2|D_i| + 2)/3$ vertices of D_i by using the inductive hypothesis of the theorem. If $\delta(D_i) \geq 3$, then the above statement holds by induction. Without loss of generality, we may assume $\delta(D_1) = 2$, which implies $\deg_{D_1}(u_1) = 2$ as $\delta(G) \geq 3$ (see Fig. 2.35). Let a_1 and a_2 be the two vertices adjacent to u_1. If $\delta(D_1 - u_1) \geq 3$, then by induction, $D_1 - u_1$ has a matching that covers at least

$$\frac{2(|D_1| - 1) + 4}{3} = \frac{2|D_1| + 2}{3} \quad \text{vertices.}$$

Note that if $D_1 - u_1$ is disconnected, we apply the inductive hypothesis to each component and obtain the above desired matching of $D_1 - u_1$. Thus we may assume that $\delta(D_1 - u_1) = 2$. If a_1 and a_2 are not adjacent, then $D_1 - u_1 + a_1 a_2$ has minimum degree three, and so by induction it has a matching M_1 covering at least $(2(|D_1| - 1) + 4)/3$ vertices. If M_1 contains an edge $a_1 a_2$, then by considering $M_1 - a_1 a_2 + a_1 u_1$ we can get the desired matching of D_1. So we may assume that a_1 and a_2 are adjacent. If $\deg_{D_1 - u_1}(a_i) \geq 3$ for $i = 1, 2$, then we can apply the inductive hypothesis to $D_1 - u_1$ and obtain the desired statement as above. Thus we may assume that $\deg_{D_1 - u_1}(a_1) = 2$.

If $\delta(D_1 - u_1 - a_1) \geq 3$, then by induction it has a matching M_2 covering at least $(2(|D_1| - 2) + 4)/3$ vertices of $D_1 - u_1 - a_1$. Hence $M_2 + a_1 u_1$ is the desired matching in D_1. By repeating this argument, we can finally find the desired matching of D_1. As an example, in Fig. 2.35, $\delta(D_1 - \{u_1, a_1, a_2, a_3\} + a_4 a_5) \geq 3$ and M_3 is its matching covering at least $2(|D_1| - 4) + 4)/3$ vertices, and if $a_4 a_5 \in M_3$, then $M_3 - a_4 a_5 + a_5 a_3 + a_1 u_1 + a_4 a_2$ is the desired matching in D_1.

Let $G = G_1 \cup G_2$, $V(G_1) \cap V(G_2) = \{w\}$, where w is a cut vertex of G. We shall show that each G_i has a matching that covers at least $(2|G_i| + 4)/3$ vertices of G_i by using the inductive hypothesis of the theorem. We can do this by a similar argument as above, Consequently the proof is complete. $\square$

A surface (a compact orientable 2-manifold) is a sphere on which a number of handles have been placed. The number of handles is referred to as the **genus** of the surface (Fig. 2.36). A surface with genus one is called a **torus**, and, of course, a plane is a surface with genus zero. A graph G is said to be embedded on a surface if G can be drawn on the surface in such a way that edges intersect only at their common end-points. The genus $\gamma(G)$ of a simple graph G is defined to be smallest genus of all surfaces on which G can be embedded.

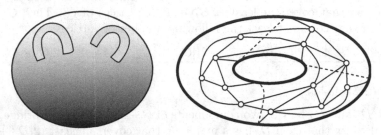

Fig. 2.36. A surface with genus two, and an embedding of a graph on the torus.

Theorem 2.45 (Nishizeki [202]). *Let G be an n-connected simple graph of even order. If $\gamma(G) < n(n-2)/4$, then G has a 1-factor. In particular, every 4-connected simple graph of even order which is embedded on the torus has a 1-factor.*

2.6 Structure theorem

We now consider the structure of a graph without 1-factors, and characterize such a graph by using vertex-decomposition together with certain properties. The resulting theorem is called the Gallai-Edmonds structure theorem and gives us much information about maximum matchings.

Let G be a graph. For a subset $X \subseteq V(G)$, $odd(G-X) - |X|$ is called the **deficiency** of X, and a subset $S \subseteq V(G)$ is called a **barrier** if

$$odd(G-S) - |S| = \mathrm{def}(G) = \max_{X \subseteq V(G)} \{odd(G-X) - |X|\}. \qquad (2.28)$$

That is, S is a barrier if its **deficiency** is equal to that of G. A barrier S is said to be **minimal** if no proper subset of S is a barrier.

Theorem 2.46. *Suppose that a connected simple graph G with even order has no 1-factor. Let S be a minimal barrier of G. Then every vertex $x \in S$ is joined to at least three odd components of $G-S$. In particular, x is the center of a certain induced claw subgraph of G.*

Proof. Since G has no 1-factor and has even order, we have $def(G) \geq 2$ by the 1-factor theorem and Lemma 2.26. If $S = \{x\}$, then $odd(G - S) \geq 3$ since $odd(G - S) \equiv |S| \pmod 2$. Hence $x \in S$ is joined to at least three odd components of $G - S$ and the theorem holds. Thus we may assume that $|S| \geq 2$.

Suppose that a vertex $x \in S$ is joined to at most two odd components of $G - S$ (Fig. 2.37). Then $odd(G - (S - x)) \geq odd(G - S) - 2$, and thus

$$odd(G - (S - x)) - |S - x| \geq odd(G - S) - |S| - 1.$$

By Lemma 2.26, we have

$$odd(G - (S - x)) - |S - x| \equiv odd(G - S) - |S| \equiv |G| \pmod 2.$$

Hence

$$odd(G - (S - x)) - |S - x| \geq odd(G - S) - |S|,$$

which implies that $S - x$ is also a barrier. This contradicts the minimality of S. Therefore every $x \in S$ is joined to at least three odd components of $G - S$.

By taking three vertices adjacent to x from each of the three odd components of $G - S$, we can obtain an induced subgraph $K_{1,3}$ with center x. $\square$

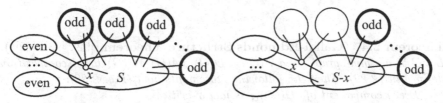

Fig. 2.37. Components of $G - S$ and those of $G - (S - x)$.

It is clear that Theorem 2.43, which says that every connected claw-free simple graph of even order has a 1-factor, is an immediate consequence of Theorem 2.46.

Consider a simple graph G. Let $D(G)$ denote the set of all vertices v of G such that v is not saturated by at least one maximum matching of G. Let $A(G)$ be the set of vertices of $V(G) - D(G)$ that are adjacent to at least one vertex in $D(G)$. Finally, define $C(G) = V(G) - D(G) - A(G)$. Then $V(G)$ is decomposed into three disjoint subsets

$$V(G) = D(G) \cup A(G) \cup C(G), \tag{2.29}$$

where

$$D(G) = \{x \in V(G) \text{ : There exists a maximum matching}$$
$$\text{that does not saturate } x\}$$
$$A(G) = N_G(D(G)) \setminus D(G)$$
$$C(G) = V(G) - A(G) - D(G).$$

Some properties of the above decomposition are given in the following Gallai-Edmonds structure theorem. This theorem was obtained by Gallai [92], [93] and Edmonds [62] independently and in different ways. The proof presented here is based on [120].

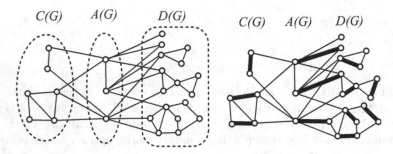

Fig. 2.38. The decomposition $V(G) = D(G) \cup A(G) \cup C(G)$ and a maximum matching of G.

Theorem 2.47 (Gallai-Edmonds Structure Theorem, [92], [93], [62]).
Let G be a simple graph, and $V(G) = D(G) \cup A(G) \cup C(G)$ the decomposition defined in (2.29). Then the following statements hold (Fig. 2.38):
(i) Every component of $\langle D(G)\rangle_G$ is factor-critical.
(ii) $\langle C(G)\rangle_G$ has a 1-factor.
(iii) Every maximum matching M in G saturates $C(G) \cup A(G)$, and every edge of M incident with $A(G)$ joins a vertex in $A(G)$ to a vertex in $D(G)$.
(iv) The number $|M|$ of vertices saturated by a maximum matching M is given by

$$|M| = |G| + \omega(\langle D(G)\rangle_G) - |A(G)|, \qquad (2.30)$$

where $\omega(\langle D(G)\rangle_G)$ denotes the number of components of $\langle D(G)\rangle_G$.

Proof. The proof of the theorem is by induction on $|G|$. We may assume that G is connected since otherwise each component of G satisfies the statements of the theorem and so does G. Moreover we may assume that G has no 1-factor since otherwise $D(G) = \emptyset$, $A(G) = \emptyset$ and $C(G) = V(G)$, and thus the theorem holds.

Let S be a maximal barrier of G, that is, S is a subset of $V(G)$ such that

$$odd(G - S) - |S| = \max_{X \subset V(G)} \{odd(G - X) - |X|\} = \text{def}(G) > 0 \qquad (2.31)$$

and

$$odd(G - Y) - |Y| < odd(G - S) - |S| \quad \text{for all} \quad S \subset Y \subseteq V(G). \quad (2.32)$$

The following claim can be proved in the same way as in the proof of the 1-factor theorem (see Problem 2.9).

Claim 1. *Every component of $G - S$ is factor-critical.*

Let $\{C_1, C_2, \ldots, C_m\}$ be the set of components of $G - S$, where $m = odd(G - S)$. We define a bipartite graph B with bipartition $S \cup \{C_1, C_2, \ldots, C_m\}$ as follows: a vertex $x \in S$ and C_i is joined by an edge of B if and only if x and C_i are joined by at least one edge of G (Fig. 2.39). Then B satisfies the following claim.

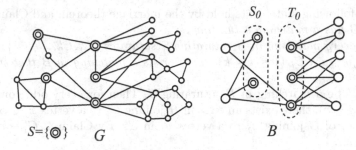

Fig. 2.39. A graph G with a maximal barrier S, and the bipartite graph B with subsets S_0 and T_0.

Claim 2. $|N_B(X)| \geq |X|$ *for all* $X \subseteq S$.

Assume that $|N_B(Y)| < |Y|$ for some $\emptyset \neq Y \subseteq S$. Then

$$odd(G - (S - Y)) - |S - Y|$$
$$\geq |\{C_1, C_2, \ldots, C_m\} - N_B(Y)| - |S - Y|$$
$$> m - |Y| - |S - Y| = m - |S| = odd(G - S) - |S|,$$

which contradicts (2.31). Hence Claim 2 holds.

Claim 3. *There exists a unique maximum proper subset* $S_0 \subset S$ *such that* $|N_B(S_0)| = |S_0|$. *Furthermore,* $|N_B(Y) \setminus N_B(S_0)| > |Y|$ *for every* $\emptyset \neq Y \subseteq S - S_0$.

By $|N_B(S)| = m > |S|$, Claim 2 and by Theorem 2.9, there exists a unique maximum proper subset $S_0 \subset S$ such that $|N_B(S_0)| = |S_0|$.

Let $\emptyset \neq Y \subseteq S - S_0$. Then it follows from the maximality of S_0 and $S_0 \subset Y \cup S_0$ that

$$|N_B(Y) \setminus N_B(S_0)| = |N_B(Y \cup S_0) - N_B(S_0)| > |Y \cup S_0| - |S_0| = |Y|.$$

Therefore the claim is proved.

Let $T_0 = S - S_0$. Then

$$odd(G - T_0) = odd(G - S) - |N_B(S_0)| = odd(G - S) - |S_0|$$
$$= odd(G - S) - (|S| - |T_0|),$$

and so $odd(G - T_0) - |T_0| = odd(G - S) - |S| = \text{def}(G)$.

Let $\{C_1', C_2', \ldots, C_r'\}$ be the set of odd components of $G - S$ corresponding to $N_B(S_0)$, where $r = |N_B(S_0)| = |S_0|$, and let $\{C_1, C_2, \ldots, C_k\}$ be the set of odd components of $G - T_0$. Then

$$Odd(G - S) = \{C_1', C_2', \ldots, C_r'\} \cup \{C_1, C_2, \ldots, C_k\},$$

and the following statements hold by the marriage theorem and Claim 3:

(i) B has a matching saturating S;

(ii) every matching in B saturating S saturates $\{C_1', C_2', \ldots, C_r'\}$; and

(iii) for each C_i $(1 \le i \le k)$, there exists a matching in B that saturates S but not C_i.

Let H be a matching in B saturating S. Then for every odd component C_j' $(1 \le j \le r)$, there exists an edge in H joining C_j' to a vertex $x_j \in S_0$. Take an edge e_j of G joining x_j to a vertex v_j in C_j'. By Claim 4, $C_j' - v_j$ has a 1-factor R_j'.

Similarly, for an odd component C_i $(1 \le i \le k)$, if H has an edge joining C_i to $x_i \in S$, then $x_i \in T_0$ and we can find an edge e_i of G joining x_i to a vertex w_i of C_i and a 1-factor R_i of $C_i - w_i$. If H has no edge joining C_i to S, then take a maximum matching R_i in C_i. Define

$$M = \bigcup_{1 \le j \le r} (R_j' + e_j) + \bigcup_{1 \le i \le k} \{(R_i + e_i) \text{ or } R_i\}. \qquad (2.33)$$

Then by Theorem 2.34, M is a maximum matching of G since the number of unsaturated vertices in M is $k - |T_0| = odd(G - T_0) - |T_0| = \text{def}(G)$.

Conversely, every maximum matching in G is obtained in this way since every matching in G cannot saturate at least $k - |T_0|$ odd components in $\{C_1, C_2, \ldots, C_k\}$, and every maximum matching M' does not saturate exactly $k - |T_0| = \text{def}(G)$ components of $\{C_1, C_2, \ldots, C_k\}$. Therefore M' induces a matching H' in B that saturates S and $\{C_1', C_2', \ldots, C_r'\}$, and thus M' can be constructed from H' as above.

Claim 5. $D(G) = V(C_1) \cup V(C_2) \cup \cdots \cup V(C_k)$ and $A(G) = T_0$.

It is clear that for every vertex v_i of any C_i $(1 \le i \le k)$, B has a matching that saturates S but not C_i, and $C_i - v_i$ has a 1-factor by Claim 1. Hence by (2.33), we can find a maximum matching in G that does not saturate v_i. Thus $V(C_1) \cup V(C_2) \cup \cdots \cup V(C_k) \subseteq D(G)$. Since every maximum matching in G is obtained in the manner mentioned above, $D(G) \subseteq V(C_1) \cup V(C_2) \cup \cdots \cup V(C_k)$. Consequently, $D(G) = V(C_1) \cup V(C_2) \cup \cdots \cup V(C_k)$.

Since $N_B(S_0) = \{C_1', C_2', \ldots, C_r'\}$, it follows that $N_B(\{C_1, C_2, \ldots, C_k\}) = S - S_0 = T_0$. Therefore

$$A(D(G)) = N_G(D(G)) \setminus D(G) = T_0.$$

It is easy to see that $\langle V(C_1') \cup \cdots \cup V(C_r') \cup S_0 \rangle$ has a 1-factor and forms the even components of $G - T_0$. Consequently the proof is complete. $\square$

We mention one application of the Gallai-Edmonds structure theorem. Since every component of $\langle D(G) \rangle_G$ is factor-critical, Theorem 2.28 is an easy consequence of the Gallai-Edmonds structure theorem.

The following lemma is interesting its own right, but it is also useful for proving the Gallai-Edmonds structure theorem; we can first prove the following stability lemma without using Gallai-Edmonds structure theorem, and then apply the lemma to prove the structure theorem ([182] section 3.2). However, we shall prove the lemma using Gallai-Edmonds structure theorem since it is shorter.

Lemma 2.48 (The Stability Lemma). *Let G be a simple graph and $V(G) = C(G) \cup A(G) \cup D(G)$. Then for every vertex $u \in A(G)$, we have $A(G - u) = A(G) - u$, $C(G - u) = C(G)$ and $D(G - u) = D(G)$.*

Proof. Let $u \in A(G)$. Then every maximum matching M in G has an edge e incident with u. Thus $M - e$ is a maximum matching in $G - u$ since the size of a maximum matching in $G - u$ must less than or equal to $\|M\| - 1$ and $\|M - e\| = \|M\| - 1$. Therefore $D(G) \subseteq D(G - u)$, and the size of a maximum matching in $G - u$ is $\|M\| - 1$.

Assume that a maximum matching H in $G - u$ does not saturate a vertex $x \in A(G) - u$. Then H is a matching in G, and H does not saturate at least $\omega(\langle D(G) \rangle_G) - (|A(G)| - 2)$ vertices in $D(G)$ and two more vertices $x, u \in A(G)$. Therefore $|V(H)| \leq |V(M)| - 4$, and so $\|H\| \leq \|M\| - 2$. This contradicts the fact that a maximum matching in $G - u$ has size $\|M\| - 1$. Hence $(A(G) - u) \cap D(G - u) = \emptyset$. We can similarly show that H saturates $C(G)$, which implies $C(G) \cap D(G - u) = \emptyset$. Consequently, $D(G - u) = D(G)$.

The other equalities $A(G - u) = A(G) - u$ and $C(G - u) = C(G)$ follow immediately from $D(G - u) = D(G)$. $\square$

Since a factor-critical graph with order at least three is not a bipartite graph (Problem 2.8), if G is a bipartite graph, then every component of $\langle D(G) \rangle_G$ must be a single vertex. Thus the following theorem holds.

Theorem 2.49. *Let G be a connected bipartite simple graph with bipartition (X, Y), and let $C_X = C(G) \cap X$ and $C_Y = C(G) \cap Y$. Then the following statements hold (Fig. 2.40):*
(i) $\langle D(G) \rangle_G$ is a set of independent vertices of G.

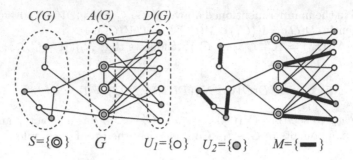

Fig. 2.40. The decomposition $V(G) = D(G) \cup A(G) \cup C(G)$ of a bipartite graph G with bipartition $U_1 \cup U_2$; and it maximum matching Mx.

(ii) $\langle C(G) \rangle_G$ *has a 1-factor, and* $|C_X| = |C_Y|$.
(iii) *Every maximum matching* M *in* G *consists of a 1-factor of* $\langle C(G) \rangle_G$ *and a matching in* $\langle A(G) \cup D(G) \rangle_G$ *saturating* $A(G)$.
(iv) *Both* $A(G) \cup C_X$ *and* $A(G) \cup C_Y$ *are minimum vertex covers of* G.
(v) *Both* $D(G) \cup C_X$ *and* $D(G) \cup C_Y$ *are maximum independent vertex subsets of* G.

2.7 Algorithms for maximum matchings

We gave an algorithm for finding a maximum matching in a bipartite graph in Section 2.3. In this section we shall propose an algorithm for finding a maximum matching in a simple graph, which was obtained by Edmonds [62].

Before stating the algorithm, let us recall Theorem 2.20, which says that "a matching M in a graph G is maximum if and only if G has no M-augmenting path". Therefore, to find a maximum matching, we should find M-augmenting paths or determine the non-existence of such paths. In order to effectively explore M-augmenting paths in a graph, we introduce some new concepts and notation.

We shall first explain the algorithm and new definitions by using examples. Let G be a graph and M a matching in G, and let v be a vertex unsaturated by M. We call v a **root**, and explore all the M-alternating paths starting with v. If $P = (v, x_1, x_2, x_3, x_4, \ldots, x_k)$ is an M-alternating path, where each x_i is a vertex of G, then we call $x_1, x_3, \ldots$ **inner vertices** and $v, x_2, x_4, \ldots$ **outer vertices** (Fig. 2.41)

Consider the graph G and the matching M given in Fig. 2.41. We try to find all the M-alternating paths starting with v as follows:

$$\{v\} \quad \{a, g\}, \quad \{b, h\}, \quad \{c, e, i\}, \quad \{d, f, j\} \quad \text{and} \quad \{f, k\}.$$

In the last step, we find that f is simultaneously an outer and inner vertex since (u, a, b, e, f) and (u, a, b, c, d, f) are both M-alternating paths. Then we

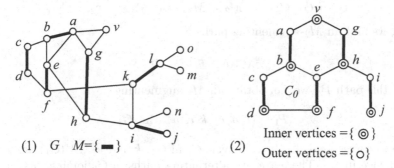

Fig. 2.41. (1) A matching M and an M-unsaturated vertex v; (2) The root v, inner vertices and outer vertices.

find an odd cycle $C_0 = (b, c, d, f, e, b)$ containing f, i.e., if we find a vertex that is simultaneously outer and inner, then there exists an odd cycle containing it, which can be easily found.

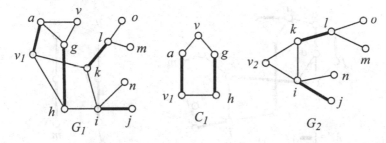

Fig. 2.42. The graph $G_1 = G/C_0$ with the matching M_1; an odd cycle $C_1 = (v, a, v_1, h, g, a)$; and the graph $G_2 = G_1/C_1$ with the matching M_2.

Next we contract C_0 into a single vertex v_1, which also implies that we delete loops and replace every multiple edge by a single edge, and denote the resulting graph by G_1. The matching of G_1 corresponding to M is obtained by deleting the edges in $C_0 \cap M$ (Fig. 2.42), i.e., we obtain

$$G_1 = G/C_0, \qquad M_1 = M - (E(C_0) \cap M) = M \cap E(G_1), \qquad v_1 = C_0.$$

In Fig. 2.42, we find an odd cycle $C_1 = (v, a, v_1, h, g)$ since h is simultaneously an outer and inner vertex in G_1. Then we obtain the new graph G_2 from G_1 by contracting C_1, where v_2 is the new root since C_1 contains the root v. In general, if the odd cycle C_i contains the root, then C_i corresponds to the new root in G_i/C_i. Otherwise, C_i corresponds to a saturated vertex in G_i/C_i.

$$G_2 = G_1/C_1, \qquad M_2 = M_1 \cap E(G_2), \qquad v_2 = C_1.$$

Next, we find an M_2-augmenting path

$$P_2 = (v_2, k, l, m) \quad \text{in} \quad G_2.$$

From this path P_2, we can obtain an M_1-augmenting path

$$P_1 = (v, a, v_1, k, l, m) \quad \text{in} \quad G_1.$$

Since v_2 corresponds to the odd cycle C_1 in G_1, k and $v_1 \in V(C_1)$ are joined by an edge in G_1. There are two alternating paths in C_1 joining v_1 to v and one of them can be added to (k, l, m). Since v_1 corresponds to C_0 in G and v_1 can be replaced by the alternating path (a, b, e, f) in C_0, we obtain the desired M-augmenting path

$$P = (v, a, b, e, f, k, l, m) \quad \text{in} \quad G.$$

Therefore we obtain a larger matching $M' = M \triangle E(P)$ in G (Fig. 2.43).

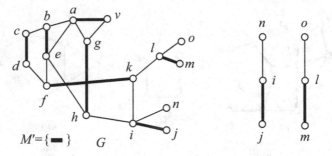

Fig. 2.43. A matching M' in G; and non-existence of M'-augmenting path in G.

There are two M'-unsaturated vertices n and o, and by the same argument as above, we can easily determine that G has no M'-augmenting paths starting with n or o, which implies that M' is a maximum matching in G.

We conclude this section with an algorithm for finding a maximum matching in a graph.

Algorithm 2.50 (Algorithm for maximum matchings) *Let G be a connected simple graph. Then a maximum matching of G can be obtained by repeating the following procedure. Let $i = 0$, $G_0 = G$, M_0 be any matching of G_0 and let v_0 be any M_0-unsaturated vertex of G_0. Initially, v_0 is the root.*

We explore all the M_i-alternating paths starting with v_i as explained above. If we find a vertex x_i that is both inner and outer, then we can find an odd cycle C_i containing x_i, and obtain a graph G_{i+1} from G_i by contracting C_i.

The contraction of C_i is the vertex v_{i+1} of G_{i+1}. If C_i contains the root of G_i, then v_{i+1} is the new root of G_{i+1} and is unsaturated by a matching $M_{i+1} = M_i - (E(C_i) \cap M_i)$. Otherwise, the vertex v_{i+1} of G_{i+1} is a vertex saturated by M_{i+1}. Set $i = i + 1$, and repeat the procedure.

If we find an M_i-augmenting path in G_i connecting the root and another M_i-unsaturated vertex y, then we can find an M_{i-1}-augmenting path in G_{i-1} connecting the root of G_{i-1} and another M_{i-1}-unsaturated vertex. Set $i = i-1$, and repeating the above procedure until $i = 1$, we get the desired M-augmenting path starting with v_0. Moreover, if G_i has no M_i-augmenting path starting with the root, then G has no M-augmenting path starting with the root v_0.

Some improvements on Algorithm 2.50 results in an algorithm that finds a maximum matching in $O(|G|^3)$ time ([182], Section 9).

2.8 Perfect matchings in cubic graphs

We conclude this chapter with some problems on perfect matchings in cubic graphs. Notice that matchings of a graph are considered as edge subsets of the graph, and that for a cubic simple graph, the edge connectivity is equal to the connectivity. The following conjecture due to Berge and Fulkerson was appeared first in [89] ([217]).

Conjecture 2.51 (Berge and Fulkerson [89]). Every 2-connected cubic simple graph G has six perfect matchings with the property that every edge of G is contained in precisely two of these perfect matchings.

Notice that if the edge set of a cubic graph G is decomposed into three perfect matchings M_1, M_2 and M_3, then every edge is contained in precisely one of these perfect matchings, and so by letting $M_4 = M_1$, $M_5 = M_2$ and $M_6 = M_3$, these six perfect matchings $\{M_i\}$ satisfy the above conjecture. On the other hand, Petersen graph of order 10 has perfect matchings but its edge set cannot be decomposed into three perfect matchings. However, Petersen graph has six perfect matchings having the property given in Conjecture 2.51 (see Fig. 2.44). Thus the conjecture says that every 2-connected cubic graph possesses a certain property closed to the decomposition of edge set into three perfect matchings.

If a cubic graph has six perfect matchings given in Conjecture 2.51, then any three of these perfect matchings have empty intersection. Thus the next conjecture holds if the above conjecture is true.

Conjecture 2.52 (Fan and Raspaud [83]). Every 2-connected cubic simple graph has three perfect matchings M_1, M_2, M_3 such that $M_1 \cap M_2 \cap M_3 = \emptyset$.

Related problems concerning the above conjecture are found in [217]. The following conjecture says that the number of 1-factors of 2-connected cubic graph is large.

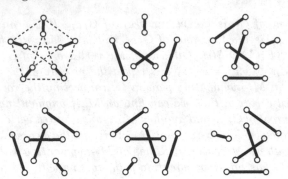

Fig. 2.44. Petersen graph has six perfect matchings such that every edge is contained in precisely two of these perfect matchings.

Conjecture 2.53 (Lovász and Plummer [182] Conjecture 8.1.8). The number of 1-factors of a 2-connected cubic simple graph is exponential in the number of vertices.

Voorhoeve [239] showned that the conjecture holds for bipartite graphs, and Chudnovsky and Seymour showed that it holds for planar graphs. Recently it is announced by Esperet, Kardos, King, Kral and Norine that the conjecture is settled.

Problems

2.1. Prove the marriage theorem by using Theorem 2.15.

2.2. Prove that if a tree has a 1-factor, then it has the unique 1-factor.

2.3. Prove that for every bipartite simple graph G with maximum degree Δ, there exists a Δ-regular bipartite simple graph H which contains G as a subgraph. Note that the two bipartite sets of H might be bigger than those of G.

2.4. Prove the following theorem: Let G be a bipartite simple graph with bipartition (A, B), and let $k \geq 1$ be an integer. Then G has a spanning subgraph H such that

$$\deg_H(x) = 1 \quad \text{for all} \quad x \in A, \quad \text{and}$$
$$\deg_H(y) \leq k \quad \text{for all} \quad y \in B$$

if and only if

$$|N_G(S)| \geq \frac{|S|}{k} \quad \text{for all} \quad S \subseteq A.$$

2.5. Let G be a bipartite simple graph with bipartition (A, B). Prove that if $|A| = |B|$ and $\delta(G) \geq |G|/4$, then G has a 1-factor.

2.6. Verify Theorem 2.16 for the following matrix:

$$M = \begin{pmatrix} 0\,1\,0\,1\,1 \\ 0\,1\,1\,0\,0 \\ 0\,0\,0\,0\,0 \\ 1\,0\,0\,1\,1 \end{pmatrix}$$

2.7. Describe many graphs that have no 1-factors and satisfy $|N_G(S)| \geq |S|$ for all $S \subset V(G)$.

2.8. Prove that a factor-critical graph is connected, of odd order and is not a bipartite graph.

2.9. Let S be a maximal barrier of a graph G. Prove that every component of $G - S$ is factor-critical.

2.10. Prove Theorem 2.31

2.11. For every even integer $r \geq 4$, find an $(r - 2)$-edge connected r-regular simple graph of even order that has no 1-factor.

2.12. Prove statement (ii) of Theorem 2.40.

2.13. Let G be a simple graph, v a vertex of G and M a maximum matching in G, and let M' be a maximum matching in $G - v$. Prove that $||M|| - 1 \leq ||M'|| \leq ||M||$ and $||M'|| = ||M||$ holds if and only if $v \in D(G)$.

2.14. Prove statements (iv) and (v) in Theorem 2.49.

2.15. Prove the following part of Algorithm 2.50: if G_1 has an M_1-augmenting path starting at the root in G_1, then G has an M-augmenting path starting at the root v.

2.16. Let M be a matching in a connected simple graph G and C be an odd cycle of G containing $(|C| - 1)/2$ edges of M, and let $G' = G/C$ be the graph obtained from G by contracting C. Prove that M is a maximum matching in G if and only if $M' = M \cap E(G')$ is a maximum matching in G'.

3

Regular Factors and f-Factors

In this chapter we consider regular factors and f-factors mainly in general graphs. For a positive integer k, a regular spanning subgraph, each of whose vertices has constant degree k, is called a k-**regular factor** or simply a k-**factor** (Fig. 3.1). In order to avoid confusion, we use "k-regular factors" in theorems, but often use "k-factors" in proofs or elsewhere.

3.1 The f-factor theorem

For a general graph G and an integer-valued function

$$f : V(G) \to \mathbb{Z}^+ = \{0, 1, 2, 3, \ldots\},$$

a spanning subgraph F satisfying

$$\deg_F(x) = f(x) \qquad \text{for all} \quad x \in V(G)$$

is called an f-**factor** (Fig. 3.1). For an integer k, if the function f is defined as $f(x) = k$ for all $x \in V(G)$, then an f-factor is equivalent to a k-regular factor, and so an f-factor is a natural generalization of a regular factor. A k-regular factor or an f-factor F is occasionally regarded as its edge set $E(F)$ since the vertex set of F is always $V(G)$ and F is determined by its edge set.

We shall give criteria for a graph to have a k-regular factor and an f-factor, and present results on regular factors and f-factors by using these criteria.

If the edge set $E(G)$ of a graph G is decomposed into edge disjoint k-regular factors $\{F_i\}$ as

$$E(G) = F_1 \cup F_2 \cup \cdots \cup F_m, \qquad F_i \cap F_j = \emptyset \ (i \neq j)$$

then G is said to be k-**factorable** and the decomposition is called a k-**factorization** of G.

The following theorem, which was obtained by Petersen in 1891, is one of the oldest results on factors.

J. Akiyama and M. Kano, *Factors and Factorizations of Graphs*,
Lecture Notes in Mathematics 2031, DOI 10.1007/978-3-642-21919-1_3,
© Springer-Verlag Berlin Heidelberg 2011

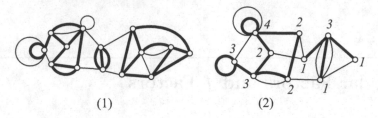

(1)   (2)

Fig. 3.1. (1) A general graph G with a 3-regular factor; (2) A general graph G with an f-factor, where numbers denote $f(v)$.

Theorem 3.1 (The 2-Factorable Theorem, Petersen [209]**).** *For every integer $r \geq 1$, every $2r$-regular general graph is 2-factorable. In particular, for every integer k, $1 \leq k \leq r$, every $2r$-regular general graph has a $2k$-regular factor.*

Proof. The union of k edge-disjoint 2-factors of G forms a $2k$-factor of G, and so the latter part follows immediately from the existence of a 2-factorization.

Let G be a $2r$-regular general graph. Since every vertex of G has even degree, G has an Euler circuit C. Traversing this Euler circuit, we get an orientation of G (Fig. 3.2). It is clear that in the oriented graph $\overrightarrow{G}$, every vertex has in-degree r and out-degree r.

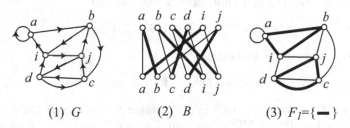

(1) G (2) B (3) F_1={━}

Fig. 3.2. (1) A graph G with Euler circuit $(a, a, b, c, d, c, j, b, i, j, d, i, a)$; (2) The bipartite graph B with a 1-factor H_1; and (3) The 2-factor F_1 corresponding to H_1.

We construct an r-regular bipartite multigraph B with bipartition (X, Y) as follows: two vertices $x \in X = V(G)$ and $y \in Y = V(G)$ are joined by an edge of B if and only if $\overrightarrow{G}$ has a directed edge (xy) (Fig. 3.2). By Theorem 2.2, B can be decomposed into r 1-factors $H_1, H_2, \ldots, H_r$. From each H_i, we obtain a spanning subgraph $F_i = (V(G), H_i)$ of G by regarding each edge of H_i as an edge of G. Since for every vertex v of $\overrightarrow{G}$, H_i has two directed edges expressed as (vx) and (yv), F_i is a 2-factor of the given graph G. Consequently the theorem is proved. □

We now present the most important theorem in this chapter, which is called the f-factor theorem and was obtained by Tutte in 1952. His original proof uses alternating trails and is interesting, but is rather long. In 1954 he gave a new, considerably simpler, proof by making use of the 1-factor theorem. Namely, he deduced an f-factor problem in a graph into a 1-factor problem in a certain large graph, and applied the 1-factor theorem to this large graph. Here we adopt this new proof. Notice that a different form of the criterion for a graph to have a k-regular factor was obtained by Ore ([205]).

Theorem 3.2 (The f-factor Theorem, Tutte [226], [228]). *Let G be a general graph and $f : V(G) \to \mathbb{Z}^+$ be a function. Then G has an f-factor if and only if for all disjoint subsets S and T of $V(G)$,*

$$\delta(S,T) = \sum_{x \in S} f(x) + \sum_{x \in T} (\deg_G(x) - f(x)) - e_G(S,T) - q(S,T) \geq 0, (3.1)$$

where $q(S,T)$ denotes the number of components C of $G - (S \cup T)$ such that

$$\sum_{x \in V(C)} f(x) + e_G(C,T) \equiv 1 \pmod 2, \tag{3.2}$$

and the function $\delta(S,T)$ is defined by (3.1).

Some remarks are in order. For convenience, we call a component C of $G - (S \cup T)$ satisfying (3.2) an **f-odd component** of $G - (S \cup T)$. It is easy to see that

$$\sum_{x \in T} \deg_G(x) - e_G(S,T) = \sum_{x \in T} \deg_{G-S}(x),$$

and thus (3.1) can be expressed as

$$\delta(S,T) = \sum_{x \in S} f(x) + \sum_{x \in T} (\deg_{G-S}(x) - f(x)) - q(S,T) \geq 0. \tag{3.3}$$

Another useful relation, due to Tutte, is the following:

$$\delta(S,T) \equiv \sum_{x \in V(G)} f(x) \pmod 2. \tag{3.4}$$

Thus if $\sum_{x \in V(G)} f(x)$ is even, then

$$\delta(S,T) \equiv 0 \pmod 2. \tag{3.5}$$

Note that the f-factor theorem holds even if $f(v) > \deg_G(v)$ or $f(v) < 0$ for some vertices v of G, since inequality (3.1) includes the trivially necessary condition that $0 \leq f(x) \leq \deg_G(x)$ for every $x \in V(G)$. This will be shown in the proof of sufficiency of the f-factor theorem.

For a constant integer $k \geq 1$, by setting $f(x) = k$ for all $x \in V(G)$, we have the following k-regular factor theorem, which was obtained by Belck and Tutte independently.

Theorem 3.3 (The Regular Factor Theorem, Belck [24] and Tutte [226]**).** *Let $k \geq 1$ be an integer and G be a general graph. Then G has a k-regular factor if and only if for all disjoint subsets S and T of $V(G)$,*

$$\delta(S,T) = k|S| + \sum_{x \in T} \deg_G(x) - k|T| - e_G(S,T) - q(S,T) \qquad (3.6)$$

$$= k|S| + \sum_{x \in T} \deg_{G-S}(x) - k|T| - q(S,T) \geq 0, \qquad (3.7)$$

where $q(S,T)$ denotes the number of components C, called k-odd components, of $G - (S \cup T)$ such that

$$k|C| + e_G(C,T) \equiv 1 \pmod{2}. \qquad (3.8)$$

Proof of the necessity of the f-factor theorem.   Suppose that a general graph G has an f-factor F. We may assume that G is connected since if each component of G satisfies (3.1), then G itself satisfies (3.1). Therefore we have

$$\sum_{x \in V(G)} f(x) = \sum_{x \in V(G)} \deg_F(x) = 2\|F\|,$$

and so G itself is not an f-odd component of $G = G - (\emptyset \cup \emptyset)$. Thus we have $\delta(\emptyset, \emptyset) = -q(\emptyset, \emptyset) = 0$.

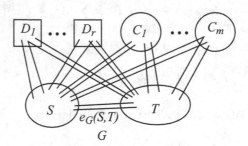

Fig. 3.3. A graph G with disjoint vertex subsets $S \cup T$, the f-odd components $C_1, \ldots, C_m$ and the other components $D_1, \ldots, D_r$.

Let S and T be two disjoint subsets of $V(G)$ such that $S \cup T \neq \emptyset$. Let $G - F$ denote the spanning subgraph of G with edge set $E(G) - E(F)$, and let $C_1, C_2, \ldots, C_m$, where $m = q(S,T)$, be the f-odd components of $G - (S \cup T)$ (Fig. 3.3). By the equation

$$\deg_G(x) - f(x) = \deg_G(x) - \deg_F(x) = \deg_{G-F}(x),$$

we have

$$\sum_{x \in T}(\deg_G(x) - f(x)) = \sum_{x \in T}\deg_{G-F}(x)$$

$$\geq e_{G-F}(T, S) + \sum_{i=1}^{m} e_{G-F}(T, C_i).$$

Furthermore, it follows that

$$\sum_{x \in S} f(x) = \sum_{x \in S}\deg_F(x) \geq e_F(S, T) + \sum_{i=1}^{m} e_F(S, C_i).$$

By the two previous inequalities, we have

$$\delta(S, T) \geq e_F(S, T) + \sum_{i=1}^{m} e_F(S, C_i) + e_{G-F}(T, S) + \sum_{i=1}^{m} e_{G-F}(T, C_i)$$

$$-e_G(S, T) - m$$

$$= \sum_{i=1}^{m}\Big(e_F(S, C_i) + e_{G-F}(T, C_i) - 1\Big).$$

Therefore, in order to prove (3.1) it is sufficient to show that for every $C = C_i$,

$$e_F(S, C) + e_{G-F}(T, C) - 1 \geq 0. \qquad (3.9)$$

If $e_{G-F}(T, C) \geq 1$, then (3.9) holds, and so we may assume that $e_{G-F}(T, C) = 0$, which implies

$$e_G(T, C) = e_F(T, C).$$

Hence

$$\sum_{x \in V(C)} f(x) + e_G(C, T) = \sum_{x \in V(C)} \deg_F(x) + e_F(C, T)$$

$$= 2||\langle V(C)\rangle_F|| + e_F(C, S \cup T) + e_F(C, T)$$

$$\equiv e_F(C, S) \pmod 2.$$

By (3.2) and the above equation, we have $e_F(C, S) \geq 1$, which implies that (3.9) holds. Consequently the necessity is proved. $\square$

Proof of (3.4). Let $C_1, C_2, \ldots, C_m$, $m = q(S, T)$, be the f-odd components of $G - (S \cup T)$, and $D_1, D_2, \ldots, D_r$ be the other components of $G - (S \cup T)$ (Fig. 3.3). Then

$$\sum_{x \in V(D_j)} f(x) + e_G(D_j, T) \equiv 0 \pmod 2 \qquad \text{for every} \quad 1 \leq j \leq r.$$

Hence

$$m \equiv \sum_{i=1}^{m} \left(\sum_{x \in V(C_i)} f(x) + e_G(C_i, T) \right)$$

$$\equiv \sum_{i=1}^{m} \left(\sum_{x \in V(C_i)} f(x) + e_G(C_i, T) \right) + \sum_{j=1}^{r} \left(\sum_{x \in V(D_j)} f(x) + e_G(D_j, T) \right)$$

$$= \sum_{x \in V(G) - (S \cup T)} f(x) + e_G(V(G) - (S \cup T), T) \pmod 2.$$

Therefore

$$\delta(S, T) \equiv \sum_{x \in S} f(x) + \sum_{x \in T} (\deg_G(x) + f(x)) + e_G(S, T) + m$$

$$\equiv \sum_{x \in S} f(x) + \sum_{x \in T} (\deg_G(x) + f(x)) + e_G(S, T)$$

$$+ \sum_{x \in V(G) - (S \cup T)} f(x) + e_G(V(G) - (S \cup T), T)$$

$$= \sum_{x \in V(G)} f(x) + 2\|\langle T \rangle_G\| + e_G(T, V(G) - T) + e_G(T, S)$$

$$+ e_G(T, V(G) - (S \cup T))$$

$$\equiv \sum_{x \in V(G)} f(x) \pmod 2.$$

Consequently (3.4) holds. □

We now prove the sufficiency of the f-factor theorem based on Tutte [228], ([32], Section 2.3).

Proof of the sufficiency of the f-factor theorem. If a general graph G satisfies condition (3.1), then each of its components satisfies (3.1), and if each component of G has the desired f-factor, then their union is the desired f-factor of G. Hence we may assume that G is connected.

For each vertex v of G, we have by (3.1)

$$\delta(\emptyset, \{v\}) = \deg_G(v) - f(v) - q(\emptyset, \{v\}) \geq 0, \quad \text{and}$$
$$\delta(\{v\}, \emptyset) = f(v) - q(\{v\}, \emptyset) \geq 0,$$

and so $0 \leq f(v) \leq \deg_G(v)$.

Since $\delta(\emptyset, \emptyset) = -q(\emptyset, \emptyset) \geq 0$, we have $q(\emptyset, \emptyset) = 0$, which implies

$$\sum_{x \in V(G)} f(x) \equiv 0 \pmod 2. \tag{3.10}$$

We now construct a new simple graph G^* from G as follows: for every edge $e = xy \in E(G)$, where we allow $x = y$, we introduce two vertices $e(x)$ and

$e(y)$ of G^*, and join them by an edge e of G^*. Thus for every vertex v of G, if $\{e_1, e_2, \ldots, e_r\}$ is the set of edges incident with v in G, then G^* has a set

$$S(v) = \{e_1(v), e_2(v), \ldots, e_r(v)\}$$

of vertices corresponding to v (Fig. 3.4).

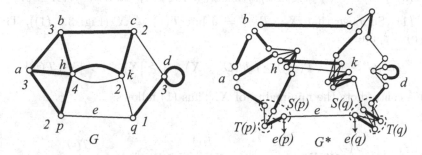

Fig. 3.4. A general graph G with an f-factor; and the corresponding simple graph G^* with a 1-factor.

Next, for every vertex v of G, we add a set

$$T(v) = \{v_1, \ldots, v_s\}, \qquad s = \deg_G(v) - f(v)$$

of vertices to G^*, and join every vertex of $T(v)$ to every vertex of $S(v)$ (Fig. 3.4). Of course, if $f(v) = \deg_G(v)$, then $T(v) = \emptyset$. Hence we have

$$V(G^*) = S^* \cup T^*, \qquad \text{where}$$
$$S^* = \bigcup_{v \in V(G)} S(v) \quad \text{and} \quad T^* = \bigcup_{v \in V(G)} T(v); \quad \text{and} \qquad (3.11)$$
$$E(G^*) = E(G) \cup \{xy : x \in S(v) \text{ and } y \in T(v)\}. \qquad (3.12)$$

We next show that G has an f-factor if and only if G^* has a 1-factor. Assume that G has an f-factor F. Then by regarding $E(F)$ as a subset of $E(G^*)$, we can get a matching that covers $f(v)$ vertices of $S(v)$ for every $v \in V(G)$. The remaining uncovered vertices of $S(v)$ can be covered by a matching of G^* that joins the remaining uncovered vertices of $S(v)$ to all the vertices of $T(v)$. Therefore G^* has a 1-factor. Conversely, if G^* has a 1-factor K, then $E(G) \cap E(K) \subseteq E(G^*)$ induces an f-factor of G by regarding it as a set of edges of G. Therefore G has an f-factor if and only if G^* has a 1-factor (Fig. 3.4).

Suppose that G satisfies condition (3.1) but has no f-factor. Then G^* has no 1-factor. By the 1-factor theorem (Theorem 2.27), there exists a subset $Y \subset V(G^*)$ such that $odd(G^* - Y) > |Y|$. Choose a minimal set X among such subsets Y. Then $X \neq \emptyset$, and by Lemma 2.26

$$odd(G^* - X) \geq |X| + 2.$$

Moreover, the following claim holds.

Claim 1. *Every vertex v of G satisfies the following statements.*
(1) *If $X \cap T(v) \neq \emptyset$, then $T(v) \subseteq X$.*
(2) *If $X \cap S(v) \neq \emptyset$, then $S(v) \subseteq X$.*
(3) *$S(v) \cup T(v) \nsubseteq X$. In particular, if $T(v) = \emptyset$ then $S(v) \cap X = \emptyset$.*

(1) Suppose that $X \cap T(v) \neq \emptyset$ but $T(v) \nsubseteq X$ (Fig. 3.5 (1)). Then $T(v) \neq \emptyset$, and

$$odd(G^* - (X \setminus T(v))) \geq odd(G^* - X) - 1 \geq |X| + 1 > |X \setminus T(v)|.$$

This contradicts the minimality of X. Thus (1) follows.

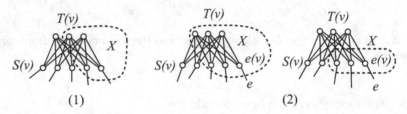

Fig. 3.5. (1) $X \cap T(v) \neq \emptyset$ but $T(v) \nsubseteq X$; (2) $X \cap S(v) \neq \emptyset$ but $S(v) \nsubseteq X$.

(2) Suppose that $X \cap S(v) \neq \emptyset$ but $S(v) \nsubseteq X$ (Fig. 3.5 (2)). Let $e(v) \in S(v) \cap X$, where $e = vu \in E(G)$. Then $e(v)$ is adjacent to at most two components of $G^* - X$ by $S(v) \setminus X \neq \emptyset$ (Fig. 3.5 (2)). Hence

$$odd\big(G^* - (X - \{e(v)\})\big) \geq odd(G^* - X) - 2 \geq |X| > |X - \{e(v)\}|.$$

This contradicts the choice of X. Hence (2) holds.

(3) Assume $S(v) \cup T(v) \subseteq X$ for some $v \in V(G)$. Choose a vertex $e(v) \in S(v)$. Then

$$odd(G^* - (X - \{e(v)\})) \geq odd(G^* - X) - 1 \geq |X| + 1 > |X - \{e(v)\}|,$$

a contradiction. Hence (3) holds.

We define

$$X(S) = \{v \in V(G) : S(v) \subseteq X\},$$
$$X(T) = \{v \in V(G) : T(v) \subseteq X\}.$$

Therefore, $T(v) = \emptyset$ implies $v \in X(T)$, and the next claim holds.

Claim 2. *The following statements hold (see (3.11)).*

(4)   The number of isolated vertices of $G^* - X$ contained in S^* is $e_G(X(T), X(S))$ (Fig. 3.6).

(5)   The number of isolated vertices of $G^* - X$ contained in T^* is $\sum_{x \in X(S)}(\deg_G(x) - f(x))$ (Fig. 3.6).

(6)   The number of odd components of $G^* - X$, whose order is at least three, is equal to $q(X(S), X(T))$.

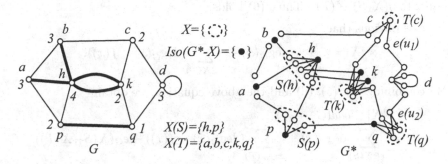

Fig. 3.6. G: G with $X(S)$ and $X(T)$, where bold lines denote $E_G(X(S), X(T))$; G^*: G^* with X. The isolated vertices of $G^* - X$ are denoted by black circles. C^* with $V(C^*) = S(d) \cup T(d) \cup \{e(u_1), e(u_2)\}$ is an odd component of $G^* - X$, and $V(C) = \{d\}$.

Let $e(v)$ be an isolated vertex of $G^* - X$ contained in S^*, and let $e = vw \in E(G) \subset E(G^*)$. Then $T(v) \subseteq X$ and $S(w) \subseteq X$, and thus $v \in X(T)$ and $w \in X(S)$, which implies $e \in E_G(X(T), X(S))$. Conversely, if $e = vw \in E_G(X(T), X(S))$, then $T(v) \subseteq X$ and $S(w) \subseteq X$, and so $e(v)$ is an isolated vertex of $G^* - X$ contained in S^*. Hence (4) holds.

(5) This follows immediately from (2) and the definition of G^*.

(6) Let C^* be an odd component of $G^* - X$ with order at least three (Fig. 3.6). It is clear that if C^* contains an edge joining a vertex of $S(x)$ to a vertex of $T(x)$, then C^* must contain $S(x) \cup T(x)$. Hence we can write

$$V(C^*) = \bigcup_x (S(x) \cup T(x)) \cup \{e(u_1), \dots, e(u_r)\},$$

where $e(u_i)$ are the end-vertices of C^* such that $T(u_i) \subseteq X$. Let

$$C = \langle\{x \in V(G) : S(x) \cup T(x) \subseteq V(C^*)\}\rangle_G.$$

Then C is a component $G - (X(S) \cup X(T))$, and

$$|C^*| = \sum_{x \in V(C)} (2\deg_G(x) - f(x)) + |\{e(u_1), \dots, e(u_r)\}|$$

$$\equiv \sum_{x \in V(C)} f(x) + e_G(C, X(T)) \tag{3.13}$$

$$\equiv 1 \pmod 2 \qquad \text{(since } C^* \text{ is an odd component of } G^* - X.)$$

By (3.13), C is an f-odd component of $G - (X(S) \cup X(T))$. Therefore, the number of those odd components of $G^* - X$ whose order is at least three is equal to $q(X(S), X(T))$. Thus, (6) holds.

It is obvious that

$$|X| = \sum_{x \in X(S)} \deg_G(x) + \sum_{x \in X(T)} (\deg_G(x) - f(x)).$$

By combining (4), (5), (6) and the above equation, we obtain

$$|X| - odd(G^* - X)$$
$$= \sum_{x \in X(S)} \deg_G(x) + \sum_{x \in X(T)} (\deg_G(x) - f(x)) - e_G(X(S), X(T))$$
$$\quad - \sum_{x \in X(S)} (\deg_G(x) - f(x)) - q(X(S), X(T))$$
$$= \sum_{x \in X(S)} f(x) + \sum_{x \in X(T)} (\deg_G(x) - f(x))$$
$$\quad - e_G(X(S), X(T)) - q(X(S), X(T))$$
$$= \delta(X(S), X(T)).$$

By (3.1), we can derive a contradiction since

$$-2 \geq |X| - odd(G^* - X) = \delta(X(S), X(T)) \geq 0.$$

Consequently the proof of the sufficiency is complete.   □

We have now proved the f-factor theorem and k-regular factor theorem. However, it is not so easy to apply these theorems to problems on f-factors or k-regular factors because $\delta(S,T)$ is fairly complicated and the number $q(S,T)$ of f-odd components of $G - (S \cup T)$ may be difficult to estimate. Consequently, many results in the following sections were obtained in the 1980's, although the k-regular factor theorem was proved in the early 1950's.

We conclude this section with a remark on the f-factor theorem in bipartite graphs. In particular, the f-factor theorem can be simplified in such a setting, as we shall see in Section 3.4.

3.2 Regular factors in regular graphs

In this section, we first consider regular factors in regular graphs. We already showed that every $2r$-regular graph has a $2k$-regular factor for every k,

$1 \leq k \leq r$. Furthermore, some results on 1-factors in an r-regular graph show the existence of $(r-1)$-regular factors in the r-regular graph by taking the complement of the 1-factor. For example, the existence of a 1-factor in every 2-edge connected cubic graph implies the existence of a 2-factor in such a graph. The proof of the next theorem is based on Kano [115]. Its proof technique is called a θ-**method**, which is a technique we will often resort to in later sections.

Theorem 3.4 (Regular Factors in Regular Graphs). *Let r, k and λ be integers such that $1 \leq k < r$ and $1 \leq \lambda$, and let G be a λ-edge connected r-regular general graph. If one of the following conditions holds, then G has a k-regular factor.*
(i) r and k are both even. (Petersen [209] (1891))
(ii) r is even, k is odd, $|G|$ is even, and $r/\lambda \leq k \leq r(1-1/\lambda)$. (Gallai [91] (1950))
(iii) r is odd, k is even and $2 \leq k \leq r(1-1/\lambda)$. (Gallai [91] (1950))
(iv) r and k are both odd and $r/\lambda \leq k$. (Gallai [91], Bäbler [22])
(v)   Let λ be an integer such that $\lambda^* \in \{\lambda, \lambda+1\}$ and $\lambda^* \equiv 1 \pmod 2$. Either r is odd, k is even and $k \leq r(1-1/\lambda^*)$; or both r and k are odd and $r/\lambda^* \leq k$. (Bollobás, Saito and Wormald [33])

Proof. We use the regular factor theorem (Theorem 3.3). It is immediate that $\delta(\emptyset, \emptyset) = 0$ since $k|G|$ is even. Let S and T be disjoint subsets of $V(G)$ such that $S \cup T \neq \emptyset$, and let $C_1, C_2, \ldots, C_m$ be the k-odd components of $G-(S \cup T)$. Set $\theta = k/r$. Then $0 < \theta < 1$, and we have

$$\delta(S,T) = k|S| + \sum_{x \in T} \deg_G(x) - k|T| - e_G(S,T) - q(S,T)$$
$$= k|S| + (r-k)|T| - e_G(S,T) - q(S,T)$$
$$= \theta r|S| + (1-\theta)r|T| - e_G(S,T) - m$$
$$= \theta \sum_{x \in S} \deg_G(x) + (1-\theta) \sum_{x \in T} \deg_G(x) - e_G(S,T) - m$$
$$\geq \theta \Big(e_G(S,T) + \sum_{i=1}^{m} e_G(S,C_i) \Big)$$
$$+ (1-\theta) \Big(e_G(T,S) + \sum_{i=1}^{m} e_G(T,C_i) \Big)$$
$$- e_G(S,T) - m$$
$$= \sum_{i=1}^{m} (\theta e_G(S,C_i) + (1-\theta)e_G(T,C_i) - 1).$$

Hence it suffices to show that for every $C = C_i$, $1 \leq i \leq m$,

$$\theta e_G(S,C) + (1-\theta)e_G(T,C) \geq 1. \tag{3.14}$$

Since C is a k-odd component of $G - (S \cup T)$, we have

$$k|C| + e_G(T, C) \equiv 1 \pmod 2. \tag{3.15}$$

Moreover, since G is λ-edge connected and

$$r|C| = \sum_{x \in V(C)} \deg_G(x) = e_G(S \cup T, C) + 2\|C\|,$$

we have

$$r|C| \equiv e_G(S \cup T, C) \pmod 2 \quad \text{and} \quad e_G(S \cup T, C) \geq \lambda. \tag{3.16}$$

It is obvious that the two inequalities $e_G(S, C) \geq 1$ and $e_G(T, C) \geq 1$ imply

$$\theta e_G(S, C) + (1 - \theta) e_G(T, C) \geq \theta + (1 - \theta) = 1.$$

Hence we may assume $e_G(S, C) = 0$ or $e_G(T, C) = 0$. We consider the following two cases under the assumption that one of conditions (i)-(iv) is satisfied. Condition (v) will be considered later.

Case 1. $e_G(S, C) = 0$.

If both r and k are even, then by (3.15) and (3.16), we have

$$e_G(T, C) \equiv 1 \pmod 2 \quad \text{and} \quad 0 \equiv e_G(T, C) \pmod 2.$$

This is a contradiction. If $k \leq r(1 - 1/\lambda)$, then $\theta \leq 1 - 1/\lambda$ and so $1 \leq (1 - \theta)\lambda$. By substituting (3.16) and $e_G(S, C) = 0$ into the left side of (3.14), we have

$$(1 - \theta) e_G(T, C) \geq (1 - \theta)\lambda \geq 1.$$

If both r and k are odd, then by (3.15) and (3.16), we have

$$|C| + e_G(T, C) \equiv 1 \pmod 2 \quad \text{and} \quad |C| \equiv e_G(T, C) \pmod 2.$$

This is again a contradiction. Therefore (3.14) follows in this case.

Case 2. $e_G(T, C) = 0$.

By (3.15) we have $k|C| \equiv 1 \pmod 2$, which implies that k must be odd. Hence we may assume that k is odd. Then $r/\lambda \leq k$ by (ii) and (iv). Hence $1 \leq \theta\lambda$, and so by (3.16) we obtain

$$\theta e_G(S, C) \geq \theta\lambda \geq 1.$$

Consequently, (3.14) holds, and thus every condition (i)-(iv) guarantees the existence of k-factors.

In order to prove the case of (v), it suffices to show that (3.14) holds under the assumption that $e_G(S, C) = 0$ or $e_G(T, C) = 0$. We first consider the case where r is odd, k is even and $k \leq r(1 - 1/\lambda^*)$, which implies $(1 - \theta)\lambda^* \geq 1$. If $e_G(S, C) = 0$, then by (3.15) and (3.16), we have

$$e_G(T, C) \equiv 1 \pmod 2 \quad \text{and} \quad e_G(T, C) \geq \lambda.$$

Hence $e_G(T, C) \geq \lambda^*$, and thus

$$(1 - \theta)e_G(T, C) \geq (1 - \theta)\lambda^* \geq 1.$$

If $e_G(T, C) = 0$, then by (3.15), we have $k|C| \equiv 1 \pmod 2$, which contradicts the assumption that k is even.

We next consider the second part of (v), i.e., we assume that both r and k are odd and $r/\lambda^* \leq k$, which implies $\theta\lambda^* \geq 1$. If $e_G(S, C) = 0$, then by (3.15) and (3.16), we have

$$|C| + e_G(T, C) \equiv 1 \pmod 2 \quad \text{and} \quad |C| \equiv e_G(T, C) \pmod 2.$$

This is a contradiction. If $e_G(T, C) = 0$, then by (3.15) and (3.16), we have

$$|C| \equiv 1 \pmod 2 \quad \text{and} \quad |C| \equiv e_G(S, C) \pmod 2,$$

which implies $e_G(S, C) \geq \lambda^*$. Thus

$$\theta e_G(S, C) \geq \theta\lambda^* \geq 1.$$

Consequently (3.14) holds under condition (v), and the proof is complete.   □

We now show that every $(r - 1)$-edge connected r-regular multigraph G has a 1-factor by making use of Theorem 3.4. First assume that r is even. Then an $(r - 1)$-edge connected r-regular graph must be r-edge connected because G is an Euler graph. Hence

$$\frac{r}{\lambda} = \frac{r}{r} = 1 = k \leq r\left(1 - \frac{1}{r}\right) = r - 1,$$

and so G has a 1-factor by (ii) of Theorem 3.4. We next consider the case where r is odd. We use (v) of Theorem 3.4. Since $\lambda = r - 1$ is even, we have $\lambda^* = r$. Hence

$$\frac{r}{\lambda^*} = \frac{r}{r} = 1 = k,$$

which implies that G has a 1-factor.

The following theorem is an immediate consequence of Theorem 3.4, and is a nice generalization of Petersen's theorem, which says that every 2-connected cubic graph has a 2-factor.

Theorem 3.5 (Bäbler [22] (1938)). *(i) Let $r \geq 3$ be an odd integer. Then every 2-edge connected r-regular general graph has a 2-regular factor.*
(ii) Let $r \geq 3$ be an odd integer and k an even integer such that $2 \leq k < r$. Then every k-edge connected r-regular general graph has a k-regular factor.

Proof. We shall prove only (ii). We apply (v) of Theorem 3.4 to an odd integer r and an even integer k such that $2 \leq k \leq r - 1$. Since $\lambda = k$ and $\lambda^* = k + 1$ in (v), we have

$$r\left(1 - \frac{1}{\lambda^*}\right) = r\left(1 - \frac{1}{k+1}\right) \geq (k+1)\left(1 - \frac{1}{k+1}\right) = k.$$

Hence the graph has a k-factor. □

Theorem 3.6 (Katerinis [144]). *Let $r \geq 2$ be an even integer and $1 \leq k \leq r/2$ be an integer. Let G be an r-edge connected r-regular multigraph of odd order. Then for every vertex v of G, $G - v$ has a k-regular factor.*

We now show that some of the conditions in Theorem 3.4 are sharp (Bollobás, Saito and Wormald, [33]). Let $r \geq 2$ be an even integer, $k \geq 1$ an odd integer and $\lambda \geq 1$ an even integer such that $k < r/\lambda$ or $r(1 - 1/\lambda) < k$.

For integers r, t, λ such that $1 \leq \lambda \leq t \leq r$, let $H(r, t; \lambda)$ be a λ-edge connected bipartite simple graph with bipartition (U, W) such that every vertex of U has degree r and every vertex of W has degree t, and let $J(r, t; \lambda)$ be a λ-edge connected simple graph of even order in which one vertex, say v_0, has degree t and all the other vertices have degree r (Fig. 3.7). We now construct our graph $G(r, \lambda)$ from $H = H(r, \lambda; \lambda)$ and $|W|$ copies of $J = J(r, \lambda; \lambda)$'s as follows: for every vertex $w \in W$ of H, replace w by J, i.e., remove w from H, remove v_0 from J and join every $u \in U$ adjacent to w in H to a vertex of J adjacent to v_0 in such a way that the resulting graph is r-regular (Fig. 3.7). It is immediate that $G(r, \lambda)$ is a λ-edge connected r-regular simple graph.

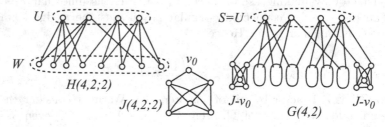

Fig. 3.7. Graphs $H(4, 2, ; 2)$, $J(4, 2)$ and $G(4, 2)$.

We now show that the graph $G = G(r, \lambda)$ has no k-regular factor for every odd integer k such that $k < r/\lambda$ or $r(1 - 1/\lambda) < k$. Let $S = U$ and $T = \emptyset$. Then each component $J - v_0$ of $G - S$ is a k-odd component since

$$k|J - v_0| + e_G(V(J - v_0), \emptyset) = k|J - v_0| \equiv 1 \pmod 2.$$

Thus if $k < r/\lambda$, then

$$\delta(U, \emptyset) = k|U| - q(U, \emptyset) < r/\lambda|U| - |W| = 0.$$

Hence G has no k-factor by the regular factor theorem. Moreover, G has no $(r-k)$-factor, which implies $k' = r - k$ satisfies $r(1 - 1/\lambda) < k'$ and an r-regular graph G has no k'-factor.

We consider graphs $H(r, \lambda^*; \lambda)$ and $J(r, \lambda^*; \lambda)$ when r is odd and $\lambda \geq 1$ and obtain a graph $G(r, \lambda)$. Then we can show that $G = G(r, \lambda)$ has no k-regular factor if either k is even and $k > r(1 - 1/\lambda^*)$, or k is odd and $k < r/\lambda^*$. The proof is left to the reader (Problem 3.1).

The next theorem says that every regular graph of small order has a regular factor. Of course, a connected 2-regular simple graph is a cycle, and so if $r = 2$ and $k = 1$, the statement (i) in the following theorem holds for every G. Avoiding this trivial case, we assume $r \geq 3$.

Theorem 3.7 (Niessen and Randerath [197]). *Let k and r be integers such that $1 \leq k < r$ and $3 \leq r$, and let G be a connected r-regular simple graph. If $k|G|$ is even and one of the following conditions holds, then G has a k-regular factor.*
(i) r is even, k is odd, and $|G| \leq 3r + 3$.
(ii) r and k are both odd, and $|G| \leq (r+2)(k+2)$.
(iii) r is odd, k is even, and $|G| \leq (r+2)(r-k+2)$.

Proof. Let S and T be disjoint subsets of $V(G)$ such that $S \cup T \neq \emptyset$, and let $C_1, C_2, \ldots, C_m$ be the k-odd components of $G - (S \cup T)$. Set $\theta = k/r$. Then we have

$$\delta(S,T) = k|S| + \sum_{x \in T} \deg_G(x) - k|T| - e_G(S,T) - q(S,T)$$

$$= k|S| + (r-k)|T| - e_G(S,T) - q(S,T)$$

$$= \theta r|S| + (1-\theta)r|T| - e_G(S,T) - m$$

$$= \theta \sum_{x \in S} \deg_G(x) + (1-\theta) \sum_{x \in T} \deg_G(x) - e_G(S,T) - m$$

$$\geq \theta \Big(e_G(S,T) + \sum_{i=1}^{m} e_G(S,C_i) \Big)$$

$$+ (1-\theta)\Big(e_G(T,S) + \sum_{i=1}^{m} e_G(T,C_i) \Big) - e_G(S,T) - m$$

$$= \sum_{i=1}^{m} \big(\theta e_G(S,C_i) + (1-\theta)e_G(T,C_i) - 1 \big).$$

We first prove the theorem in case (i). We may assume that $k \leq r/2$ since if G has a k-factor F for an odd integer $1 \leq k \leq r/2$, then $G - F$ is an $(r-k)$-factor and $r-k$ is an odd integer with $r/2 \leq r-k < r$. Thus $\theta \leq 1/2$, and so $1 - \theta \geq \theta$ and

$$\theta e_G(S,C_i) + (1-\theta)e_G(T,C_i) \geq \theta e_G(C_i, S \cup T).$$

We may assume that $1 \leq |C_i| \leq r$ for $1 \leq i \leq a$ and $|C_i| \geq r+1$ for $a+1 \leq i \leq a+b = m$.

Suppose $1 \leq |C_j| \leq r$. Since G is a simple graph, for every vertex $x \in V(C_j)$, $|N_G(x) \cap (S \cup T)| \geq r - |C_j| + 1$. Hence

$$e_G(C_j, S \cup T) \geq |C_j|(r - |C_j| + 1) \geq r$$

because the last inequality holds when $|C_j| = 1$ and $|C_j| = r$ and $|C_j|(r - |C_j| + 1)$ is an upward-convex function of $|C_j|$. Therefore

$$\theta e_G(C_j, S \cup T) - 1 \geq \theta r - 1 = k - 1 \geq 0.$$

There exist at most two components C_i with $|C_i| \geq r+1$ as $|G| \leq 3r+3$ and $S \cup T \neq \emptyset$. Thus $b \leq 2$, and we obtain

$$\delta(S,T) \geq \sum_{i=1}^{m} \left(\theta e_G(C_i, S \cup T) - 1 \right)$$

$$\geq \sum_{i=1}^{b} \left(\theta e_G(C_{a+i}, S \cup T) - 1 \right) \geq 2(\theta - 1) > -2,$$

which implies $\delta(S,T) \geq 0$ by $\delta(S,T) \equiv 0 \pmod 2$ (see (3.5)). Consequently, if G satisfies (i), then G has a k-factor.

We next consider case (ii). Let $C_1, C_2, \ldots, C_m$ be the k-odd components of $G - (S \cup T)$ such that

$|C_i| \leq r$ for $1 \leq i \leq a$;
$|C_i| \geq r+1$ and $e_G(T, C_i) \geq 1$ for $a < i \leq a+b$; and
$|C_i| \geq r+1$ and $e_G(T, C_i) = 0$ for $a+b < i \leq a+b+c = m$,
where some of a, b, c are possibly 0.

Since C_i is a k-odd component, we have

$$k|C_i| + e_G(T, C_i) \equiv |C_i| + e_G(T, C_i) \equiv 1 \pmod 2.$$

If $e_G(S, C_i) = 0$, then

$$|C_i| \equiv r|C_i| = \sum_{x \in V(X_i)} \deg_G(x) = e_G(S \cup T, C_i) + 2\|C_i\|$$

$$\equiv e_G(T, C_i) \pmod 2.$$

The two congruences above result in a contradiction. Therefore we may assume

$$e_G(S, C_i) \geq 1 \qquad \text{for all} \quad 1 \leq i \leq m. \tag{3.17}$$

If $a \geq 1$, then for every C_i, $1 \leq i \leq a$, by the same argument as before, we have $e_G(S \cup T, C_i) \geq r$, and so

$$\theta e_G(S, C_i) + (1 - \theta)e_G(T, C_i) \geq \min\{\theta, 1 - \theta\}e_G(S \cup T, C_i)$$
$$\geq \min\{\theta, 1 - \theta\}r = \min\{k, r - k\}. \tag{3.18}$$

For every C_i, $a < i \leq a + b$, we have

$$\theta e_G(S, C_i) + (1 - \theta)e_G(T, C_i) - 1 \geq \theta + 1 - \theta - 1 = 0.$$

If $a + b < i \leq a + b + c$, then $k|C_i| + e_G(T, C_i) \equiv |C_i| \equiv 1 \pmod 2$, and so $|C_i| \geq r + 2$, which implies $c \leq k$ since $|G| \leq (r + 2)(k + 1)$ and $S \cup T \neq \emptyset$.

Claim. *If $a \geq 1$, then $\delta(S, T) \geq 0$. Hence we may assume $a = 0$.*

Suppose $a \geq 1$. If $k \leq r/2$, then

$$\delta(S, T) \geq \sum_{i=1}^{a} \left(\theta e_G(S, C_i) + (1 - \theta)e_G(T, C_i) - 1\right)$$

$$+ \sum_{i=a+b+1}^{a+b+c} \left(\theta e_G(S, C_i) - 1\right)$$

$$\geq (k - 1)a + (\theta - 1)c \qquad \text{(by (3.18))}$$

$$\geq (k - 1) + (\theta - 1)k = k\theta - 1 > -2. \quad \text{(by } c \leq k)$$

This implies $\delta(S, T) \geq 0$.
    If $k > r/2$, then

$$\delta(S, T) \geq \sum_{i=1}^{a} \left(\theta e_G(S, C_i) + (1 - \theta)e_G(T, C_i) - 1\right)$$

$$+ \sum_{i=a+b+1}^{a+b+c} \left(\theta e_G(S, C_i) - 1\right)$$

$$\geq (r - k - 1)a + (\theta - 1)c \geq r - k - 1 + (\theta - 1)k$$

$$= r - 2k - 1 + \frac{k^2}{r} > r - 2r - 1 + \frac{r^2}{r} = -1. \quad \text{(by } r/2 < k < r)$$

This implies $\delta(S, T) \geq 0$. Therefore G has a k-factor, and so we may assume that $a = 0$. We consider the following two cases.

Case 1. $|S| > |T|$.

In this case, $|S| - |T| \geq 1$ and $m = b + c \leq k + 1$ as $|G| \leq (r + 2)(k + 1) < (r + 1)(k + 2)$. Hence

$$\delta(S, T) \geq \theta r|S| + (1 - \theta)r|T| - e_G(S, T) - m$$

$$\geq \theta r + r|T| - e_G(S, T) - m$$

$$\geq k + \sum_{x \in T} \deg_G(x) - e_G(S, T) - (k + 1) \geq -1.$$

Thus $\delta(S,T) \geq 0$.

Case 2. $|S| \leq |T|$.

It follows that

$$\begin{aligned}
\delta(S,T) &\geq \theta r|S| + (1-\theta)r|T| - e_G(S,T) - m \\
&\geq \theta r|S| + (1-\theta)r|S| - e_G(S,T) - m \\
&= \sum_{x \in S} \deg_G(x) - e_G(S,T) - m \\
&\geq e_G(S,T) + \sum_{i=1}^{m} e_G(S,C_i) - e_G(S,T) - m \\
&\geq 0. \qquad\qquad\qquad \text{(by (3.17))}
\end{aligned}$$

Therefore $\delta(S,T) \geq 0$. Consequently, G has a k-factor under condition (ii).

(iii) is an easy corollary of (ii) since if r is odd and k is even, then $r - k$ is odd and so G has an $(r-k)$-factor F by (ii), and then $G - F$ is a k-factor. □

Theorem 3.8 (Katerinis [137]). *Let a, b, k be odd integers such that $1 \leq a < k < b$. If a general graph G has both an a-regular factor and a b-regular factor, then G has a k-regular factor. In particular, if an r-regular general graph H has a 1-factor, then for every integer h, $1 \leq h \leq r$, H has an h-regular factor.*

Proof. We first prove that G has a k-factor. Assume that G has no k-factor. Then by the regular factor theorem, there exist two disjoint subsets S and T of $V(G)$ such that

$$\delta(S,T) = k|S| - \sum_{x \in T} \deg_G(x) - k|T| + e_G(S,T) - q(S,T;k) < 0, \qquad (3.19)$$

where $q(S,T;k)$ denote the number of k-odd components C of $G - (S \cup T)$ that satisfy

$$k|C| + e_G(C,T) \equiv 1 \pmod 2. \qquad (3.20)$$

Since G has both an a-factor and a b-factor, we have

$$a|S| - \sum_{x \in T} \deg_G(x) - a|T| + e_G(S,T) - q(S,T;a) \geq 0$$

$$b|S| - \sum_{x \in T} \deg_G(x) - b|T| + e_G(S,T) - q(S,T;b) \geq 0.$$

Since a, b and k are all odd integers, by (3.20) we can easily see that $q(S,T;k) = q(S,T;a) = q(S,T;b)$. By the above two inequalities and (3.19), we obtain

$$(a - k)(|S| - |T|) > 0 \quad \text{and} \quad (b - k)(|S| - |T|) > 0.$$

If $|S| \geq |T|$, then $(a - k)(|S| - |T|) \leq 0$ since $a < k$, which is a contradiction. Similarly, if $|S| < |T|$, then $(b - k)(|S| - |T|) \leq 0$ since $k < b$, which is again a contradiction. Consequently G has a k-factor.

Suppose that an r-regular general graph H has a 1-factor F_1. If r is odd, then by the statement proved above, for every odd integer $c, 1 < c < r$, H has a c-factor F_c, and also an $(r - c)$-factor $H - E(F_c)$, where $r - c$ is an even number. Therefore for every integer $h, 1 \leq h \leq r$, H has an h-factor.

If r is even, then $H - F_1$ is an $(r - 1)$-factor. Hence for every odd integer $c, 1 \leq c \leq r - 1$, H has a c-factor by the statement proved above. By the 2-factorable theorem (Theorem 3.1), it is obvious that for every even integer $d, 2 \leq d \leq r$, H has a d-factor. Hence the latter part of this theorem is proved. □

By a similar argument to the one used in the proof of Theorem 3.8 above, the following can be proved. Note that its short proof was given by Volkmann [238].

Theorem 3.9 (Kano and Yu [136]). *Let G be a connected r-regular general graph of even order, where $r \geq 2$. Assume that for every edge of G, G has a 1-factor containing it. Then for every edge e of G and every integer k, $1 \leq k \leq r - 1$, G has a k-regular factor containing e and another k-regular factor excluding e.*

The next theorem shows the existence of a k-regular factor that contains a given edge or excludes such an edge. The proof given here is based on [114].

Theorem 3.10. *Let r, k and λ be integers such that $1 \leq k < r$ and $2 \leq \lambda$, and let G be a λ-edge connected r-regular general graph. If $k|G|$ is even, and G satisfies one the following conditions, which are the same as in Theorem 3.4, then for any given edge e, G has a k-regular factor containing e and another k-regular factor excluding e.*
(i) r and k are both even.
(ii) r is even, k is odd, and $r/\lambda \leq k \leq r(1 - 1/\lambda)$.
(iii) r is odd, k is even, and $2 \leq k \leq r(1 - 1/\lambda)$.
(iv) r and k are both odd, and $r/\lambda \leq k$.
(v)   Let λ be an integer such that $\lambda^* \in \{\lambda, \lambda + 1\}$ and $\lambda^* \equiv 1 \pmod 2$. Either r is odd, k is even and $k \leq r(1 - 1/\lambda^*)$; or r is odd, k is odd and $r/\lambda^* \leq k$.

Proof. Let $e = uv$ be a given edge of G. Here we assume that e is not a loop since otherwise the theorem follows easily. Let H be the graph obtained from G by inserting a new vertex w of degree two into e (Fig. 3.8). Then $V(H) = V(G) \cup \{w\}$, and define two functions f_1, $f_2 : V(H) \to \mathbb{Z}^+$ as

$$f_1(x) = \begin{cases} 0 & \text{if } x = w \\ k & \text{otherwise,} \end{cases} \quad \text{and} \quad f_2(x) = \begin{cases} 2 & \text{if } x = w \\ k & \text{otherwise.} \end{cases} \quad (3.21)$$

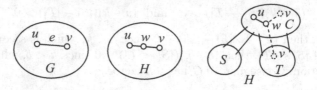

Fig. 3.8. A graph G with an edge e; a new graph H with a new vertex w of degree two; and an f-odd component C of $H - (S \cup T)$.

Then it is obvious that G has a k-factor excluding e or containing e if and only if H has an f_1-factor or an f_2-factor, respectively. So we use the f-factor theorem. For convenience, we write f for f_1 or f_2 because most parts of the proof of the existence of an f_i-factor, $i \in \{1, 2\}$, are the same.

Since both $\sum_{x \in V(H)} f_1(x) = k|G|$ and $\sum_{x \in V(H)} f_2(x) = k|G| + 2$ are even and G is connected, $\delta_H(\emptyset, \emptyset) = 0$. Let S and T be disjoint subsets of $V(H)$ such that $S \cup T \neq \emptyset$. We shall show that $\delta(S, T) \geq 0$, which implies the existence of the desired f-factor. Set $\theta = k/r$. Then $0 < \theta < 1$, and

$$f(x) = k = \theta \deg_H(x) \qquad \text{for all} \quad x \in V(H) - w,$$
$$f_1(w) = 0 = \theta \deg_H(w) - 2\theta,$$
$$f_2(w) = 2 = \theta \deg_H(w) + 2(1 - \theta).$$

Hence we have

$$\delta(S, T) = \sum_{x \in S} f(x) + \sum_{x \in T} (\deg_H(x) - f(x)) - e_H(S, T) - q(S, T)$$
$$\geq \theta \sum_{x \in S} \deg_H(x) + (1 - \theta) \sum_{x \in T} \deg_H(x)$$
$$- e_H(S, T) - q(S, T) - \epsilon,$$

where

$$\epsilon = \begin{cases} 2\theta & \text{if } w \in S \text{ and } f = f_1 \\ 2(1 - \theta) & \text{if } w \in T \text{ and } f = f_2 \\ 0 & \text{otherwise.} \end{cases} \tag{3.22}$$

Since $\delta(S, T) \equiv 0 \pmod 2$ by (3.5), it suffices to show that $\delta(S, T) \geq -\epsilon$ since $\epsilon < 2$.

Let $C_1, C_2, \ldots, C_m$ be the f-odd components of $H - (S \cup T)$, where $m = q(S, T)$. Then

$$\delta(S,T) \geq \theta \sum_{x \in S} \deg_H(x) + (1-\theta) \sum_{x \in T} \deg_H(x) - e_H(S,T) - m - \epsilon$$

$$\geq \theta \Big(e_H(S,T) + \sum_{i=1}^{m} e_H(S,C_i) \Big)$$

$$+ (1-\theta) \Big(e_H(T,S) + \sum_{i=1}^{m} e_H(T,C_i) \Big) - e_H(S,T) - m - \epsilon$$

$$= \sum_{i=1}^{m} \big(\theta e_H(S,C_i) + (1-\theta) e_H(T,C_i) - 1 \big) - \epsilon.$$

We shall show that every $C = C_i$, $1 \leq i \leq m$, satisfies

$$\theta e_H(S,C) + (1-\theta) e_H(T,C) \geq 1, \tag{3.23}$$

which implies $\delta(S,T) \geq -\epsilon$, and so the proof is complete.

Since C is an f-odd component of $H - (S \cup T)$, we have

$$\sum_{x \in V(C)} f(x) + e_H(T,C) \equiv 1 \pmod 2. \tag{3.24}$$

Moreover,

$$\sum_{x \in V(C)} \deg_H(x) = e_H(S \cup T, C) + 2\|C\|, \tag{3.25}$$

and so

$$\text{if } w \notin C \text{ then } r|C| \equiv \sum_{x \in V(C)} \deg_H(x) \equiv e_H(S \cup T, C) \pmod 2. \tag{3.26}$$

If $V(C) = \{w\}$, then $f(w) + e_H(\{w\}, T) = 1 \pmod 2$ by (3.24). So $e_H(\{w\}, T) = 1$, which implies $|\{u,v\} \cap T| = 1$ and $|\{u,v\} \cap S| = 1$. Hence

$$\theta e_H(S,C) + (1-\theta) e_H(T,C) = \theta + (1-\theta) = 1,$$

and thus (3.23) holds.

If C does not contain w or if C contains w and at least one of $\{u,v\}$ (see Fig. 3.8), then

$$e_H(S \cup T, C) \geq \lambda. \tag{3.27}$$

It is obvious that the two inequalities $e_H(S,C) \geq 1$ and $e_H(T,C) \geq 1$ imply

$$\theta e_H(S,C) + (1-\theta) e_H(T,C) \geq \theta + (1-\theta) = 1.$$

Hence we may assume $e_H(S,C) = 0$ or $e_H(T,C) = 0$. We consider two cases under the assumption that one of conditions (i)-(iv) holds. Condition (v) will be considered later.

Case 1. $e_H(S,C) = 0$.

If both r and k are even, then by (3.24) and (3.25), we have

$$e_H(T,C) \equiv 1 \pmod 2 \quad \text{and} \quad 0 \equiv e_H(T,C) \pmod 2,$$

which is a contradiction. If $k \le r(1 - 1/\lambda)$, then $1 \le (1 - \theta)\lambda$. By (3.27) and $e_H(S,C) = 0$, we have $e_H(T,C) \ge \lambda$, and hence

$$(1 - \theta)e_H(T,C) \ge (1 - \theta)\lambda \ge 1.$$

Therefore we may assume that both r and k are odd.

If $w \notin C$, then by (3.24) and (3.25), we have

$$|C| + e_G(T,C) \equiv 1 \pmod 2 \quad \text{and} \quad |C| \equiv e_G(T,C) \pmod 2.$$

This is a contradiction. If $w \in C$, then by (3.24) and (3.25), we have

$$|C| - 1 + e_G(T,C) \equiv 1 \pmod 2 \quad \text{and} \quad |C| - 1 \equiv e_G(T,C) \pmod 2.$$

This is again a contradiction.

Case 2. $e_H(T,C) = 0$.

By (3.24) we have $\sum_{x \in V(C)} f(x) \equiv 1 \pmod 2$, which implies that k must be odd. Hence we may assume that k is odd. Then $r/\lambda \le k$ by (ii) and (iv). Hence $1 \le \theta\lambda$, and so by (3.27) we obtain

$$\theta e_H(S,C) \ge \theta\lambda \ge 1.$$

Consequently, statements (i)-(iv) in the theorem guarantee the existence of a k-factor excluding or including the edge $e = uv$.

We next prove the first part of (v), i.e., we assume that r is odd, k is even and $k \le r(1 - 1/\lambda^*)$. Suppose first that $e_H(S,C) = 0$. Then by (3.24) and (3.25), we have

$$e_H(T,C) \equiv 1 \pmod 2 \quad \text{and} \quad e_H(T,C) \ge \lambda.$$

Hence $e_H(T,C) \ge \lambda^*$, and thus

$$(1 - \theta)e_H(T,C) \ge (1 - \theta)\lambda^* \ge 1.$$

Next, assume that $e_H(T,C) = 0$. Then by (3.24), we have $k|C| \equiv 1 \pmod 2$, which is a contradiction.

The remaining part of (v) can be proved similarly (Problem 3.2). □

The next theorem deals with a regular factor of an $(r - 1)$-edge connected r-regular graph that contains or excludes more than one edge.

Theorem 3.11 (Katerinis [143]). *Let* $1 \leq k \leq r - 1$ *be integers, and* G *be an* $(r-1)$-*edge connected* r-*regular multigraph of even order. Then for any* $r - k$ *edges of* G, G *has a* k-*regular factor excluding these edges. Moreover, for any* k *edges of* G, G *has a* k-*regular factor containing these edges.*

Note that the latter part of the theorem is an immediate consequence of the first part because for any k edges, where $k = r - (r - k)$, if G has a $(r-k)$-regular factor F excluding these k edges, then G has a k-factor $G - E(F)$, which contains the given k edges.

When we deal with factors in an almost regular graph, most of whose vertices have constant degree r but some vertices have not degree r, the following theorem is useful, and it can be regarded as a generalization of Theorem 3.4. For a vertex subset X of a graph G, let $\partial_G(X)$ denote $e_G(X, V(G) - X)$.

Theorem 3.12 (Kano [118]). *Let* G *be a connected general graph,* θ *be a real number such that* $0 < \theta < 1$, A *and* B *be disjoint subsets of* $E(G)$, *and* $f : V(G) \to \mathbb{Z}^+$ *be a function. If the following four conditions hold, then* G *has an* f-*factor* F *such that* $E(F) \supset A$ *and* $E(F) \cap B = \emptyset$.
(i) $\sum_{x \in V(G)} f(x)$ *is even.*
(ii) $\sum_{x \in V(G)} |f(x) - \theta \deg_G(x)| + 2(1 - \theta)|A| + 2\theta|B| < 2$.
(iii) $\theta \partial_G(X) \geq 1$ *for all* $X \subset V(G)$ *such that* $\langle X \rangle_G$ *is connected and* $\sum_{x \in X} f(x)$ *is odd.*
(iv) $(1 - \theta)\partial_G(X) \geq 1$ *for all* $X \subset V(G)$ *such that* $\langle X \rangle_G$ *is connected and* $\sum_{x \in X} f(x) + \partial_G(X)$ *is odd.*

3.3 Regular factors and f-factors in graphs

In this section we consider regular factors in simple graphs. In particular, we mainly deal with k-regular factors, where $k \geq 2$, since 1-factors have already been treated.

We showed that every 1-tough graph of even order has a 1-factor (Theorem 2.42). Here we consider regular factors and f-factors in t-tough graphs. The following theorem was conjectured by Chvátal in 1973, and proved by Enomoto, Jackson, Katerinis and Saito in 1985. Chvátal also proposed the conjecture that every 2-tough graph has an Hamiltonian cycle. By the following theorem, such a graph has a 2-factor. However a counterexample to this conjecture was found [23].

Theorem 3.13 (Enomoto, Jackson, Katerinis and Saito [78] (1985)). *Let* $k \geq 2$ *be an integer. Then every* k-*tough simple graph* G *with* $|G| \geq k + 1$ *and* $k|G|$ *even has a* k-*regular factor.*

Suppose that a simple graph G has no k-factor. Then there exists a pair (S, T) of disjoint subsets of $V(G)$ such that $\delta(S, T) < 0$. We say that (S, T) is a **minimal pair** of G if $\delta(S, T) < 0$ and

$$\text{either} \quad T = \emptyset \quad \text{or} \quad \delta(S, T') \geq 0 \quad \text{for all} \quad T' \subset T.$$

The following lemma will be used in the proof of Theorem 3.13.

Lemma 3.14. [78] *Let $k \geq 2$ be an integer, and G be a simple graph with $k|G|$ even and having no k-regular factor. Let (S, T) be a minimal pair of G. If $T \neq \emptyset$, then the maximum degree of $\langle T \rangle_G$ is at most $k - 2$.*

Proof. Let $v \in T$ and $V = V(G)$. Since $\delta(S, T) < 0$ and $k|G|$ is even, we have $\delta(S, T) \leq -2$ since $\delta(S, T) \equiv 0 \pmod{2}$ by (3.5). It follows from $\delta(S, T - v) \geq 0$ that

$$
\begin{aligned}
-2 &\geq \delta(S, T) - \delta(S, T - v) \\
&= k|S| + \sum_{x \in T} \deg_G(x) - k|T| - e_G(S, T) - q(S, T) \\
&\quad - \left(k|S| + \sum_{x \in T - v} \deg_G(x) - k|T - v| - e_G(S, T - v) - q(S, T - v) \right) \\
&= \deg_G(v) - k - e_G(S, v) - q(S, T) + q(S, T - v).
\end{aligned}
$$

Since $q(S, T - v) \geq q(S, T) - e_G(v, V - (S \cup T))$, we have

$$
\begin{aligned}
-2 &\geq \deg_G(v) - k - e_G(S, v) - e_G(v, V - (S \cup T)) \\
&= e_G(v, T) - k.
\end{aligned}
$$

Hence $e_G(v, T) \leq k - 2$, which implies $\Delta(\langle T \rangle_G) \leq k - 2$. $\quad\square$

Proof of Theorem 3.13. Suppose that G has no k-factor. Then G is not a complete graph since a complete graph with order at least $k+1$ and $k|G|$ even has a k-factor. Since $k|G|$ is even and G is connected, $\delta(\emptyset, \emptyset) = -q(\emptyset, \emptyset) = 0$, and thus every minimal pair (S, T) of G satisfies $S \cup T \neq \emptyset$.

Let (S, T) be a minimal pair of G, and let $C_1, C_2, \ldots, C_m$ be the components of $G - (S \cup T)$, where

$$m = \omega(G - (S \cup T)) \geq q(S, T).$$

For any two non-adjacent vertices $u \in S$ and $x \in V(G) - u$, $\delta(S, T)$ does not change if a new edge ux is added to G, and so the resulting graph still does not have a k-factor. Hence we may assume that $N_G(u) = V(G) - u$ for every $u \in S$ (Fig. 3.9). Similarly, joining two non-adjacent vertices in any component C_i of $G - (S \cup T)$ results in a graph that has no k-factor. Therefore, we may assume that the following claim holds.

Claim 1. $N_G(u) = V(G) - u$ *for all $u \in S$, and every $\langle V(C_i) \rangle_G$ is a complete graph (Fig. 3.9).*

For convenience, let $U = V(G) - (S \cup T)$. Then

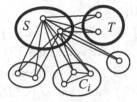

Fig. 3.9. $G - (S \cup T)$, where $N_G(u) = V(G) - u$ for all $u \in S$ and every $\langle V(C_i) \rangle_G$ is a complete graph.

$$0 > \delta(S,T) = k|S| + \sum_{x \in T} \deg_G(x) - k|T| - e_G(S,T) - q(S,T)$$

$$\geq k|S| + e_G(T,U) + e_G(T,S) + 2||\langle T \rangle_G|| - k|T| - e_G(S,T) - m$$

$$= k|S| + e_G(T,U) + 2||\langle T \rangle_G|| - k|T| - m. \qquad (3.28)$$

Let B_1 be a maximal independent set of $\langle T \rangle_G$ and $T_1 = T - B_1$. For every i, $2 \leq i \leq k-1$, we recursively define B_i and T_i as follows: let B_i be a maximal independent set of $\langle T_{i-1} \rangle_G$ and $T_i = T_{i-1} - B_i$ (Fig. 3.10). By Lemma 3.14, and since $N_G(B_i) \supseteq T_i$, it follows that $\langle T_{k-2} \rangle_G$ is an independent set, if it exists, and thus $B_{k-1} = T_{k-2}$ and $T_{k-1} = \emptyset$.

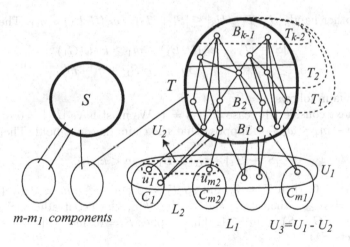

Fig. 3.10. A minimal pair (S,T) and the structure of G.

Claim 2. $|S| + |T_1| + e_G(U, B_1) - m \geq k|B_1|$.

Put $U = V(G) - (S \cup T)$, and let

$$U_1 = \{x \in U \mid e_G(x, B_1) \geq 1\},$$
$$L_1 = \{C_i \mid e_G(C_i, B_1) \geq 1\} = \{C_1, \ldots, C_{m_1}\},$$
$$L_2 = \{C_i \mid e_G(u, B_1) = 1 \text{ for some } u \in V(C_i)\} = \{C_1, \ldots, C_{m_2}\},$$
$$U_2 = \{u_i \mid \text{For each } C_i, \; 1 \leq i \leq m_2, \text{ choose one vertex}$$
$$u_i \in V(C_i) \text{ such that } e_G(u_i, B_1) = 1\}, \quad \text{and}$$
$$U_3 = U_1 - U_2. \qquad \text{(see Fig. 3.10)}$$

Then $m_2 \leq m_1 \leq m$ and $|U_2| = m_2$, and

$$L_1 - L_2 = \{C_i \mid e_G(u, B_1) \geq 2 \text{ for all } u \in V(C_i)\},$$
$$|U_1| \leq e_G(U_1, B_1) - (m_1 - m_2) = e_G(U, B_1) - (m_1 - m_2).$$

Hence

$$|U_3| \leq e_G(U, B_1) - m_1.$$

Let $G_1 = G - (S \cup T_1 \cup U_3)$. Since B_1 is an independent set of $\langle T \rangle_G$ and $e_G(B_1, U - U_1) = 0$, each vertex of B_1 belongs to a different component of G_1. Hence $\omega(G_1) \geq |B_1| + m - m_1$.

We first assume $w(G_1) \geq 2$. Then since G is k-tough, we have

$$|S \cup T_1 \cup U_3| \geq k \cdot \omega(G_1) \geq k(|B_1| + m - m_1).$$

On the other hand, $|S \cup T_1 \cup U_3| \leq |S| + |T_1| + e_G(U, B_1) - m_1$. Therefore

$$|S| + |T_1| + e_G(U, B_1) - m_1 \geq k \cdot \omega(G_1)$$
$$\geq k(|B_1| + m - m_1) \geq k|B_1| + m - m_1,$$

and the claim follows.

We next consider the case $\omega(G_1) = 1$. We must have $|B_1| = 0$ or 1, since $|B_1| + m - m_1 \leq \omega(G_1)$. Suppose that the claim does not hold. Then

$$|S| + |T_1| + e_G(U, B_1) - m < k|B_1|. \qquad (3.29)$$

If $|B_1| = 0$, then $T = \emptyset$, and so $T_1 = \emptyset$ and $|S| - m < 0$ by (3.29). If $\omega(G - S) = m \geq 2$, then $|S| \geq k\omega(G - S) = km$, which contradicts $|S| - m < 0$. Therefore $\omega(G - S) = m \leq 1$. This implies $S = \emptyset$ since $|S| - m < 0$, and contradicts $S \cup T \neq \emptyset$.

Hence we may assume that $|B_1| = 1$, and this holds for any choice of B_1. Therefore $\langle T \rangle_G$ is a complete graph. Furthermore, $\omega(G_1) = 1$ means that $m = m_1$ and for every component C_i, there exists at least one edge joining C_i to B_1. Thus, $e_G(U, B_1) \geq m$. Then by (3.29), we have $|S| + |T_1| < k|B_1| = k$. Hence $|S \cup T| = |S| + |T_1| + |B_1| < k + 1$. On the other hand, we shall show that $m \geq 2$ and $|S \cup T| \geq 2k$, which contradicts the above inequality $|S \cup T| < k + 1$. Assume $m = 1$. Since G is not a complete graph, by Claim 1 there exist two non-adjacent vertices $x_1 \in V(C_1)$ and $y_1 \in T$. Setting $B_1 = \{y_1\}$, then

$V(C_1) - U_3$ contains x_1 and so $\omega(G_1) \geq 2$, which gives us a contradiction since in the previous case $\omega(G_1) \geq 2$. Hence $m \geq 2$, which implies

$$|S \cup T| \geq k \cdot \omega(G - (S \cup T)) = km \geq 2k.$$

Therefore Claim 2 is proved.

Claim 3. *For every i, $2 \leq i \leq k-1$, it follows that*

$$|S| + e_G(T - T_{i-1}, B_i) + |T_i| + e_G(U, B_i) \geq k|B_i|. \tag{3.30}$$

We may assume $B_i \neq \emptyset$ since otherwise (3.30) trivially holds. Since B_i is a set of isolated vertices of $G - N_G(B_i)$, if $\omega(G - N_G(B_i)) \geq 2$, then

$$|N_G(B_i)| \geq k \cdot \omega(G - N_G(B_i)) \geq k|B_i|.$$

It is clear that

$$|N_G(B_i)| \leq |S| + e_G(T - T_{i-1}, B_i) + |T_i| + e_G(U, B_i).$$

Hence (3.30) holds. Suppose $\omega(G - N_G(B_i)) = 1$, which implies $|B_i| = 1$ and $V(G) = N_G(B_i) \cup B_i$. Then

$$\begin{aligned} k|B_i| + 1 = k + 1 \leq |V(G)| &= |N_G(B_i) \cup B_i| \\ &\leq |S| + e_G(T - T_{i-1}, B_i) + |T_i| + e_G(U, B_i) + 1 \end{aligned}$$

Consequently (3.30) follows.

By Claim 2, (3.30) and since $T_{k-1} = \emptyset$, we have

$$\begin{aligned} k|T| = k \sum_{i=1}^{k-1} |B_i| \qquad &\text{(by } T = \bigcup_{i=1}^{k-1} B_i\text{)} \\ \leq\ & |S| + |T_1| + e_G(U, B_1) - m \\ &+ \sum_{i=2}^{k-1} \Big(|S| + e_G(T - T_{i-1}, B_i) + |T_i| + e_G(U, B_i) \Big) \\ \leq\ & (k-1)|S| + \sum_{i=1}^{k-1} |T_i| + \sum_{i=2}^{k-1} e_G(T - T_{i-1}, B_i) + \sum_{i=1}^{k-1} e_G(U, B_i) - m \\ \leq\ & (k-1)|S| + \sum_{i=1}^{k-2} |T_i| + \|\langle T \rangle_G\| + e_G(U, T) - m. \tag{3.31} \end{aligned}$$

By (3.28) and (3.31), we have

$$\begin{aligned} (k-1)|S| &+ \sum_{i=1}^{k-2} |T_i| + \|\langle T \rangle_G\| + e_G(T, U) - m \\ &\geq k|T| > k|S| + e_G(T, U) + 2\|\langle T \rangle_G\| - m, \end{aligned}$$

and thus

$$\sum_{i=1}^{k-2} |T_i| > |S| + ||\langle T \rangle_G|| \geq ||\langle T \rangle_G||. \tag{3.32}$$

On the other hand, since B_i is a maximal independent set of $\langle T_{i-1} \rangle_G$ and $T_i = T_{i-1} - B_i$, we have $e_G(x, B_i) \geq 1$ for every vertex $x \in T_i$, and thus we obtain

$$||\langle T \rangle_G|| = \sum_{i=1}^{k-2} e_G(B_i, T_i) \geq \sum_{i=1}^{k-2} |T_i|.$$

This contradicts (3.32). Consequently the theorem is proved. □

Theorem 3.15. ([78]) *Let $k \geq 1$ be integer. For every real number $\epsilon > 0$, there exists a $(k - \epsilon)$-tough simple graph G with $k|G|$ even and $|G| \geq k + 1$ that has no k-factor.*

Proof. We construct the desired graph $G = G(m, k)$ as follows (Fig. 3.11).

$$V(G) = A \cup B \cup C = A \cup \left(\overset{k+2}{\underset{i=1}{\bigcup}} B_i \cup \overset{k+2}{\underset{i=1}{\bigcup}} B_i' \right) \cup \overset{k(k+2)}{\underset{i=1}{\bigcup}} C_i,$$

where

$$A = \{a_1, a_2, \ldots, a_k\},$$
$$B_i = \{b_{i,1}, b_{i,2}, \ldots, b_{i,m}\} \qquad (1 \leq i \leq k + 2),$$
$$B_i' = \{b_{i,1}', b_{i,2}', \ldots, b_{i,m}', b_{i,m+1}'\} \qquad (1 \leq i \leq k + 2),$$
$$C_i = \{x_{i,1}, \ldots, x_{i,m}, y_{i,1}, \ldots, y_{i,m+1}, z_i\} \qquad (1 \leq i \leq k(k + 2)).$$

The adjacency of G is defined as follows:

1. $N_G(a_i) = V(G) - a_i$,
2. $\langle C_i \rangle_G = K_{2m+2}$.
3. $N_G(b_{i,j}) = A \cup \{x_{(i-1)k+1,j}, \ldots, x_{ik,j}\}$ $(1 \leq i \leq k + 2, \ 1 \leq j \leq m)$,
4. $N_G(b_{i,j}') = A \cup \{y_{(i-1)k+2,j}, \ldots, x_{ik+1,j}\}$ $(1 \leq i \leq k + 1, \ 1 \leq j \leq m)$,
5. $N_G(b_{k+2,j}') = A \cup \{y_{k(k+1)+2,j}, \ldots, y_{k(k+2),j}, y_{1,j}\}$,
6. $e_G(C_i, C_j) = 0$ $(1 \leq i < j \leq m + 1)$.

Sketch of the proof of Theorem 3.15   Let $G = G(m, k)$ be the graph defined above. It is clear that

$$|G| = |A| + (k + 2)|B_i| + (k + 2)|B_i'| + k(k + 2)|C_i|$$
$$= k + (k + 2)m + (k + 2)(m + 1) + k(k + 2)(2m + 2)$$
$$\equiv k + km + k(m + 1) \equiv 0 \pmod 2.$$

Hence $|G| \geq k + 1$ and $k|G|$ is even.

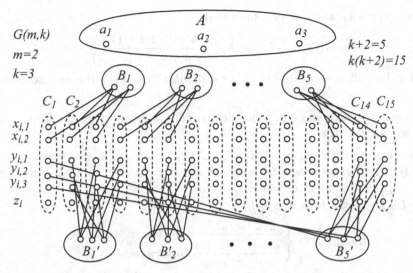

Fig. 3.11. The graph $G(m,k)$ with $k = 3$ and $m = 2$, where $|A| = k, |B_i| = m, |B_i'| = m + 1, |C_i| = 2m + 2$, $N_G(x) = V(G) - x$ for $x \in A$, and $\langle C_i \rangle_G$ is a complete graph.

Let $S = A$ and $T = B$. Then every $C = C_i$ is a component of $G - (S \cup T)$, and
$$k|C| + e_G(C,T) = k(2m+2) + 2m + 1 \equiv 1 \pmod{2}.$$
Hence C is a k-odd component of $G - (S,T)$. Thus we have
$$\begin{aligned}
\delta(S,T) &= k|S| + \sum_{x \in T} \deg_{G-S}(x) - k|T| - q(S,T) \\
&= k^2 + \sum_{x \in T}(\deg_{G-S}(x) - k) - k(k+2) \\
&= k^2 - k(k+2) = -2k.
\end{aligned}$$

Hence G has no k-factor.
We can show that
$$k - 1 < tough(G) < k.$$
Let
$$X = \bigcup_{i=1}^{k(k+2)} (C_i - z_i) \ \cup A.$$
Then $|X| = k(k+2)(2m+1) + k$ and
$$\omega(G - S_0) = \left| \bigcup_{i=1}^{k+2}(B_i \cup B_i') \right| + \left| \bigcup_{i=1}^{k(k+2)}\{z_i\} \right| = (2m + k + 1)(k+2).$$

If $m \geq k(k+1)$, then we can show that

$$tough(G) = \frac{\omega(G-X)}{|X|} = \frac{k(k+2)(2m+1)+k}{(2m+k+1)(k+2)} < k,$$

and so for sufficiently large m, $k - \epsilon < tough(G)$, and the theorem is proved.
□

The next theorem gives a sufficient condition using toughness for a graph to have an f-factor.

Theorem 3.16 (Katerinis [141]). *Let G be a connected simple graph.*
(i) If G is 2-tough, then for every function $f : V(G) \rightarrow \{1,2\}$ such that $\sum_{x \in V(G)} f(x)$ is even, G has an f-factor.
(ii) Let $1 \leq a \leq b$ be integers. If

$$tough(G) \geq \begin{cases} \frac{(a+b)^2 + 2(b-a)}{4a} & if \ \ b \equiv a \pmod 2 \\ \frac{(a+b)^2 + 2(b-a)+1}{4a} & otherwise, \end{cases}$$

then for every function $f : V(G) \rightarrow \{a, a+1, \ldots, b\}$ such that $\sum_{x \in V(G)} f(x)$ is even, G has an f-factor.

Proof. We shall prove only statement (i) here. Since G is 2-*tough*, G is connected. Suppose that the theorem does not hold. Then G is not a complete graph, and so

$$|G| \geq 6 \quad \text{and} \quad \delta(G) \geq 4$$

since $tough(G) \geq 2$ (see Problem 3.3). By Theorem 4.5, there exists a pair (S, T) of disjoint subsets S and T of $V(G)$ such that

$$\rho(S,T) = |S| + \sum_{x \in T} \deg_{G-S}(x) - 2|T| - \omega(G - (S \cup T)) \leq -2, \quad (3.33)$$

where $\omega(G - (S \cup T))$ denotes the number of components of $G - (S \cup T)$. We choose such a pair (S, T) so that T is minimal. Then $S \cup T \neq \emptyset$. It follows that $T \neq \emptyset$, since if $\omega(G - S) = 1$ then $\rho(S, \emptyset) = |S| - \omega(G - S) \geq 0$; otherwise, $|S| \geq 2\omega(G - S)$ and thus $\rho(S, \emptyset) = |S| - \omega(G-S) \geq \omega(G-S) \geq 2$, which are contradictions.

Claim 1. *For every $x \in T$, it follows that $e_G(x, T) = 0$ and $\deg_{G-S}(x) \leq 2$. In particular, $|S| \geq 2$.*

Let $v \in T$. By the choice of (S, T), it follows that

$$1 \leq \rho(S, T - v) - \rho(S, T)$$
$$= |S| + \sum_{x \in T-v} \deg_{G-S}(x) - 2|T - v| - \omega(G - (S \cup (T - v)))$$
$$\quad - \left(|S| + \sum_{x \in T} \deg_{G-S}(x) - 2|T| - \omega(G - (S \cup T)) \right)$$
$$\leq -\deg_{G-S}(v) + 2 + e_G(v, V(G) - (S \cup T)) - 1$$
$$= -e_G(v, T) + 1.$$

Hence $e_G(v, T) = 0$.

Similarly, it follows from $\omega(G - (S \cup T)) = \omega(G - ((S+v) \cup (T-v)))$ that

$$1 \le \rho(S+v, T-v) - \rho(S, T)$$
$$= |S| + 1 + \sum_{x \in T-v} \deg_G(x) - e_G(S+v, T-v) - 2|T-v|$$
$$- \left(|S| + \sum_{x \in T} \deg_G(x) - e_G(S, T) - 2|T| \right)$$
$$= 1 - \deg_G(v) + e_G(v, S) - e_G(v, T-v) + 2$$
$$= - \deg_{G-S}(v) + 3. \qquad (\text{by } e_G(v, T) = 0)$$

Hence $\deg_{G-S}(v) \le 2$. Since $\delta(G) \ge 4$, we have

$$4 \le \deg_G(v) \le |S| + \deg_{G-S}(v) \le |S| + 2,$$

which implies $|S| \ge 2$.

Claim 2. $|T| \ge 2$.

Assume $T = \{v\}$. If $\omega(G - (S \cup \{v\})) = 1$, then it follows from $|S| \ge 2$ that

$$\rho(S, \{v\}) \ge |S| - 2|\{v\}| - \omega(G - (S \cup \{v\})) \ge -1,$$

which contradicts (3.33). If $\omega(G - (S \cup \{v\})) \ge 2$, then $|S \cup \{v\}| \ge 2\omega(G - (S \cup \{v\}))$, and so

$$\rho(S, \{v\}) \ge |S| - 2|\{v\}| - \omega(G - (S \cup \{v\}))$$
$$\ge \omega(G - (S \cup \{v\})) - 1 - 2|\{v\}| \ge -1,$$

a contradiction. Therefore the claim holds.

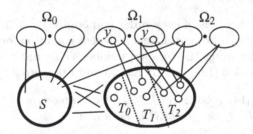

Fig. 3.12. A graphs G with a pair (S, T).

By Claim 1, we can partition T into the following three disjoint subsets.

$$T_0 = \{x \in T \mid \deg_{G-S}(x) = 0\}; \quad T_1 = \{x \in T : \deg_{G-S}(x) = 1\};$$
$$T_2 = \{x \in T \mid \deg_{G-S}(x) = 2\}.$$

We divide the set Ω of components of $G - (S \cup T)$ into the following three subsets.

$$\Omega_0 = \{C \in \Omega \mid e_G(C,T) = 0\},$$
$$\Omega_1 = \{C \in \Omega \mid e_G(C,T) \geq 1 \text{ and } e_G(x,T) = 1 \text{ for some vertex } x \in V(C)\},$$
$$\Omega_2 = \{C \in \Omega \mid e_G(C,T) \geq 1 \text{ and } e_G(x,T) \geq 2 \text{ for all } x \in V(C)\}.$$

For each $C \in \Omega_1$, choose one vertex y of C such that $e_G(y,T) = 1$, and let Y denote the set of these vertices, that is,

$$Y = \{y \in V(C) \mid e_G(y,T) = 1, \ C \in \Omega_1, \ |Y \cap V(C)| = 1\}.$$

Then by Claim 2, $G - (S \cup (N_G(T) - Y))$ has at least $|T| \geq 2$ components, and so

$$|S| + |T_1| + 2|T_2| - |\Omega_1| - |\Omega_2| \geq |S \cup (N_G(T) - Y))|$$
$$\geq 2\omega\big(G - (S \cup (N_G(T) - Y))\big)$$
$$\geq 2(|\Omega_0| + |T_0| + |T_1| + |T_2|).$$

Hence

$$|S| \geq 2|\Omega_0| + |\Omega_1| + |\Omega_2| + 2|T_0| + |T_1|.$$

Therefore

$$\rho(S,T) = |S| + \sum_{x \in T} \deg_{G-S}(x) - 2|T| - \omega(G - (S \cup T))$$
$$\geq |S| + |T_1| + 2|T_2| - 2(|T_0| + |T_1| + |T_2|) - (|\Omega_0| + |\Omega_1| + |\Omega_2|)$$
$$\geq |S| - (2|T_0| + |T_1| + |\Omega_0| + |\Omega_1| + |\Omega_2|) \geq 0.$$

This contradicts (3.33). Consequently statement (i) is proved. $\square$

We showed that if G is of even order and $bind(G) \geq 4/3$, then G has a 1-factor (Theorem 2.41). We now give a sufficient condition for a graph to have a regular factor by using the binding number. This result was obtained by Katerinis and Woodall [147] and by Egawa and Enomoto [66], independently.

Theorem 3.17 ([147], [66]). *Let $k \geq 2$ be an integer, and G be a connected simple graph. If $|G| \geq 4k - 6$, $k|G|$ is even and*

$$bind(G) > \frac{(2k-1)(|G|-1)}{k(|G|-2)+3}, \tag{3.34}$$

then G has a k-regular factor. In particular, (i) if $bind(G) \geq 3/2$, then G has a 2-factor; (ii) if $bind(G) \geq 5/3$, then G has a 3-factor; and (iii) if $bind(G) \geq 2$, then G has a k-regular factor for every $k \geq 2$ such that $|G| \geq 4k - 6$.

We remark that Woodall [249] proved that every graph G with $bind(G) \geq 3/2$ has a Hamiltonian cycle; therefore, G has a 2-factor.

In order to prove the above theorem, we need to introduce a new concept, which will be used in the proofs of other theorems as well. Suppose that a connected graph G has no k-factor. Then by the regular factor theorem, there exists a pair (X, Y) of disjoint subsets $V(G)$ such that $\delta(X, Y) < 0$. We say that (S, T) is a **maximal pair** if $S \cup T$ is maximal among all pairs (X, Y) with $\delta(X, Y) < 0$. In other words, if (S, T) is a maximal pair, then $\delta(S, T) < 0$ and for every pair (X, Y) of disjoint subsets $V(G)$ with $X \cup Y \supset S \cup T$, it follows that $\delta(X, Y) \geq 0$.

Lemma 3.18. *Let $k \geq 1$ be an integer and G be a connected simple graph with $k|G|$ even. Suppose that G has no k-factor. Let (S, T) be a maximal pair of G. Then the following statements hold.*
(i) For every $u \in V(G) - (S \cup T)$, $\deg_{G-S}(u) \geq k+1$ and $|N_G(u) \cap T| \leq k-1$.
(ii) For every component C of $G - (S \cup T)$, we have $|V(C)| \geq 3$.

Proof. (i) Let $u \in V(G) - (S \cup T)$. Note that $\delta(S, T) \equiv 0 \pmod 2$ implies $\delta(S, T) \leq -2$. Combining this with the fact that $\delta(S, T \cup \{u\}) \geq 0$, we have

$$
\begin{aligned}
2 &\leq \delta(S, T \cup \{u\}) - \delta(S, T) \\
&= k|S| + \sum_{x \in T \cup \{u\}} \deg_{G-S}(x) - k|T \cup \{u\}| - q(S, T \cup \{u\}) \\
&\quad - \left(k|S| + \sum_{x \in T} \deg_{G-S}(x) - k|T| - q(S, T) \right) \\
&\leq \deg_{G-S}(u) - k - q(S, T \cup \{u\}) + q(S, T) \\
&\leq \deg_{G-S}(u) - k + 1. \qquad (\text{by } q(S, T) - 1 \leq q(S, T \cup \{u\}))
\end{aligned}
$$

Hence

$$k + 1 \leq \deg_{G-S}(u).$$

Similarly, it follows that

$$
\begin{aligned}
2 &\leq \delta(S \cup \{u\}, T) - \delta(S, T) \\
&= k|S \cup \{u\}| + \sum_{x \in T} \deg_G(x) - k|T| \\
&\quad - e_G(S \cup \{u\}, T) - q(S \cup \{u\}, T) \\
&\quad - \left(k|S| + \sum_{x \in T} \deg_G(x) - k|T| - e_G(S, T) - q(S, T) \right) \\
&\leq k - e_G(u, T) - q(S \cup \{u\}, T) + q(S, T) \\
&\leq k - e_G(u, T) + 1. \qquad (\text{by } q(S, T) - 1 \leq q(S \cup \{u\}, T))
\end{aligned}
$$

Hence

$$|N_G(u) \cap T| = e_G(u, T) \leq k - 1.$$

Therefore statement (i) holds.

Let us now prove (ii). Let C be a component of $G - (S \cup T)$ and $u \in V(C)$ (Fig. 3.13). By (i), we have

$$|V(C) - \{u\}| \geq |N_G(u) \cap V(C)| = \deg_{G-S}(u) - |N_G(u) \cap T|$$
$$\geq k + 1 - (k - 1) = 2.$$

Hence $|V(C)| \geq 3$. Consequently the lemma is proved. $\square$

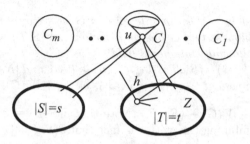

Fig. 3.13. $G - (S \cup T)$, $u \in V(C)$ and $M = V(G) - (S \cup T) - V(D)$.

Lemma 3.19. *Let G be a connected simple graph with* $\text{bind}(G) > b$. *Then*

$$\delta(G) > \frac{(b-1)|G| + 1}{b}.$$

Moreover, if X is a non-empty independent set of G, then

$$|N_G(X)| > \frac{(b-1)|G| + |X|}{b}.$$

Proof. Let X be a non-empty independent set of G, and let $Y = V(G) - N_G(X)$. Then $X \subseteq Y$ and $N_G(Y) \subseteq V(G) - X$, and thus

$$|G| - |X| \geq |N_G(Y)| > b|Y| = b(|G| - |N_G(X)|).$$

Hence

$$|N_G(X)| > \frac{(b-1)|G| + |X|}{b}.$$

For any vertex v of G, since $\{v\}$ is an independent set of G, we have

$$\deg_G(v) = |N_G(\{v\})| > \frac{(b-1)|G| + 1}{b},$$

and the lemma is proved. $\square$

In the proof of [147], the authors assume $k \geq 3$ since they use a result of [249] for $k = 2$. Our proof includes the case $k = 2$ and is based on [66]. Moreover, the cases considered in the proof often appears in proofs of some other f-factor and k-factor theorems.

Proof of Theorem 3.17. Let $n = |G|$, and b be the right-hand side of (3.34). By Lemma 3.19,

$$\delta(G) > \frac{(b-1)n+1}{b} = \frac{\left(\frac{(2k-1)(n-1)}{k(n-2)+3} - 1\right)n + 1}{\frac{(2k-1)(n-1)}{k(n-2)+3}}$$

$$= \frac{(k-1)n+2k-3}{(2k-1)} = \frac{(k-1)n-2}{(2k-1)} + 1.$$

Hence

$$\delta(G) \geq \frac{(k-1)n-1}{(2k-1)} + 1 > \frac{(k-1)n}{(2k-1)}. \qquad (3.35)$$

Assume that G has no k-factor. Let (S,T) be a maximal pair of G, where $S \cup T \neq \emptyset$ since $\delta(\emptyset, \emptyset) = 0$ is implied by $kn \equiv 0 \pmod 2$. Put

$$s = |S| \qquad \text{and} \qquad t = |T|.$$

Let m be the number of components of $G - (S \cup T)$, and c denote the minimum order of components of $G - (S \cup T)$. Then by the regular factor theorem and Lemma 3.18, we have

$$-2 \geq \delta(S,T) = k|S| + \sum_{x \in T} \deg_{G-S}(x) - k|T| - q(S,T)$$

$$\geq ks + \sum_{x \in T} \deg_{G-S}(x) - kt - m, \qquad (3.36)$$

$$m \leq n - s - t, \qquad (3.37)$$

$$3 \leq c \leq \frac{n-s-t}{m} \qquad \text{(when } m \geq 1\text{), and} \qquad (3.38)$$

$$\delta(G) \leq c - 1 + s + l. \qquad (3.39)$$

Claim. $T \neq \emptyset$.

Suppose $T = \emptyset$. Then $t = 0$ and $s \geq 1$, and by (3.36) and (3.37),

$$ks + 2 \leq m \leq n - s. \qquad (3.40)$$

Hence by (3.35), (3.39), (3.38) and by (3.40), we have

$$\frac{(k-1)n}{(2k-1)} < \delta(G) \leq c - 1 + s \leq \frac{n-s}{m} - 1 + s$$

$$\leq \frac{n-s}{ks+2} - 1 + s = \frac{n-2}{k+1} - \frac{(n-2-ks-s)(ks-k+1)}{(k+1)(ks+2)}.$$

Since $n - 2 - ks - s \geq 0$ by (3.40), it follows that

$$\frac{(k-1)n}{2k-1} < \frac{n-2}{k+1}.$$

This is equivalent to $(k^2 - 1)n < (2k - 1)(n - 2)$, which is a contradiction since $k^2 - 1 > 2k - 1$ when $k \geq 3$ and $3n > 3(n - 2)$ when $k = 2$. Therefore the claim is proved.

By the above claim, we define

$$h = \min_{x \in T} \deg_{G-S}(x).$$

Clearly

$$\delta(G) \leq h + s. \tag{3.41}$$

We shall consider the following three cases.

Case 1. $h = 0$.

Let $Z = \{x \in T \mid \deg_{G-S}(x) = 0\} \neq \emptyset$. Since Z is an independent set, we have by Lemma 3.19,

$$s = |S| \geq |N_G(Z)| > \frac{(b-1)n + |Z|}{b}$$
$$= \frac{\left(\frac{(2k-1)(n-1)}{k(n-2)+3} - 1\right)n + |Z|}{\frac{(2k-1)(n-1)}{k(n-2)+3}} = \frac{(k-1)n + k|Z| + \frac{(k-3)(n-|Z|)}{n-1}}{2k-1}$$
$$\geq \frac{(k-1)n + |Z| - 1}{2k-1}. \tag{3.42}$$

On the other hand, by (3.36) and (3.37), we obtain

$$-2 \geq ks - k|Z| + \sum_{x \in T - Z}(\deg_{G-S}(x) - k) - m$$
$$\geq ks - k|Z| + (1 - k)(t - |Z|) - (k - 1)(n - s - t),$$

where $m \leq (k-1)(n - s - t)$. Hence

$$s \leq \frac{(k-1)n + |Z| - 2}{2k-1}.$$

This contradicts (3.42).

Case 2. $1 \leq h \leq k - 1$.

By (3.36), (3.37) and by the assumption $k - h \geq 1$ of this case, we have

$$ks + (h - k)t - (k - h)(n - s - t) \leq -2.$$

Thus

$$s \leq \frac{(k-h)n-2}{2k-h}.$$ (3.43)

On the other hand, we obtain

$$\frac{(k-1)n-1}{2k-1}+1 \leq \delta(G) \leq s+h$$

by (3.35) and (3.41). By combining the above inequality with (3.43), we have

$$\frac{(k-1)n-1}{2k-1} \leq h-1+\frac{(k-h)n-2}{2k-h}.$$

Hence

$$(h-1)kn \leq (h-1)(2k-1)(2k-h)-(2k+h-2).$$

This implies $h-1 \geq 1$, that is, the above inequality does not hold if $h=1$. Thus

$$kn \leq (2k-1)(2k-h)-\frac{2k+h-2}{h-1}$$
$$= 4k^2-(2k-1)h-2k-1-\frac{2k-1}{h-1}$$
$$\leq 4k^2-8k+2 < 4k^2-6k. \qquad \text{(by } 2 \leq h)$$

This contradicts the assumption $n \geq 4k-6$.

Case 3. $h=k$.

It is clear from (3.36) that $m \geq ks+2$. So by (3.38),

$$3 \leq c \leq \frac{n-s-t}{m} \leq \frac{n-s-1}{ks+2},$$

and thus

$$ks+2 \leq \frac{n-s-t}{3} \leq \frac{n-1}{3}.$$ (3.44)

If $k \geq 3$, then by (3.35) and (3.41), we have

$$\frac{(k-1)n}{2k} < \frac{(k-1)n+2k-3}{2k-1} \leq \delta(G) \leq h+s=k+s.$$

Since $k+s \leq sk+2$, we have

$$\frac{(k-1)n}{2k} < \frac{n-1}{3},$$

which is equivalent to $(k-3)n+2k < 0$, a contradiction. If $k=2$, then by (3.35), (3.41) and (3.44), we have

$$\frac{n-1}{3} < \frac{(k-1)n-1}{2k-1} + 1 \le \delta(G) \le h+s = k+s \le sk+2 \le \frac{n-1}{3}.$$

This is also a contradiction.

Case 4. $h > k$.

By (3.36), we have $-2 \ge ks + (h-k)t - m$, and so

$$m \ge ks + t + 2 \ge s + t + 2 \tag{3.45}$$

Then $m \ge 3$, and by (3.45), (3.38) and (3.39), we obtain

$$\delta(G) \le c - 1 + s + t \le c + m - 3$$
$$\le c + m - 3 + \frac{(c-3)(m-3)}{3} = \frac{cm}{3} \le \frac{n}{3}.$$

This contradicts (3.35). Consequently the theorem is proved. $\square$

We now show that the conditions in Theorem 3.17 are sharp. Let us define the graph $G_{n,k}$ as follows:

$$G_{n,k} = K_m + (nk-1)K_2, \quad \text{where } m = 2(nk - n - 1).$$

Then we can show that $G_{n,k}$ has no k-factor, kn is even, and

$$bind(G_{n,k}) = 2 - \frac{2n-1}{2nk-3} = \frac{(2k-1)(n-1)}{k(n-2)+3}.$$

Moreover, it follows that (i) when $k = 2$, $bind(G) \to \frac{3}{2}$ as $n \to \infty$; (ii) when $k = 3$, $bind(G) \to \frac{5}{3}$ as $n \to \infty$; and (iii) when $k \ge 4$, $bind(G) \to 2 - \frac{1}{k}$ as $n \to \infty$.

Theorem 3.20 (Kano and Tokushige [135]). *Let a and b be integers such that $1 \le a \le b$ and $2 \le b$, and let G be a connected simple graph with order $|G| \ge (a+b)^2/2$. Let $f : V(G) \to \{a, a+1, \ldots, b\}$ be a function such that $\sum_{x \in V(G)} f(x) \equiv 0 \pmod 2$. If one of the following two conditions holds, then G has an f-factor.*

(1) $bind(G) \ge \dfrac{(a+b-1)(|G|-1)}{a|G| - (a+b) + 3}$.

(2) $\delta(G) \ge \dfrac{b|G| - 2}{a+b}$.

In order to prove the above theorem, by the f-factor theorem we should show that

$$\delta(S,T) = \sum_{x \in S} f(x) + \sum_{x \in T} \big(\deg_{G-S}(x) - f(x)\big) - q(S,T) \ge 0.$$

Since $a \leq f(x) \leq b$, it suffices to show that

$$a|S| + \sum_{x \in T} \deg_G(x) - b|T| - q(S,T) \geq 0.$$

The proof of this inequality is similar to that of Theorem 3.17.

It is well-known that if $\alpha(G) \leq \kappa(G)$, then the graph G has a Hamiltonian cycle (Theorem 1.13). The next theorem gives a sufficient condition for a graph to have a regular factor using the independence number $\alpha(G)$ and connectivity $\kappa(G)$. A similar result was obtained by Katerinis [138], which says that if $k \geq 2$, $k|G|$ is even, $|G| \geq k+1$ and

$$\frac{(k+1)^2}{4k}\alpha(G) + \frac{5k-4}{8} - \frac{2}{k} \leq \kappa(G),$$

then G has a k-regular factor.

Theorem 3.21 (Nishimura [199]). *Let $k \geq 2$ be an integer, and G be a connected simple graph. If $k|G|$ is even and one of the following two conditions holds, then G has a k-regular factor.*
(i) k is odd, $(k+1)^2/2 \leq \kappa(G)$ and $((k+1)^2/4k)\alpha(G) \leq \kappa(G)$.
(ii) k is even, $k(k+2)/2 \leq \kappa(G)$ and $((k+2)/4)\alpha(G) \leq \kappa(G)$.

Proof. Assume that k is odd and G has no k-factor. Let (S,T) be a maximal pair of G. Since $k|G|$ is even and G is connected, $S \cup T \neq \emptyset$. Set $U = V(G) - (S \cup T)$.

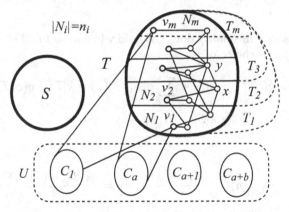

Fig. 3.14. A maximal pair (S,T), vertices v_i, closed neighborhoods N_i and components C_i of $G - (S \cup T)$.

Let $T_1 = \langle T \rangle_G$, and v_1 be a vertex of T_1 with minimum degree, and let $N_1 = N_{T_1}[v_1] = N_{T_1}(v_1) \cup \{v_1\}$ and $T_2 = T_1 - N_1$. For $i \geq 2$, we inductively define v_i, T_i and N_i until $T_{i+1} = \emptyset$ as follows (Fig. 3.14):

Let v_i be a vertex of T_i with minimum degree, and let

$$N_i = N_{T_i}[v_i] \qquad \text{and} \qquad T_{i+1} = T_i - N_i.$$

Assume $T_{m+1} = \emptyset$, and let $n_i = |N_i|$ for every $1 \le i \le m$.

Then the following three statements hold, and their proofs are given below.

$$\alpha(\langle T \rangle_G) \ge m, \tag{3.46}$$
$$|T| = n_1 + n_2 + \cdots + n_m \tag{3.47}$$
$$\sum_{1 \le i \le m} \left(\sum_{x \in N_i} \deg_{T_i}(x) \right) \ge \sum_{1 \le i \le m} (n_i^2 - n_i). \tag{3.48}$$

(3.46) is an immediate consequence of the fact that $\{v_1, \ldots, v_m\}$ is an independent set of $\langle T \rangle_G$. (3.47) follows from $T_{m+1} = \emptyset$. Since $\deg_{T_i}(x) \ge n_i - 1$ for every $x \in V(T_i)$ and $|N_i| = n_i$, (3.48) holds.

An edge xy joining $x \in N_i$ and $y \in N_j$ $(1 \le i < j \le m)$ is counted in $\deg_{G-S}(x)$, $\deg_{G-S}(y)$ and $\deg_{T_i}(x)$ but not in $\deg_{T_j}(y)$, and thus by (3.48), we have

$$\sum_{x \in T} \deg_{G-S}(x) \ge \sum_{1 \le i \le m} (n_i^2 - n_i) + \sum_{1 \le i < j \le m} e_G(N_i, N_j) + e_G(T, U). \tag{3.49}$$

Let $c = \kappa(G - S)$. Then we have

$$e_G\left(N_i, \bigcup_{j \ne i} N_j\right) + e_G(N_i, U) \ge c \qquad \text{for every } 1 \le i \le m. \tag{3.50}$$

Note that the above inequality does not hold when $m = 1$ and $U = \emptyset$. However, in this case, by the choice of v_1, we have

$$0 > \delta(S, T) = k|S| + \sum_{x \in T} \deg_{G-S}(x) - k|T| - q(S, T)$$
$$\ge k|S| + n_1(n_1 - 1) - kn_1,$$

which implies

$$|G| = |S| + n_1 < n_1 - \frac{n_1(n_1 - 1)}{k} + n_1 = \frac{n_1(2k + 1 - n_1)}{k}$$
$$\le \frac{k(k + 1)}{k} = k + 1,$$

(since $n_1(2k + 1 - n_1)$ takes a maximum value when $n_1 = k$).

This contradicts the assumption that $|G| \ge \kappa(G) \ge \frac{(k+1)^2}{2}$.

By summing up (3.50), we have

$$\sum_{1 \leq i \leq m} \left(e_G(N_i, \bigcup_{j \neq j} N_j) + e_G(N_i, U) \right)$$

$$= 2 \sum_{1 \leq i < j \leq m} e_G(N_i, N_j) + e_G(T, U) \geq cm. \qquad (3.51)$$

By (3.49) and (3.51), we get

$$\sum_{x \in T} \deg_{G-S}(x) \geq \sum_{1 \leq i \leq m} n_i(n_i - 1) + \frac{cm + e_G(T, U)}{2}. \qquad (3.52)$$

By applying (3.49) and (3.52) to $\delta(S, T) \leq -2$, we have

$$-2 \geq \delta(S, T) = k|S| + \sum_{x \in T} \deg_{G-S}(x) - k|T| - q(S, T)$$

$$\geq k|S| + \sum_{1 \leq i \leq m} n_i(n_i - 1) + \frac{mc + e_G(T, U)}{2} - \sum_{1 \leq i \leq m} kn_i - \omega(U)$$

$$= k|S| + \sum_{1 \leq i \leq m} n_i(n_i - k - 1) + \frac{mc + e_G(T, U)}{2} - \omega(U)$$

$$\geq k|S| - \frac{(k+1)^2 m}{4} + \frac{mc + e_G(T, U)}{2} - \omega(U) \qquad (3.53)$$

(since $n_i(n_i - k - 1)$ takes a minimum value when $n_i = \frac{k+1}{2}$).

Let $C_1, \ldots, C_a$ be the components of $G - (S \cup T)$ such that $V(C_i) \subseteq N_G(\{v_1, v_2, \ldots, v_m\})$, and let $C_{a+1}, \ldots, C_{a+b}$ be the other components of $G - (S \cup T)$. Let us estimate $e_G(T, U)$. If $c \geq 1$ and $T = \emptyset$, then $\omega(U) = 1$ and so $q(S, \emptyset) \leq 1$ and $\delta(S, \emptyset) = k|S| - q(S, \emptyset) \geq k - 1 \geq 0$, a contradiction. Therefore if $c \geq 1$, then $T \neq \emptyset$, and thus

$$e_G(C_i, T) \geq c \quad \text{for all} \quad 1 \leq i \leq a + b.$$

By Lemma 3.18, $e_G(C_i, T) \geq |C_i| \geq 3$ for every $1 \leq i \leq a$. Hence

$$e_G(T, U) \geq \sum_{1 \leq i \leq m} e_G(C_i, T) \geq 3a + cb. \qquad (3.54)$$

Since

$$\omega(U) = a + b, \quad \alpha(G) \geq |\{v_1, \ldots, v_m\}| + b = m + b, \quad \text{and} \quad \kappa(G) \leq |S| + c,$$

by (3.53) we have

$$-2 \geq k|S| - \frac{(k+1)^2 m}{4} + \frac{mc + e_G(T,U)}{2} - \omega(U)$$

$$\geq k(\kappa(G) - c) - \frac{(k+1)^2 m}{4} + \frac{mc + 3a + cb}{2} - a - b$$

$$\geq k(\kappa(G) - c) - \left(\frac{(k+1)^2}{4} - \frac{c}{2}\right)(m+b) \qquad (\text{by } \tfrac{(k+1)^2}{4} > 1)$$

$$\geq k(\kappa(G) - c) - \left(\frac{(k+1)^2}{4} - \frac{c}{2}\right)\alpha(G)$$

$$\geq k(\kappa(G) - c) - \left(\frac{(k+1)^2}{4} - \frac{c}{2}\right)\frac{4k\kappa(G)}{(k+1)^2}$$

$$(\text{by the assumption } \tfrac{(k+1)^2\alpha(G)}{4k} \leq \kappa(G))$$

$$\geq kc\left(\frac{2\kappa(G)}{(k+1)^2} - 1\right) \geq 0. \quad (\text{by the assumption } \tfrac{(k+1)^2}{2} \leq \kappa(G))$$

This contradiction completes the proof of the case where k is odd.

When k is even, instead of (3.53) we have

$$-2 \geq k|S| - \frac{k(k+2)^2 m}{4} + \frac{mt + e_G(T,U)}{2} - \omega(U),$$

since $n_i(n_i - k - 1) \geq -k(k+2)/4$. Thus we can similarly derive a contradiction. Consequently the theorem is proved. $\square$

We now show that the conditions $((k+1)^2/4k)\alpha(G) \leq \kappa(G)$ and $((k+2)/4)\alpha(G) \leq \kappa(G)$ are best possible. Let $k \geq 2$ be an integer, and

$$G = K_s + tK_{(k+1)/2} \qquad \text{or} \qquad G = K_s + tK_{k/2}$$

according to the parity of k. If s and t satisfy

$$ks < \frac{t(k+1)^2}{4} \qquad \text{or} \qquad ks < \frac{tk(k+2)}{4},$$

then G has no k-factor. Also, the above inequalities are equivalent to

$$\frac{(k+1)^2}{4k}\alpha(G) > \kappa(G) \qquad \text{or} \qquad \frac{k(k+1)}{4k}\alpha(G) > \kappa(G).$$

Hence we can choose s and t so that the difference between $((k+1)^2/4k)\alpha(G)$ and $\kappa(G)$ is as small as we want. Therefore the conditions are sharp. Note that we can easily choose s and t so that they satisfy the above conditions and so that $k|G|$ is even.

We next consider a sufficient condition for a graph to have a regular factor, using minimum degree.

Theorem 3.22 (Egawa and Enomoto [66] and Katerinis [138]). *Let $k \geq 2$ be an integer, and G be a connected simple graph. If $\delta(G) \geq |G|/2$, $|G| \geq 4k - 6$ and $k|G|$ is even, then G has a k-regular factor.*

Proof. The proof of this theorem is similar to that of Theorem 3.17. Actually, these two theorems are simultaneously proved in the paper [66]. Here we use the proof in [66].

Let $n = |G|$. Assume that G does not have a k-factor. Let (S, T) be a maximal pair of G, where $S \cup T \neq \emptyset$ since $\delta(\emptyset, \emptyset) = 0$ by noting that $kn \equiv 0 \pmod 2$. Put

$$s = |S| \qquad \text{and} \qquad t = |T|.$$

Let m be the number of components of $G - (S \cup T)$, and c denote the minimum order of components of $G - (S \cup T)$. Then by the regular factor theorem and Lemma 3.18, we have

$$-2 \geq \delta(S, T) = k|S| + \sum_{x \in T} \deg_{G-S}(x) - k|T| - q(S, T)$$

$$\geq ks + \sum_{x \in T} \deg_{G-S}(x) - kt - m, \tag{3.55}$$

$$m \leq n - s - t, \tag{3.56}$$

$$3 \leq c \leq \frac{n - s - t}{m} \qquad \text{(when } m \geq 1\text{)}, \tag{3.57}$$

$$\delta(G) \leq c - 1 + s + t, \qquad \text{and} \tag{3.58}$$

$$\delta(G) \geq \frac{n}{2} \geq \frac{(k-1)n + 2k - 3}{2k - 1}. \qquad \text{(by } n \geq 4k - 6\text{)} \tag{3.59}$$

Claim. $T \neq \emptyset$.

Suppose $T = \emptyset$. Then $t = 0$ and $s \geq 1$, and $m \geq ks + 2 \geq 4$ by (3.55). Therefore, by (3.58) and (3.57), we have

$$\frac{n}{2} \leq \delta(G) \leq c - 1 + s \leq \frac{n - s}{m} - 1 + s$$

$$\leq \frac{n - s}{4} - 1 + s = \frac{n - 3s - 4}{4}.$$

This is a contradiction. Hence the claim holds.

By the above claim, we can define

$$h = \min_{x \in T} \deg_{G-S}(x).$$

Clearly

$$\delta(G) \leq h + s. \tag{3.60}$$

We first consider the following case.

Case 1. $h = 0$.

Let $Z = \{x \in T \mid \deg_{G-S}(x) = 0\} \neq \emptyset$. Then for a vertex $v \in Z$, we have

$$s = |S| \geq |N_G(v)| = \deg_G(v) \geq \delta(G) \geq \frac{n}{2}.$$

On the other hand, by (3.55) and (3.56), we obtain

$$-2 \geq ks + \sum_{x \in T}(\deg_{G-S}(x) - k) - m$$
$$\geq ks - kt - k(n - s - t) = 2ks - kn.$$

Hence

$$\frac{n}{2} \leq s \leq \frac{kn - 2}{2k}.$$

This is a contradiction.

Case 2. $1 \leq h \leq k - 1$.

Case 3. $h = k$.

Case 4. $h > k$.

In each of the above cases, we can apply exactly the same argument as in the proof of Theorem 3.17 in order to derive a contradiction. Consequently the theorem is proved. $\square$

Theorem 3.22 can be generalized as follows.

Theorem 3.23. *Let $k \geq 2$ be an integer, and G be a connected simple graph such that $\delta(G) \geq k$ and $k|G|$ is even. If one of the following two conditions holds, then G has a k-regular factor.*
(i) $\deg_G(x) + \deg_G(y) \geq |G|$ for every pair of non-adjacent vertices x and y of G and $|G| \geq 4k - 5$.   (Iida and Nishimura [107])
(ii) $\max\{\deg_G(x), \deg_G(y)\} \geq |G|/2$ for every pair of non-adjacent vertices x and y of G and $|G| \geq 4k - 3$.   (Nishimura [201]).

For a graph G, we define $NC(G)$ as

$$NC(G) = \min\{|N_G(x) \cup N_G(y)| : x, y \in V(G),\ xy \notin E(G)\},$$

that is, $NC(G)$ is the minimum size of $N_G(x) \cup N_G(y)$ taken over all pairs of non-adjacent vertices x and y of G. For a graph G with sufficiently large order, Niessen generalizes the above theorem by using $NC(G)$ as follows.

Theorem 3.24 (Niessen [194] (1995)). *Let $k \geq 2$ be an integer, and G be a connected simple graph such that $k|G|$ is even and $|G| \geq 8k - 7$. If one of the following two conditions holds, then G has a k-regular factor.*
(i) $NC(G) \geq |G|/2$ and $\delta(G) \geq k + 1$.
(ii) $NC(G) \geq (|G| + k - 2)/2$ and $\delta(G) \geq k$.

Note that Niessen obtained a more general result, that is, he proved that if $\delta(G) \geq k$, $|G| \geq 8k - 7$ and $NC(G) \geq |G|/2$, then G has a k-regular factor or G belongs to two families of exceptional graphs ([194]), which have no k-factors. The above theorem is an easy consequence of this result, but we shall prove the above theorem since it is simple to do so. Furthermore, when $k = 2$, Broersma, van den Heuvel and Veldman obtained the next theorem, and so we assume $k \geq 3$, which makes the proof shorter. However, the proof of Theorem 3.24 is still long and complicated, and so readers may skip it.

Theorem 3.25 ([35]). *Let G be a 2-connected graph. If $NC(G) \geq |G|/2$, then G has a Hamiltonian cycle, G is the Petersen graph, or G belongs to one of three families of exceptional graphs.*

We prove only (i) of Theorem 3.24, but the proof is almost the same as the original one. In order to do so, we need the next two lemmas.

Lemma 3.26. *Let G be a connected simple graph of order n, and let $\emptyset \neq X \subset V(G)$. If $NC(G) \geq n/2$, then the following statements hold.*
(i)   If $|X| \geq 3$, then $\omega(G - X) \leq |X|$.
(ii) If $|X| \leq 3$, then $\omega(G - X) \leq 3$.
(iii) If $4 \leq |X| < (n + 20)/6$, then $\omega(G - X) \leq 4$.

Proof. (i) Let $m = \omega(G - X)$. If $m \leq 3$, then the lemma holds, and so we may assume $m \geq 4$. Let $C_1, C_2, \ldots, C_m$ be the components of $G - X$, where $|C_1| \leq |C_2| \leq \cdots \leq |C_m|$. For two vertices $x \in V(C_1)$ and $y \in V(C_2)$, we have

$$\frac{n}{2} \leq |N_G(x) \cup N_G(y)| \leq |C_1| - 1 + |C_2| - 1 + |X|.$$

Thus

$$\frac{n}{2} + 2 - |X| \leq |C_1| + |C_2| \leq 2 \cdot \frac{n - |X|}{m}.$$

This is equivalent to

$$m(n + 4 - 2|X|) \leq 4(n - |X|). \tag{3.61}$$

If $|X| \geq n/2$, then $m \leq |V(G) - X| \leq n/2 \leq |X|$, and (i) holds. Thus we may assume $3 \leq |X| < n/2$. Suppose $m \geq |X| + 1$. Then by (3.61),

$$(|X| + 1)(n + 4 - 2|X|) \leq 4(n - |X|),$$

which implies

$$(2|X| - n)(|X| - 3) \geq 4.$$

This is a contradiction. Therefore $m \leq |X|$, which implies (i).

(ii) We may assume $1 \leq |X| \leq 3$ since if $X = \emptyset$, (ii) clearly holds. Assume $m \geq 4$. By (3.61), we have

$$4(n + 4 - 2|X|) \leq 4(n - |X|),$$

which implies $4 \leq |X|$, a contradiction. Thus (ii) follows.

(iii) Assume that $4 \leq |X| < (n + 20)/6$ and $m \geq 5$. Then $n + 4 - 2|X| > (2n - 8)/3 \geq 0$. Thus, by (3.61) we have

$$5(n + 4 - 2|X|) \leq 4(n - |X|),$$

which implies $n + 20 \leq 6|X|$. This contradicts $|X| < (n + 20)/6$, and thus (iii) holds. $\square$

Lemma 3.27. *Let $k \geq 2$ be an integer. Suppose that a connected simple graph G with $k|G|$ even has no k-regular factor. Let (S, T) be a maximal pair of G. Assume $T \neq \emptyset$. Choose a subset $T^* \subseteq T$ so that $\delta(S, T^*) < 0$, but $\delta(S, T') \geq 0$ for every proper subset $T' \subset T^*$. If $T^* \neq \emptyset$, then for every $u \in T^*$ we have*

$$\deg_{G-S}(u) \leq k - 2 + c(u) \leq k - 2 + q(S, T^*),$$

where $c(u)$ denotes the number of k-odd components of $G - (S \cup T^*)$ joined to u by edges of G, and $q(S, T^*)$ is the number of k-odd components of $G - (S \cup T^*)$. In particular, $|N_G(u) \cap T^*| \leq k - 2$ for every $u \in T^*$.

Proof. Since $\delta(S, T^*) \leq -2$ and $\delta(S, T^* \setminus \{u\}) \geq 0$, we have

$$
\begin{aligned}
2 \leq {}& \delta(S, T^* \setminus \{u\}) - \delta(S, T^*) \\
= {}& k|S| + \sum_{x \in T^* \setminus \{u\}} \deg_{G-S}(x) - k|T^* \setminus \{u\}| - q(S, T^* \setminus \{u\}) \\
& - \left(k|S| + \sum_{x \in T^*} \deg_{G-S}(x) - k|T^*| - q(S, T^*) \right) \\
= {}& - \deg_{G-S}(u) + k - h_G(S, T^* \setminus \{u\}) + q(S, T^*).
\end{aligned}
$$

Since $q(S, T^* \setminus \{u\}) \geq q(S, T^*) - c(u)$, we obtain

$$\deg_{G-S}(u) \leq k - 2 + c(u) \leq k - 2 + q(S, T^*).$$

$\square$

Proof of (i) of Theorem 3.24   Suppose that $k \geq 3$ and G satisfies the conditions in (i), but has no k-factor. Let (S, T) be a maximal pair, and if $T \neq \emptyset$, choose a subset $T^* \subseteq T$ as Lemma 3.27. Then by the regular factor theorem, we have $\delta(S, T) \leq -2$, which is equivalent to

$$q(S, T) \geq k|S| + 2 - k|T| + \sum_{x \in T} \deg_{G-S}(x). \tag{3.62}$$

Since kn is even, $S \cup T \neq \emptyset$. Let

$$s = |S|, \quad t = |T| \quad \text{and} \quad U = V(G) - (S \cup T).$$

Claim 1. $T \neq \emptyset$ and $T^* \neq \emptyset$.

Assume $T = \emptyset$. Then $S \neq \emptyset$. By (3.62), we have $q(S, \emptyset) \geq k|S| + 2 \geq 5$. This implies $\omega(G - S) \geq 5$, which contradicts Lemma 3.26. Hence $T \neq \emptyset$. If $T^* = \emptyset$, then $\delta(S, \emptyset) \leq -2$, and so (3.62) holds, and we can derive a contradiction as above. Therefore $T^* \neq \emptyset$.

Claim 2. $|T^*| \leq k - 1$.

Since the proof of Claim 2 is difficult and long, we shall present it after the proof of the theorem, i.e., we continue the proof under the assumption that Claim 2 holds.

Claim 3. T^* induces a complete graph of G.

Suppose that there exist two non-adjacent vertices $v, w \in T^*$. Then $|T^*| \geq 2$. By $NC(G) \geq n/2 \geq 4k - (7/2)$, we have

$$\deg_{G-S}(v) + \deg_{G-S}(w) \geq |N_G(v) \cup N_G(w)| - |S| \geq 4k - 3 - |S|. \quad (3.63)$$

Furthermore, for every $x \in T^* - \{v, w\}$, it follows that $\deg_{G-S}(x) \geq \delta(G) - |S| \geq k + 1 - |S|$. Hence setting $|T^*| = t^*$, we have by (3.62) that

$$\begin{aligned}
q(S, T^*) &\geq k|S| + 2 - k|T^*| + \sum_{x \in T^*} \deg_{G-S}(x) \\
&\geq ks + 2 - kt^* + (t^* - 2)(k + 1 - s) + 4k - 3 - s \\
&= s(k + 1 - t^*) + t^* + 2k - 3 \\
&\geq 2s + t^* + 2k - 3 \geq 2 + 6 - 3 = 5. \qquad \text{(by Claim 2)}
\end{aligned}$$

We can derive a contradiction from Lemma 3.26 since $q(S, T^*) > |S \cup T^*| = s + t^*$ (see the above inequality). Therefore Claim 3 is proved.

By the previous claims, we know that

$$1 \leq |T^*| \leq k - 1 \qquad \text{and} \qquad T^* \text{ induces a complete graph.}$$

Let $C_1, C_2, \ldots, C_m$ be the k-odd components of $G - (S \cup T^*)$, where $m = q(S, T^*)$ and $|C_1| \leq |C_2| \leq \cdots \leq |C_m|$.

If $\omega(G - (S \cup T^*)) \geq 2$, then the edge-maximality of G implies $m = \omega(G - S \cup T^*)$, that is, if there is an even component, then we can add an edge joining an even component to a k-odd component. This is due to the fact that the resulting graph still has no k-factor, since $\delta(S, T^*)$ does not change. By the same arguments, we may assume that each component C_i induces a complete graph. We consider two cases.

Case 1. $|T^*| = 1$.

Let $T^* = \{u\}$. If $S = \emptyset$, then by (3.62) we have

$$q(\emptyset, \{u\}) \geq 2 - k + \deg_G(u) \geq 2 - k + \delta(G) \geq 3. \quad (3.64)$$

By Lemma 3.26 we have $q(\emptyset, \{u\}) \leq \omega(G - \{u\}) \leq 3$, and so $q(\emptyset, \{u\}) = 3$. By substituting $q(\emptyset, \{u\}) = 3$ into (3.64), we have

$$\deg_G(u) = \delta(G) = k + 1.$$

Let C_1, C_2, C_3 be the components of $G - T^* = G - \{u\}$. Since $1 \leq e_G(u, C_i) \leq \deg_G(u) - 2 = k - 1$ for every $1 \leq i \leq 3$ and $|C_i| \geq \delta(G) \geq k + 1$, there exists a vertex $y_i \in V(C_i)$ such that $xy_i \notin E(G)$. It follows that

$$\sum_{i=1}^{3} \Big(|C_i| + e_G(u, V(G) - V(C_i)) \Big) = n - 1 + 2\deg_G(u) = n + 2k + 1.$$

We may assume that C_1 takes the minimum value among all $|C_i| + e_G(u, V(G) - V(C_i))$, $1 \leq i \leq 3$. For two vertices u and $y_1 \in V(C_1)$, we have

$$\frac{n}{2} \leq |N_G(u) \cup N_G(y_1)| \leq |C_1| + e_G(u, V(G) - V(C_1)) - 1 \leq \frac{n + 2k - 2}{3}.$$

This is a contradiction. Therefore $S \neq \emptyset$.

It is clear that $\deg_{G-S}(u) \geq \delta(G) - |S| \geq k + 1 - s$. By (3.62), we have

$$q(S, \{u\}) \geq k|S| + 2 - k + \deg_{G-S}(u) \geq (k - 1)s + 3 \geq 5.$$

This inequality contradicts Lemma 3.26 since $q(S, \{u\}) \geq s + 1$.

Case 2. $|T^*| \geq 2$.

It follows from Claim 2 that $2 \leq |T^*| \leq k - 1$. We first assume $S = \emptyset$. Since $\delta(S, T^*) \leq -2$ and $\delta(G) \geq k + 1$, we have

$$q(\emptyset, T^*) \geq 2 - k|T^*| + \sum_{x \in T^*} \deg_G(x) \geq 2 + |T^*| \geq 4.$$

This contradicts Lemma 3.26 as $q(\emptyset, T^*) \geq |T^*| + 1$. Thus we may assume that $S \neq \emptyset$. Let

$$U^* = V(G) - (S \cup T^*).$$

Since $|S| + |T^*| \geq 1 + 2 = 3$, by Lemma 3.26 we have

$$q(S, T^*) \leq s + t^*. \tag{3.65}$$

By Claim 3 and since $\delta(G) \geq k + 1$, we have

$$\sum_{x \in T^*} \deg_{G-S}(x) = t^*(t^* - 1) + e_G(T^*, U^*)$$

$$\geq t^*(t^* - 1) + t^*(k + 1 - s - (t^* - 1)) \tag{3.66}$$

Combining (3.62), (3.65) and (3.66), we obtain

$$s + t^* \geq q(S, T^*) \geq ks + 2 - kt^* + t^*(t^* - 1) + t^*(k + 1 - s - (t^* - 1))$$
$$= (k - t^*)s + 2 + t^* \geq s + 2 + t^*. \qquad \text{(by Claim 2).}$$

This is a contradiction. Consequently Theorem 3.24 is proved.

We now prove Claim 2.

Claim 2. $|T^*| \leq k - 1$.

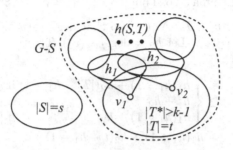

Fig. 3.15. Proof of Claim 2; $|T^*| \geq k$, v_1, v_2, $G - S$, S and T.

Proof. Suppose

$$|T^*| \geq k.$$

Then $3 \leq k \leq |T^*| \leq |T| = t$. Define

$$h_1 = \min_{x \in T} \deg_{G-S}(x),$$

and choose a vertex $v_1 \in T$ so that $\deg_{G-S}(v_1) = h_1$, and $v_1 \in T^*$, if possible (see Fig. 3.15).

Sub-Claim 2A. $h_1 \leq k$.

Assume $h_1 \geq k + 1$. Then by (3.62) we have

$$\omega(G - (S \cup T)) \geq q(S, T) \geq ks + 2 + t \geq 4.$$

This contradicts Lemma 3.26 as $|S \cup T| = s + t < ks + t + 2 \leq \omega(G - (S \cup T))$. Hence the sub-claim is proved.

Sub-Claim 2B. $T \setminus N_G[v_1] \neq \emptyset$.

Suppose $T \subseteq N_G[v_1]$, which implies $T \subseteq N_{G-S}[v_1]$. If $v_1 \in T^*$, then $|N_G(v_1) \cap T^*| = |T^* - v_1| \geq k - 1$ as $|T^*| \geq k$, which contradicts Lemma 3.27. Hence $v_1 \notin T^*$. Thus

$$|T| \geq |T^*| + 1 \geq k + 1 \quad \text{and} \quad h_1 \geq |T - v_1| \geq |T^*| \geq k.$$

On the other hand, $|T| \leq |N_{G-S}[v_1]| \leq h_1 + 1 \leq k + 1$ by Sub-Claim 2A. Hence
$$t = |T| = k + 1, \ h_1 = k, \ T = T^* \cup \{v_1\} \ \text{and} \ t^* = k.$$

By the choice of v_1, $v_1 \notin T^*$ implies $\deg_{G-S}(x) \geq h_1 + 1$ for every $x \in T^*$. By (3.62) we have

$$q(S, T) \geq ks + 2 - kt + h_1 + t^*(h_1 + 1) = ks + 2 + k > s + t,$$

which contradicts Lemma 3.26. Thus $T \setminus N_G[v_1] \neq \emptyset$, and Sub-Claim 2B holds.

Choose a vertex $v_2 \in T \setminus N_G(v_1)$ so that $\deg_{G-S}(v_2)$ is minimum among all vertices in $T \setminus N_G(v_1)$. Let $h_2 = \deg_{G-S}(v_2)$. Then

$$\sum_{x \in T} \deg_{G-S}(x) \geq \begin{cases} h_1(h_1 + 1) + h_2(|T| - h_1 - 1) & \text{if } |T| \geq h_1 + 2, \\ h_1^2 + h_2 & \text{if } |T| = h_1 + 1, \\ h_1(h_1 - 1) + h_2 & \text{if } |T| = h_1, \end{cases}$$
$$\geq h_1(h_1 + 1) + h_2(t - h_1 - 1). \qquad \text{(by } h_1 \leq h_2\text{)} (3.67)$$

Hereafter, we often use the above inequality without specifically mentioning it. By $NC(G) \geq \frac{n}{2}$, we have $\frac{n}{2} \leq |N_G(v_1) \cup N_G(v_2)| \leq |S| + h_1 + h_2$, and so

$$s = |S| \geq \left\lceil \frac{n}{2} \right\rceil - h_1 - h_2 \geq 4k - 3 - h_1 - h_2. \qquad \text{(by } n \geq 8k - 7\text{)} \quad (3.68)$$

We consider three cases.

Case A. $0 \leq h_2 \leq k - 1$.

Using (3.62), we have

$$q(S, T) \geq k|S| + 2 - k|T| + \sum_{x \in T} \deg_{G-S}(x)$$
$$\geq ks + 2 - kt + h_1(h_1 + 1) + h_2(t - h_1 - 1)$$
$$= ks + 2 + (h_2 - k)t + (h_1 + 1)(h_1 - h_2)$$

Since $q(S, T) \leq \omega(S, T) \leq |U|$, by the above inequality we have

$$(k - h_2)t + |U| \geq ks + 2 + (h_1 + 1)(h_1 - h_2)$$
$$= k(n - t - |U|) + 2 + (h_1 + 1)(h_1 - h_2)$$

and thus

$$t \geq \frac{kn + 2 + (h_1 + 1)(h_1 - h_2)}{2k - h_2} - \frac{k + 1}{2k - h_2}|U|$$
$$\geq \frac{kn + 2 + (h_1 + 1)(h_1 - h_2)}{2k - h_2} - |U| \qquad \text{(by } 2k - h_2 \geq k + 1\text{)}$$

By (3.68) we can obtain

$$n = t + |U| + s \geq \frac{kn + 2 + (h_1 + 1)(h_1 - h_2)}{2k - h_2} + \frac{n}{2} - h_1 - h_2$$

$$= \frac{h_1^2 - (2k - 1)h_1 - h_2}{2k - h_2} + \frac{kn + 2}{2k - h_2} + \frac{n}{2} - h_2$$

Here the right-hand-side takes its minimum value when $h_1 = h_2$ since $0 \leq h_1 \leq h_2 \leq k - 1$. Therefore

$$n \geq \frac{kn + 2}{2k - h_2} + \frac{n}{2} - 2h_2,$$

which is equivalent to $8h_2 k - 4h_2^2 - 4 \geq nh_2$. Therefore $h_2 \geq 1$ and hence $n \leq 8k - 4h_2 - \frac{4}{h_2} \leq 8k - 8$. This contradicts the assumption $n \geq 8k - 7$.

Case B. $h_2 = k$.

By Lemma 3.18, we have $|U| \geq 3q(S,T)$. Combining this with (3.62), we have

$$\frac{|U|}{3} \geq q(S,T) \geq k|S| + 2 + \sum_{x \in T} \deg_{G-S}(x) - k|T|$$

$$= ks + 2 + h_1(h_1 + 1) + k(t - h_1 - 1) - kt$$

$$= k(n - t - |U|) + 2 - (k - h_1)(h_1 + 1),$$

and thus

$$\left(k + \frac{1}{3}\right)(n - s) = \left(k + \frac{1}{3}\right)(|U| + t)$$

$$\geq kn + 2 - (k - h_1)(h_1 + 1) + \frac{t}{3}.$$

$$\left(k + \frac{1}{3}\right)n - kn - 2 \geq \left(k + \frac{1}{3}\right)s - (k - h_1)(h_1 + 1).$$

Therefore we have

$$\frac{n}{3} - 2 = \left(k + \frac{1}{3}\right)n - kn - 2 \geq \left(k + \frac{1}{3}\right)s - (k - h_1)(h_1 + 1)$$

$$\geq \left(k + \frac{1}{3}\right)\left(\frac{n}{2} - h_1 - k\right) - (k - h_1)(h_1 + 1) \qquad \text{(by (3.68))}$$

$$= h_1^2 - \left(2k + \frac{1}{3} - 1\right)h_1 + \left(k + \frac{1}{3}\right)\left(\frac{n}{2} - k\right) - k.$$

The right-hand-side takes the minimum value when $h_1 = k$ since $0 \leq h_1 \leq h_2 = k$. Therefore

$$\frac{n}{3} - 2 \geq \left(k + \frac{1}{3}\right)\left(\frac{n}{2} - 2k\right).$$

Thus

$$12k^2 + 4k - 12 \geq (3k - 1)n \geq (3k - 1)(8k - 7),$$

i.e., we have

$$0 \geq 12k^2 - 33k + 19k = (k - 2)(12k - 9) + 1.$$

This is a contradiction since $k \geq 3$.

Case C. $h_2 \geq k + 1$.

We shall consider six subcases.

Subcase C1. $h_1 = k$ and $t = k$.

In this case $T^* = T$ as $k = |T| \geq |T^*| \geq k$. By (3.62), we have

$$q(S, T) \geq ks + 2 - k^2 + k(k - 1) + k + 1 = ks + 3 \geq 3.$$

If $s + t \geq q(S, T)$, then $s + t \geq q(S, t) \geq ks + 3$, which implies $s = 0$ (if $s \geq 1$, then $s + k = s + t \geq ks + 3 \geq k + s + 2$, a contradiction.) Therefore

$$k + 1 \leq \delta(G) \leq \deg_G(v_1) = \deg_{G-S}(v_1) = h_1 = k,$$

a contradiction. Hence $s + t = |S \cup T| < q(S, T)$.

By Lemma 3.26 and $t = k \geq 3$, we have

$$|S \cup T| < q(S, T) \leq \omega(G - (S \cup T)) \leq |S \cup T|.$$

This is a contradiction.

Subcase C2. $h_1 = k$ and $t = k + 1$.

Since $|T| = k + 1 \geq 4$, we can apply Lemma 3.26 to $S \cup T$, and so by (3.62)

$$\begin{aligned} k + 1 + |S| = |T| + |S| &\geq q(S, T) \geq k|S| + 2 - k|T| + h_1^2 + h_2 \\ &= k|S| + 2 - k + h_2. \end{aligned}$$

Hence

$$2k - 1 \geq (k - 1)|S| + h_2.$$

Since $h_2 \geq k + 1$, we have $S = \emptyset$ and $h_2 \leq 2k - 1$. By (3.68) we obtain

$$4k - 3 \leq \left\lceil \frac{n}{2} \right\rceil \leq h_1 + h_2 \leq 3k - 1.$$

Thus $k \leq 2$, a contradiction.

Subcase C3. $h_1 = k$ and $t \geq k + 2$.

Since $|S| + |T| \geq k + 2 \geq 5$, we can apply Lemma 3.26 to $S \cup T$, and so by (3.62)

$$s + t \geq q(S,T) \geq k|S| + 2 - k|T| + k(k+1) + h_2(t - k - 1)$$
$$0 \geq (k-1)s + 2 + t(h_2 - k - 1) + (k+1)(k - h_2)$$
$$\geq (k-1)s + h_2 - 2k$$

Hence $h_2 \leq 2k - (k-1)s$. By (3.68)

$$h_2 \geq \left\lceil \frac{n}{2} \right\rceil - s - h_1 \geq 3k - 3 - s,$$

and so we have $3 \geq k + (k-2)s$. This is a contradiction if $k \geq 4$. So we may assume $k = 3$. Then $s = 0$ and the equality holds in all the above inequalities, in particular, $t = q(S,T) = 5 \geq \omega(G - (S \cup T))$. By Lemma 3.26 (3), we have $5 = t = |S \cup T| \geq (n+20)/6$. This contradicts $n \geq 8k - 7 = 17$.

Subcase C4. $h_1 = k - 1$ and $t \geq k$.

Clearly

$$s = |S| \geq \deg_G(v_1) - \deg_{G-S}(v_1) \geq \delta(G) - h_1 \geq k + 1 - h_1 = 2. \quad (3.69)$$

Since $s + t \geq 2 + k \geq 4$, we can apply Lemma 3.26 to $S \cup T$. If $t = k$, then by (3.67) and (3.69) we have

$$s + t \geq q(S,T) \geq k|S| + 2 - kt + h_1^2 + h_2$$
$$\geq s + t + (k-1)s - 2k + h_2 + 3 > s + t, \qquad \text{(by } h_2 \geq k + 1)$$

a contradiction.

Thus we may assume $t \geq k + 1$. By (3.62) and (3.67) we have

$$s + t \geq q(S,T) \geq k|S| + 2 - kt + (k-1)k + h_2(t - k).$$

Thus

$$0 \geq (k-1)s + 2 + t(h_2 - k - 1) + k(k-1) - h_2 k$$
$$\geq (k-1)s + 2 + (k+1)(h_2 - k - 1) + k(k-1) - h_2 k$$
$$= (k-1)s + h_2 - 3k + 1$$
$$= (k-2)s + s + h_2 - 3k + 1$$
$$\geq (k-2)s + k - 2 - h_1 = (k-2)s - 1 \qquad \text{(by (3.68))}$$

This is a contradiction since $k \geq 3$ and $s \geq 2$.

Subcase C5. $h_1 \leq k - 2$ and $t \geq k$.

It follows that

$$s = |S| \geq \deg_G(v_1) - \deg_{G-S}(v_1) \geq \delta(G) - h_1 \geq k + 1 - h_1 \geq 3. \quad (3.70)$$

By applying Lemma 3.26 to $S \cup T$, and by (3.62) and (3.67) we have

$$s + t \geq q(S,T) \geq k|S| + 2 - kt + h_1(h_1 + 1) + h_2(t - h_1 - 1).$$

Thus by $h_2 \geq k+1$ and $s \geq k+1 - h_1$ (by 3.70), we have

$$\begin{aligned}
0 &\geq (k-1)s + 2 + t(h_2 - k - 1) - (h_2 - h_1)(h_1 + 1) \\
&\geq (k-1)s + 2 + k(h_2 - k - 1) - (h_2 - h_1)(h_1 + 1) \quad\quad (3.71) \\
&\geq (k-1)(4k - 3 - h_1 - h_2) + 2 + k(h_2 - k - 1) \\
&\quad -h_2 h_1 - h_2 + h_1^2 + h_1 \quad\quad\quad\quad\quad\quad (\text{by } (3.68))
\end{aligned}$$

Then we have

$$\begin{aligned}
h_1(k - 2 + h_2) - h_1^2 &\geq (k-1)(4k-3) - k(k+1) + 2 \\
&= (3k-2)(k-2) + 1.
\end{aligned}$$

Since left-hand-side takes its maximum when $h_1 = k - 2$ as $0 \leq h_1 \leq k - 2$, we have

$$(k-2)h_2 \geq (3k-2)(k-2) + 1.$$

Hence $h_2 \geq 3k - 1$. By substituting $s \geq k+1 - h_1$ (by 3.70) in (3.71), we have

$$\begin{aligned}
0 &\geq (k-1)(k+1-h_1) + 2 + k(h_2 - k - 1) - (h_2 - h_1)(h_1 + 1) \\
&= 1 - k + h_2(k - h_1 - 1) + h_1(h_1 + 2 - k) \\
&= (k - h_1 - 2)(h_2 - h_1) + h_2 + 1 - k \\
&\geq 2k > 0 \quad\quad (\text{by } h_1 \leq k - 2, h_2 \geq h_1 \text{ and } h_2 \geq 3k - 1.)
\end{aligned}$$

This is a contradiction. Consequently Claim 2 is proved, and the proof of Theorem 3.24 is complete. $\square$

Two results similar to Theorem 3.20 using minimum degree, connectivity and independence number are given in the following theorem.

Theorem 3.28 (Katerinis and Tsikopoulos [146], [145]). *Let a and b be integers such that $1 \leq a \leq b$ and $2 \leq b$, and let G be a connected simple graph. Let $f : V(G) \to \{a, a+1, \ldots, b\}$ be a function such that $\sum_{x \in V(G)} f(x) \equiv 0$ (mod 2). If one of the following two conditions holds, then G has an f-factor.*

$$(i) \quad \delta(G) \geq \frac{b}{a+b}|G| \quad and \quad |G| \geq \frac{(a+b)^2 - 3(a+b)}{a}$$

$$(ii) \quad \kappa(G) \geq \frac{2(b-1)}{a}\alpha(G) \quad and \quad |G| \geq 8.$$

It is well-known that a claw-free graph possesses nice properties (Fig. 3.16). For example, we showed that a connected claw-free graph of even order has a 1-factor. Here we consider regular factors in such graphs. It is easy to see that every line graph is claw-free, and Nishimura [200] showed that for an integer $k \geq 2$, if a connected graph G with $k|E(G)|$ even satisfies $\delta(L(G)) \geq (9k + 12)/8$, then the line graph $L(G)$ of G has a k-factor. This result was extended to the following theorem by Egawa and Ota. Moreover, a similar but weaker result, which says that every claw-free graph with $k|G|$ even and minimum degree at least $2k$ has a k-regular factor, was obtained by Choudum and Paulraj [53] independently.

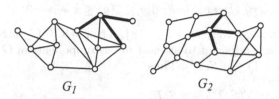

$$G_1 \qquad\qquad G_2$$

Fig. 3.16. A claw-free graph G_1 and a $K_{1,4}$-free graph G_2, which is not claw-free.

Theorem 3.29 (Egawa and Ota [73]). *Let $k \geq 1$ be an integer and G be a connected claw-free simple graph. If $k|G|$ is even and*

$$\delta(G) \geq \frac{9k + 12}{8}, \tag{3.72}$$

then G has a k-regular factor.

In the same paper, Egawa and Ota obtained the following theorem, and showed that the condition on minimum degree is almost best possible. By substituting $n = 3$ into inequality (3.73) below, we obtain $\delta(G) \geq (9/8)k + (3/2) + (1/2k)$, which is slightly weaker than (3.72).

Theorem 3.30 (Egawa and Ota [73]). *Let $n \geq 3$ and $k \geq 2$ be integers such that $k \geq n - 1$ if k is odd, and let G be a $K_{1,n}$-free connected simple graph with $k|G|$ even. If*

$$\delta(G) \geq \frac{n^2}{4(n-1)}k + \frac{3n - 6}{2} + \frac{n - 1}{4k}, \tag{3.73}$$

then G has a k-regular factor.

Notice that the condition $k \geq n-1$ for an odd integer k cannot be removed. Namely, for every $n \geq 4$ and every odd integer k such that $k < n - 1$, there exists a $K_{1,n}$-free connected graph with even order and minimum degree $2k$ that has no k-factor. We shall only prove Theorem 3.30, however, Theorem 3.29 can be proved in the same way with some careful estimation.

Proof of Theorem 3.30 Suppose that G has no k-factor. Then there exists a pair (S, T) of disjoint vertex subsets S and T of G such that

$$\delta(S,T) = k|S| + \sum_{x \in T} \deg_{G-S}(x) - k|T| - q(S,T) < 0.$$

Since G is connected and $k|G|$ is even, it follows that $\delta(\emptyset, \emptyset) = -q(\emptyset, \emptyset) = 0$ and, therefore $S \cup T \neq \emptyset$. Choose such a pair (S,T) so that T is minimal. Then for every proper subset $T' \subset T$, if any exist, $\delta(S,T') \geq 0$. This implies $\delta(S,T') \geq \delta(S,T) + 2$ by $\delta(X,Y) \equiv 0 \pmod 2$.

Claim 1. *For every vertex* $u \in T$, $\deg_{G-S}(u) \leq k + n - 3$.

For a vertex $u \in T$, let $T' = T - \{u\}$. Then $\delta(S,T') \geq \delta(S,T) + 2$. Since G is $K_{1,n}$-free, u is adjacent to at most $n - 1$ components of $G - (S \cup T)$, and so $q(S,T) - (n-1) \leq q(S,T')$. Thus

$$
\begin{aligned}
2 &\leq \delta(S,T') - \delta(S,T) \\
&= k|S| + \sum_{x \in T'} \deg_{G-S}(x) - k|T'| - q(S,T') \\
&\quad - \left(k|S| + \sum_{x \in T} \deg_{G-S}(x) - k|T| - q(S,T) \right) \\
&\leq -\deg_{G-S}(u) + k + n - 1.
\end{aligned}
$$

Hence $\deg_{G-S}(u) \leq k + n - 3$.

For a subset $Y \subseteq T$, we write $\rho(Y)$ for the number of k-odd components of $G - (S \cup T)$ that are joined to Y by edges of G. For $y \in T$, let $\rho(y) = \rho(\{y\})$. Then $\rho(y) \leq n - 1$ since G is $K_{1,n}$-free. Assume $T \neq \emptyset$. Then we define a set $\{y_1, y_2, \ldots, y_r\}$ of vertices of T as follows: let y_1 be a vertex of T such that $\deg_{G-S}(y) - \rho(y)$ is minimum among all $y \in T$. Let $T_1 = N_G[y_1] \cap T$, where $N_G[y_1]$ is the closed neighborhood of y_1. Then for $j \geq 2$, if $T - (T_1 \cup \cdots \cup T_{j-1}) \neq \emptyset$, then choose y_j so that $\deg_{G-S}(y) - \rho(y)$ is minimum among all $y \in T - (T_1 \cup \cdots \cup T_{j-1})$, and assume $T - (T_1 \cup \cdots \cup T_r) = \emptyset$ (see Fig. 3.17). When $T = \emptyset$, we define $r = 0$. Then $\{y_1, y_2, \ldots, y_r\}$ is an independent set of G.

Claim 2. $|S| \geq \dfrac{1}{n-1} \displaystyle\sum_{i=1}^{r} e_G(v_i, S) + \dfrac{1}{k}\big(q(S,T) - \rho(T)\big)$.

Let $\ell = q(S,T) - \rho(T)$. Then there exist ℓ k-odd components $C_1, C_2, \ldots, C_\ell$ such that $e_G(T, C_i) = 0$ for all $1 \leq i \leq \ell$.

Suppose $S = \emptyset$. Then $\ell = 0$ since $T \neq \emptyset$, and so the claim holds. Hence we may assume $S \neq \emptyset$. Since G is connected, $e_G(S, C_i) \geq 1$ for all $1 \leq i \leq \ell$, and thus we can choose a vertex $z_i \in V(C_i)$ so that z_i and S are adjacent. Let $X = \{y_1, \ldots, y_r, z_1, \ldots, z_\ell\}$ (see Fig. 3.17). Since X is an independent set and G is $K_{1,n}$-free, every $x \in S$ is adjacent to at most $n - 1$ vertices of X. Therefore

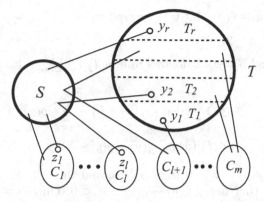

Fig. 3.17. The graph G and vertices y_i and z_j.

$$(n-1)|S| \geq e_G(X, S) = \sum_{i=1}^{r} e_G(y_i, S) + \sum_{j=1}^{\ell} e_G(z_j, S)$$

$$\geq \sum_{i=1}^{r} e_G(y_i, S) + \ell.$$

On the other hand, if k is even then $\ell = 0$ since for any k-odd component C_i of $G - (S \cup T)$, it follows that $k|C_i| + e_G(T, C_i) \equiv e_G(T, C_i) \equiv 1 \pmod 2$. If k is odd, then $k \geq n - 1$ by the assumption of the theorem. Hence

$$(n-1)|S| \geq \sum_{i=1}^{r} e_G(y_i, S) + \ell \geq \sum_{i=1}^{r} e_G(y_i, S) + \frac{n-1}{k}\ell.$$

Therefore Claim 2 holds.

By Claim 2, we have

$$\delta(S, T) \geq \frac{k}{n-1} \sum_{i=1}^{r} e_G(y_i, S) + \sum_{x \in T} \left(\deg_{G-S}(x) - k \right) - \rho(T)$$

$$\geq \sum_{i=1}^{r} \left(\frac{k}{n-1} e_G(y_i, S) + \sum_{x \in T_i} \left(\deg_{G-S}(x) - k \right) - \rho(T_i) \right).$$

Let

$$\beta_i = \sum_{x \in T_i} \left(\deg_{G-S}(x) - k \right) - \rho(T_i).$$

Then in order to prove the theorem, it suffices to show that

$$\frac{k}{n-1} e_G(y_j, S) + \beta_j \geq 0 \qquad \text{for all } 1 \leq j \leq r \tag{3.74}$$

because this implies $\delta(S,T) \geq 0$, a contradiction.

Case 1. $\rho(y_j) \leq n-2$.

Let $y = y_j$ and $\beta = \beta_j$. It follows from $\rho(T_j) \leq \sum_{x \in T_j} \rho(x)$ and the choice of y that

$$\beta \geq \sum_{x \in T_j} \left(\deg_{G-S}(x) - k \right) - \sum_{x \in T_j} \rho(x)$$

$$\geq |T_j|(\deg_{G-S}(y) - \rho(y) - k).$$

If $\deg_{G-S}(y) - \rho(y) - k \geq 0$ then $\beta \geq 0$, and thus (3.74) holds. Therefore we may assume that $\deg_{G-S}(y) - \rho(y) - k < 0$. On the other hand, since $N_{G-S}[y] \supseteq T_j$, we have

$$|T_j| \leq \deg_{G-S}(y) - \rho(y) + 1.$$

Therefore

$$\beta \geq \left(\deg_{G-S}(y) - \rho(y) + 1 \right)(\deg_{G-S}(y) - \rho(y) - k) \tag{3.75}$$

$$= \left(\deg_{G-S}(y) - \rho(y) - \frac{k-1}{2} \right)^2 - \frac{(k+1)^2}{4}. \tag{3.76}$$

Subcase 1.1 $\deg_{G-S}(y) - \dfrac{k-1}{2} \leq n-2$.

By (3.76) we have $\beta \geq -(k+1)^2/4$. Since

$$e_G(y,S) = \deg_G(y) - \deg_{G-S}(y) \geq \deg_G(y) - \left(n - 2 + \frac{k-1}{2}\right),$$

it follows that

$$\frac{k}{n-1}e_G(y,S) + \beta \geq \frac{k}{n-1}\left(\deg_G(y) - n + 2 - \frac{k-1}{2} \right) - \frac{(k+2)^2}{4}.$$

By the condition of the minimum degree of G, we have

$$\frac{k}{n-1}e_G(y,S) + \beta$$

$$\geq \frac{k}{n-1}\left(\frac{n^2}{4(n-1)}k + \frac{3n-6}{2} + \frac{n-1}{4k} - n + 2 - \frac{k-1}{2} \right) - \frac{(k+1)^2}{4}$$

$$= \frac{k}{n-1}\left(\frac{kn^2}{4(n-1)} + \frac{n-1-k}{2} + \frac{n-1}{4k} \right) - \frac{(k+1)^2}{4}$$

$$= \frac{k^2n^2 + 2k(n-1)(n-1-k) + (n-1)^2 - (k+1)^2(n-1)^2}{4(n-1)^2}$$

$$= \frac{k^2}{4(n-1)^2} \geq 0$$

Hence (3.74) holds.

Subcase 1.2 $\deg_{G-S}(y) - \dfrac{k-1}{2} > n - 2.$

Since $\alpha(y) \le n - 2$, the right-hand side of (3.76) takes the minimum value when $\rho(y) = n - 2$. Hence it follows from (3.75) that

$$\beta \ge (\deg_{G-S}(y) - n + 3)(\deg_{G-S}(y) - n - k + 2). \qquad (3.77)$$

Then

$$\frac{k}{n-1} e_G(y, S) + \beta$$
$$= \frac{k}{n-1}\left(\deg_G(y) - \deg_{G-S}(y)\right) + \beta$$
$$\ge \frac{k}{n-1}\left(\frac{n^2}{4(n-1)}k + \frac{3n-6}{2} + \frac{n-1}{4k} - \deg_{G-S}(y)\right)$$
$$\quad + \left(\deg_{G-S}(y) - n + 3\right)(\deg_{G-S}(y) - n - k + 2)$$
$$= \left(\deg_{G-S}(y) - \frac{nk + (n-1)(2n-5)}{2(n-1)}\right)^2 \ge 0$$

Consequently (3.74) holds.

Case 2. $\rho(y_j) = n - 1.$

Let $y = y_j$ and $\beta = \beta_j$. In this case, there exist $n - 1$ k-odd components $C_1, C_2, \ldots, C_{n-1}$ of $G - (S \cup T)$ such that every C_i, $1 \le i \le n - 1$, contains a vertex z_i adjacent to y. Since G is $K_{1,n}$-free, every $x \in T_j - \{y\}$ must be adjacent to some z_i. In particular, $e_G(x, C_i) \ge 1$ for some i and hence C_i is counted in $\rho(x)$. Therefore

$$\rho(T_j) \le \rho(y) + \sum_{x \in T_j - \{y\}} (\rho(x) - 1)$$
$$= \sum_{x \in T_j} \rho(x) - |T_j| + 1.$$

Thus

$$\beta = \sum_{x \in T_j} (\deg_{G-S}(x) - k) - \rho(T_j)$$
$$\ge \sum_{x \in T_j} \left(\deg_{G-S}(x) - \rho(x) - k\right) + |T_j| - 1$$
$$\ge |T_j|(\deg_{G-S}(y) - \rho(y) - k) + |T_j| - 1$$
$$= |T_j|(\deg_{G-S}(y) - n - k + 2) - 1.$$

By Claim 1, $\deg_{G-S}(y) \le n+k-3$, and so $\deg_{G-S}(y)-n-k+2 \le -1 < 0$. On the other hand, $|T_j| \le \deg_{G-S}(y) - \alpha(y) + 1 = \deg_{G-S}(y) - n + 2$. Therefore

$$\beta \geq (\deg_{G-S}(y) - n + 2)(\deg_{G-S}(y) - n - k + 2) - 1$$
$$\geq (\deg_{G-S}(y) - n + 3)(\deg_{G-S}(y) - n - k + 2)$$

This inequality is equivalent to (3.77). Hence by the same argument given above, we can show that (3.74) holds. Consequently, in each case we can derive a contradiction, and thus the proof of the theorem is complete.   □

We now give some variety of sufficient conditions for a graph to have regular factors without proofs.

Theorem 3.31 (Saito [214]). *Let G be a connected simple graph with $|G| \geq 4$ and having a 1-factor F, and let $k \geq 2$ be an integer. If $G - \{x, y\}$ has a k-regular factor for every edge $xy \in E(F)$, then G itself has a k-regular factor.*

Theorem 3.32 (Kotani [159]). *Let G be a connected simple graph, and p be an integer such that $1 \leq p < |G|$. Let $f : V(G) \rightarrow \mathbb{Z}^+$ be a function such that $2 \leq f(x) \leq \deg_G(x)$ for all $x \in V(G)$ and $\sum_{x \in V(G)} f(x)$ is even. If every connected induced subgraph H of order p has an f-factor, then G has an f-factor.*

Theorem 3.33 (Niessen [194]). *Let $k \geq 4$ be an even integer and let G be a simple graph of order at least $k + 1$. If*

$$\delta(G) > \frac{k+2}{4}\alpha(G) + \frac{5k-3}{8} - \frac{2}{k},$$

then G has a k-regular factor, where $\alpha(G)$ denotes the independence number of G.

Note that the situation of Theorem 3.33 is changed a lot when we consider an odd regular factor. Namely, for every integer $\alpha \geq 2$ and any arbitrary large number δ, there exist graphs with independence number α and minimum degree δ that have no k-factor. For example, such graphs are $K_b + \alpha K_d$, where $\alpha \geq kd + 2$, d is odd and $b + \alpha d$ is even.

For positive integers n and k with nk even, let $\rho(n, k)$ denote the maximal number of edges in a graph of order n that has a unique k-factor. Some studies on $\rho(n, k)$ were done as presented below, and for some k, n, $\rho(n, k)$ is settled.

Theorem 3.34 (Hetyei [103] (1972)). *If n is an even positive integer, then*

$$\rho(n, 1) = \frac{n^2}{4}.$$

Proof. ([105]) Assume that a simple graph G has the unique 1-factor F. Let $H_1, H_2, \ldots, H_{n/2}$ be the components of F. There can be at most two edges joining H_i to H_j in G since otherwise $\langle V(H_i) \cup V(H_j) \rangle_G$ contains two 1-factors and so does G. Hence

$$|E(G)| \leq \frac{n}{2} + 2\binom{n/2}{2} = \frac{n^2}{4}.$$

□

Theorem 3.35 (Hendry [101] (1984)). *If n is a positive integer, then*

$$\rho(n, 2) = \left\lceil \frac{n(n+1)}{4} \right\rceil.$$

Hendry [101] determined $\rho(n, 2)$ as above. Moreover, for other $\rho(n, k)$ he constructed a class of graphs G that have a unique k-factor and satisfy

$$\|G\| = \frac{n^2}{4} + \frac{(k-1)n}{4} \qquad \text{if } k \leq \frac{n}{2}, \quad \text{and}$$

$$\|G\| = \frac{nk}{2} + \binom{n-k}{2} \qquad \text{if } k > \frac{n}{2}.$$

He conjectured that these two values yield $\rho(n, k)$, that is, he conjectured that a graph in the class has a unique k-factor and has the maximum number of edges. This conjecture was partially proved by Johann in the following theorem.

Theorem 3.36 (Johann [109]). *Let n and k with nk even and $k < n$. Then*

$$\rho(n, k) \leq \frac{n^2}{4} + \frac{(k-1)n}{4} \qquad \text{if } k \leq \frac{n}{2},$$

where equality holds for $n = 2kl$, $l \geq 1$, and

$$\rho(n, k) = \frac{nk}{2} + \binom{n-k}{2} \qquad \text{if } k > \frac{n}{2}.$$

Recently $\rho(n, 3)$ and $\rho(n, k)$ for many k were determined by Volkmann as follows.

Theorem 3.37 (Volkmann [236]). *Let n and k be positive integers such that kn or $3n$ is even. Then*

$$\rho(n, k) = k^2 + 2\binom{n-k}{2} \qquad \text{for } \frac{n}{3} \leq k \leq \frac{n}{2},$$

and $\rho(n, 3) = n^2/4 + n/2$ for $n \equiv 0, 4 \pmod{6}$, $\rho(n, 3) = n^2/4 + n/2 - 1$ for $n \equiv 2 \pmod{6}$.

3.4 Regular factors and f-factors in bipartite graphs

The criterion for a graph to have an f-factor or a regular factor becomes much simpler when we consider bipartite graphs; $q(S,T)$, the most difficult term in $\delta(S,T)$ to estimate, disappears. It may also be assumed that S and T are contained in distinct partite sets. We begin with this simpler criterion, which was obtained by Ore [205], and Folkman and Fulkerson [86].

Theorem 3.38 (The f-Factor Theorem for Bipartite Graphs). *Let G be a bipartite multigraph with bipartition (A,B) and let $f : V(G) \to \mathbb{Z}^+$ be a function. Then G has an f-factor if and only if $\sum_{x \in A} f(x) = \sum_{x \in B} f(x)$ and*

$$
\begin{aligned}
\delta^*(S,T) &= \sum_{x \in S} f(x) + \sum_{x \in T} \big(\deg_{G-S}(x) - f(x) \big) \\
&= \sum_{x \in S} f(x) + \sum_{x \in T} \big(\deg_G(x) - f(x) \big) - e_G(S,T) \geq 0 \quad (3.78)
\end{aligned}
$$

for all $S \subseteq A$ and $T \subseteq B$. Moreover, condition (3.78) is equivalent to

$$
\delta^*(T,S) = \sum_{x \in T} f(x) + \sum_{x \in S} \big(\deg_{G-T}(x) - f(x) \big) \geq 0 \quad (3.79)
$$

for all $S \subseteq A$ and $T \subseteq B$.

We first show that under the assumption $\sum_{x \in A} f(x) = \sum_{x \in B} f(x)$, conditions (3.78) and (3.79) are equivalent.

Proof of (3.78) $\Leftrightarrow$ (3.79). Let $S \subseteq A$ and $T \subseteq B$. Then it follows from $\sum_{x \in A} f(x) = \sum_{x \in B} f(x)$ that

$$
\begin{aligned}
&\delta^*(A - S, B - T) \\
&= \sum_{x \in A-S} f(x) + \sum_{x \in B-T} \big(\deg_G(x) - f(x) \big) - e_G(A - S, B - T) \\
&= \sum_{x \in A} f(x) - \sum_{x \in S} f(x) + \sum_{x \in B-T} \deg_G(x) \\
&\quad - \sum_{x \in B} f(x) + \sum_{x \in T} f(x) - e_G(A - S, B - T) \\
&= -\sum_{x \in S} f(x) + e_G(A, B - T) + \sum_{x \in T} f(x) - e_G(A - S, B - T) \\
&= \sum_{x \in T} f(x) - \sum_{x \in S} f(x) + e_G(S, B) - e_G(S, T) \\
&= \sum_{x \in T} f(x) + \sum_{x \in S} (\deg_G(x) - f(x)) - e_G(S, T) \\
&= \delta^*(T, S).
\end{aligned}
$$

Therefore (3.78) holds for all S, T if and only if (3.79) holds for all S, T.   □

An elementary direct proof of Theorem 3.38 will be given in the next chapter. Here we shall prove the theorem by using the f-factor theorem. We use the following lemma.

Lemma 3.39. *Let G be a bipartite multigraph with bipartition (A, B) and let $f : V(G) \to \mathbb{Z}^+$ satisfy $\sum_{x \in A} f(x) = \sum_{x \in B} f(x)$. If $\delta^*(S, T) \geq 0$ for all $S \subseteq A$ and $T \subseteq B$, then*

$$\delta^*(X, Y) \geq 0 \qquad \text{for all} \quad X, Y \subseteq V(G), \quad X \cap Y = \emptyset.$$

Proof. By the equivalence of (3.78) and (3.79), the following also holds.

$$\delta^*(T, S) \geq 0 \qquad \text{for all} \quad S \subseteq A \text{ and } T \subseteq B.$$

Let X and Y be two disjoint subsets of $V(G)$. Let $X_A = X \cap A$, $X_B = X \cap B$, $Y_A = Y \cap A$ and $Y_B = Y \cap B$. Then

$$
\begin{aligned}
\delta^*(X, Y) &= \sum_{x \in X} f(x) + \sum_{x \in Y} \left(\deg_G(x) - f(x) \right) - e_G(X, Y) \\
&= \sum_{x \in X_A} f(x) + \sum_{x \in X_B} f(x) + \sum_{x \in Y_A} (\deg_G(x) - f(x)) \\
&\quad + \sum_{x \in Y_B} (\deg_G(x) - f(x)) - e_G(X_A, Y_B) - e_G(X_B, Y_A) \\
&= \delta^*(X_A, Y_B) + \delta^*(X_B, Y_A) \\
&\geq 0
\end{aligned}
$$

Hence the lemma holds.   □

Proof of Theorem 3.38. First assume that G has an f-factor F. Let $S \subseteq A$ and $T \subseteq B$. Then

$$
\begin{aligned}
\delta^*(S, T) &= \sum_{x \in S} f(x) + \sum_{x \in T} \left(\deg_{G-S}(x) - f(x) \right) \\
&= \sum_{x \in S} \deg_F(x) + \sum_{x \in T} \left(\deg_{G-S}(x) - \deg_F(x) \right) \\
&\geq \sum_{x \in T} \deg_{G-S}(x) - \left(\sum_{x \in T} \deg_F(x) - e_F(S, T) \right) \\
&\geq \sum_{x \in T} \deg_{G-S}(x) - \sum_{x \in T} \deg_{F-S}(x) \\
&\geq 0.
\end{aligned}
$$

We next prove the sufficiency by using the f-factor theorem. Assume that G does not have an f-factor. Then by the f-factor theorem, there exist two disjoint subsets $X, Y \subseteq V(G)$ such that

$$\delta(X,Y) = \delta^*(X,Y) + q(X,Y) < 0.$$

Since $\delta(\emptyset, \emptyset) = 0$ by the fact that $\sum_{x \in A} f(x) = \sum_{x \in B} f(x)$, we have $X \cup Y \neq \emptyset$. Take X, Y so that $\delta(X,Y) < 0$ and $X \cup Y$ is maximal. If $q(X,Y) = 0$, then by Lemma 3.39, $\delta(X,Y) = \delta^*(X,Y) \geq 0$, a contradiction. Hence $q(X,Y) \geq 1$.

Let D be an f-odd component of $G - (X \cup Y)$, and let $S = V(D) \cap A$ and $T = V(D) \cap B$. Since there does not exist any edges between D and any other components of $G - (X \cup Y)$, it follows from (3.4) that

$$q(X \cup S, Y \cup T) = q(X \cup T, Y \cup S) = q(X,Y) - 1. \tag{3.80}$$

Then we obtain

$$
\begin{aligned}
&\delta(X \cup S, Y \cup T) + \delta(X \cup T, Y \cup S) - 2\delta(X,Y) \\
={} & \sum_{x \in X \cup S} f(x) + \sum_{x \in Y \cup T} (\deg_G(x) - f(x)) \\
& - e_G(X \cup S, Y \cup T) - q(X \cup S, Y \cup T) \\
& + \sum_{x \in X \cup T} f(x) + \sum_{x \in Y \cup S} (\deg_G(x) - f(x)) \\
& - e_G(X \cup T, Y \cup S) - q(X \cup T, Y \cup S) \\
& - 2\sum_{x \in X} f(x) - 2\sum_{x \in Y} (\deg_G(x) - f(x)) + 2e_G(X,Y) + 2q(X,Y) \\
={} & \sum_{x \in S \cup T} f(x) + \sum_{x \in S \cup T} (\deg_G(x) - f(x)) \\
& - 2e_G(S,T) - e_G(S \cup T, X \cup Y) + 2 \\
={} & \sum_{x \in S \cup T} \deg_G(x) - \sum_{x \in S \cup T} \deg_G(x) + 2 \\
={} & 2.
\end{aligned}
$$

Since

$$\delta(X,Y) \equiv \delta(X \cup S, Y \cup T) \equiv \delta(X \cup T, Y \cup S) \pmod 2,$$

it follows from the above equation that

$$\delta(X,Y) \geq \delta(X \cup S, Y \cup T) \quad \text{or} \quad \delta(X \cup T, Y \cup S).$$

In each case, this inequality contradicts the maximality of $X \cup Y$. Consequently the theorem is proved. □

A generalization of a regular factor in a bipartite graph is given in the next theorem.

Theorem 3.40. *Let $a \geq 1$ and $b \geq 1$ be integers, and G be a bipartite multigraph with bipartition (A, B). Then G has a factor F such that*

$$\deg_F(x) = a \qquad \text{for all} \quad x \in A, \quad \text{and}$$
$$\deg_F(y) \leq b \qquad \text{for all} \quad y \in B \qquad\qquad (3.81)$$

if and only if

$$\delta^*(T, S) = b|T| + \sum_{x \in S} \deg_{G-T}(x) - a|S| \geq 0 \qquad\qquad (3.82)$$

for all $S \subseteq A$ and $T \subseteq B$ (see 3.79)).

Proof. Suppose that G has an factor F satisfying the condition (3.81). Then for every $S \subseteq A$ and $T \subseteq B$, we have

$$\delta^*(T, S) = b|T| + \sum_{x \in S} \deg_{G-T}(x) - a|S|$$

$$\geq \sum_{x \in T} \deg_F(x) + \sum_{x \in S} \deg_{G-T}(x) - \sum_{x \in S} \deg_F(x)$$

$$\geq \sum_{x \in S} \deg_{G-T}(x) - \left(\sum_{x \in S} \deg_F(x) - e_F(S, T) \right)$$

$$\geq \sum_{x \in S} \deg_{G-S}(x) - \sum_{x \in S} \deg_{F-S}(x) \geq 0.$$

Hence (3.82) holds.

We next prove the sufficiency. We use Theorem 3.38 with (3.79). Since $\delta^*(B, A) = b|B| - a|A| \geq 0$, we have $b|B| \geq a|A|$. If $b|B| = a|A|$, then G has the desired factor by Theorem 3.38 (see (3.79)). Thus we may assume that $t = b|B| - a|A| \geq 1$.

By adding a new vertex w to A and by joining w to every vertex of B by t multiple edges, we obtain a new bipartite multigraph G^*. Define a function $f : V(G^*) \to \mathbb{Z}^+$ by

$$f(x) = \begin{cases} a & \text{if } x \in A \\ t & \text{if } x = w \\ b & \text{if } x \in B. \end{cases}$$

Then $\sum_{x \in A \cup \{w\}} f(x) = \sum_{x \in B} f(x)$. Let $S \subseteq A$ and $T \subseteq B$, where $S, T \subset V(G^*)$. Then

$$\delta^*(T, S) = \sum_{x \in T} f(x) + \sum_{x \in S} \left(\deg_{G^*-T}(x) - f(x) \right)$$

$$= b|T| + \sum_{x \in S} \deg_{G-T}(x) - a|S| \geq 0.$$

It is clear that $\delta^*(B, S \cup \{w\}) = b|B| - a|S| - t = 0$. Thus we may assume $T \subset B$. Since $\deg_{G^*-T}(w) = t(|B| - |T|)$ and $|T| < |B|$, we have

$$\delta^*(T, S \cup \{w\}) = \sum_{x \in T} f(x) + \sum_{x \in S \cup \{w\}} \left(\deg_{G^*-T}(x) - f(x) \right)$$

$$= b|T| + \sum_{x \in S} \deg_{G-T}(x) + t(|B| - |T|) - a|S| - t$$

$$\geq b|T| + \sum_{x \in S} \deg_{G-T}(x) - a|S|$$

$$\geq 0.$$

Therefore, G^* has an f-factor F by Theorem 3.38. Hence $F \cap E(G) = F - w$ is the desired factor of G, which satisfies condition (3.81). $\quad\square$

We now give some sufficient conditions for a bipartite graph to have regular factors or f-factors, which are proved using Theorem 3.38.

Theorem 3.41. *Let G be a bipartite multigraph and $k \geq 1$ be an integer. Suppose that the degree of every vertex of G is divisible by k. Then for a function f defined by*

$$f(x) = \frac{\deg_G(x)}{k} \qquad \text{for all} \quad x \in V(G), \tag{3.83}$$

G is decomposed into k f-factors. In particular, G has an f-factor.

Proof. We first show that G has an f-factor. Let X and Y be disjoint subsets of $V(G)$. Then

$$\delta^*(X, Y) = \sum_{x \in X} f(x) + \sum_{x \in Y} \left(\deg_G(x) - f(x) \right) - e_G(X, Y)$$

$$= \frac{1}{k} \sum_{x \in X} \deg_G(x) + \left(1 - \frac{1}{k} \right) \sum_{x \in Y} \deg_G(x) - e_G(X, Y)$$

$$\geq \frac{1}{k} e_G(X, Y) + \left(1 - \frac{1}{k} \right) e_G(Y, X) - e_G(X, Y)$$

$$= 0.$$

Hence G has an f-factor F by Theorem 3.38. It is obvious that $G - F$ satisfies

$$f(x) = \frac{\deg_{G-F}(x)}{k - 1} \qquad \text{for all} \quad x \in V(G).$$

Hence by the induction on k, $G - F$ is decomposed into $k - 1$ f-factors, and thus the theorem follows. $\quad\square$

Theorem 3.42 (Katerinis [140]). *Let G be a bipartite simple graph of order at least four. If $\text{tough}(G) \geq 1$, then G has a 2-regular factor.*

Lemma 3.43. *Let G be a bipartite simple graph of order at least three and with bipartition (A, B). Then $tough(G) \leq 1$, and $tough(G) = 1$ implies $|A| = |B|$.*

Proof. We may assume $|A| \geq |B|$. Then $|A| = \omega(G - B) \geq 2$ by $|G| \geq 3$, and so

$$tough(G) \leq \frac{|B|}{\omega(G - B)} = \frac{|B|}{|A|} \leq 1.$$

In particular, $tough(G) = 1$ implies $|A| = |B|$. □

Proof of Theorem 3.42 By the above lemma, $|A| = |B|$. Hence $2|A| = 2|B|$. Suppose that G does not have a 2-factor. Then by Theorem 3.38 there exist subsets $S \subseteq A$ and $T \subseteq B$ such that

$$\delta^*(S, T) = 2|S| + \sum_{x \in T} \deg_{G-S}(x) - 2|T| < 0.$$

Choose such a pair (S, T) so that T is minimal. Since $\delta^*(S, \emptyset) = 2|S| \geq 0$ and $\delta^*(A, T) = 2|A| - 2|T| \geq 0$, we have $T \neq \emptyset$ and $S \neq A$.

Claim *For every $v \in T$, $\deg_{G-S}(v) \leq 1$.*

Let $T' = T - \{v\}$. Then

$$
\begin{aligned}
1 &\leq \delta^*(S, T') - \delta^*(S, T) \\
&= 2|S| + \sum_{x \in T'} \deg_{G-S}(x) - 2|T'| \\
&\quad - \left(2|S| + \sum_{x \in T} \deg_{G-S}(x) - 2|T|\right) \\
&= -\deg_{G-S}(v) + 2
\end{aligned}
$$

Hence $\deg_{G-S}(v) \leq 1$.

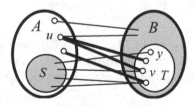

Fig. 3.18. A bipartite graph G with $tough(G) = 1$.

Let $R = \{y \in T \mid \deg_{G-S}(y) = 0\}$ and $X = S \cup (B - T)$. By the above claim, if $u \in A - S$ and $v \in T$ are adjacent, then the component D of

$G - X = \langle (A - S) \cup T \rangle_G$ containing u satisfies $V(D) \cap (A - S) = \{u\}$ (see Fig. 3.18). Thus

$$\omega(G - X) = |A| - |S| + |R|.$$

Assume first $\omega(G - X) \geq 2$. By $tough(G) = 1$, we have

$$|A| - |S| + |R| = \omega(G - X) \leq |X| = |S| + |B| - |T|,$$

which implies $0 \leq 2|S| - |T| - |R|$. On the other hand, by the above claim we obtain

$$\delta^*(S, T) = 2|S| + |T - R| - 2|T| = 2|S| - |R| - |T| < 0.$$

This contradicts the above inequality $0 \leq 2|S| - |T| - |R|$. Therefore we may assume $\omega(G - X) = 1$. Then $|A - S| = 1$ and $R = \emptyset$, and so by $|A| \geq 2$ and the above claim, we have

$$\delta^*(S, T) = 2(|A| - 1) + |T| - 2|T| \geq 2|A| - 2 - |T| \geq 0$$

since $|A| = |B| \geq 2$. This is a contradiction. Consequently the proof is complete. $\square$

Katerinis showed that a similar result does not hold for a 3-factor. Namely, he showed that for any large integer $\lambda \geq 2$, there exist 1-tough λ-edge connected bipartite graphs that have no 3-factor. In the same paper, he obtained another sufficient condition for a bipartite graph to have a 2-factor, which is given in the next theorem.

Theorem 3.44 (Katerinis [140]). *Let G be a bipartite simple graph with bipartition (A, B). If the following two conditions hold, then G has a 2-regular factor.*
(i) $|A| = |B|$, *and*
(ii) for every subset $S \subseteq A$, it follows that

$$|N_G(S)| \geq \frac{3}{2}|S| \qquad if \ |S| < \left\lfloor \frac{2}{3}|A| \right\rfloor, \quad and$$
$$N_G(S) = B \qquad otherwise. \tag{3.84}$$

Proof. By the symmetry of (3.78) and (3.79), we may assume that the following condition (3.85) holds instead of (3.84). Namely, we shall prove the existence of a 2-factor under the following condition (3.85) instead of condition (3.84).

$$|N_G(T)| \geq \frac{3}{2}|T| \qquad if \ T \subseteq B, \ |T| < \left\lfloor \frac{2}{3}|B| \right\rfloor, \quad and$$
$$N_G(T) = A \qquad otherwise. \tag{3.85}$$

Suppose that G has no 2-factor. Since $2|A| = 2|B|$ by (i), there exists a pair (S, T) of subsets $S \subseteq A$ and $T \subseteq B$ such that

$$\delta^*(S,T) = 2|S| + \sum_{x \in T} \deg_{G-S}(x) - 2|T| < 0. \quad \text{(see (3.79))}$$

Choose such a pair (S,T) so that T is minimal. Since $\delta^*(S,T) \geq 2(|S|-|T|) \geq 0$ if $|S| \geq |T|$, we have $|S| < |T|$. By the same arguments used in the proof of Theorem 3.42, we have

$$|S| < |T|, \quad S \neq A, \quad T \neq \emptyset \quad \text{and}$$
$$\deg_{G-S}(v) \leq 1 \quad \text{for every } v \in T.$$

Let $R_0 = \{x \in T \mid \deg_{G-S}(x) = 0\}$ and $R_1 = \{x \in T \mid \deg_{G-S}(x) = 1\}$. Then $T = R_0 \cup R_1$. Since

$$\delta^*(S,T) = 2|S| + |R_1| - 2|T| < 0 \quad \text{and} \quad |N_G(T)| \leq |S| + |R_1|,$$

we have

$$2|N_G(T)| \leq 2|S| + 2|R_1| < 2|T| + |R_1| \leq 3|T|.$$

Hence

$$|T| > \frac{2}{3}|N_G(T)|.$$

Therefore it follows from (3.85) that

$$N_G(T) = A \quad \text{and} \quad |T| \geq \left\lfloor \frac{2}{3}|B| \right\rfloor + 1,$$

which implies $e_G(x,T) \geq 1$ for all $x \in A$. If $e_G(x,T) \geq 2$ for all $x \in A - S$, then

$$\delta^*(S,T) = 2|S| + \sum_{x \in T} \deg_{G-S}(x) - 2|T|$$
$$\geq 2|S| + 2|A - S| - 2|T| = 2|A| - 2|T| \geq 0,$$

a contradiction. Hence there exists a vertex $u \in A - S$ with $e_G(u,T) = 1$. Let $w \in T$ be a vertex adjacent to u. Then

$$u \notin N_G(T - \{w\}), \quad \text{but} \quad |T - \{w\}| \geq \left\lfloor \frac{2}{3}|B| \right\rfloor.$$

This contradicts (3.85). Consequently the theorem is proved. $\quad \square$

Note that Theorem 3.44 is best possible in the sense that there exist bipartite graphs G that have no 2-factor and satisfy the requirement that for every subset $S \subseteq A$, $|N_G(S)| \geq \lfloor (3/2)|S| \rfloor$ if $|S| < \lfloor (2/3)|A| \rfloor$, and $N_G(S) = B$ otherwise.

Theorem 3.44 can be generalized as follows.

Theorem 3.45 (Enomoto, Ota and Kano [80]). *Let $a \geq 2$ and $b \geq 2$ be integers, and let $\lambda = a - 1 + (1/b)$. Let G be a bipartite simple graph with bipartition (A, B). Suppose that G satisfies the following two conditions:*
(i) $a|A| \leq b|B|$ and $|B| \geq a$,   and
(ii) for every $S \subseteq A$, it follows that

$$|N_G(S)| \geq \lambda|S| \quad if \;\; |S| < \left\lfloor \frac{|B|}{\lambda} \right\rfloor, \; and$$
$$N_G(S) = B \quad otherwise. \tag{3.86}$$

Then G has a spanning subgraph F such that

$$\deg_F(x) = a \quad for \; all \;\; x \in A, \; and \tag{3.87}$$
$$\deg_F(y) \leq b \quad for \; all \;\; y \in B. \tag{3.88}$$

If $a|A| = b|B|$, then equality holds in (3.88), and by substituting $a = b = 2$ in the above theorem, we have $\lambda = 3/2$ and obtain Theorem 3.44. Hence Theorem 3.45 is a generalization of Theorem 3.44.

Proof of Theorem 3.45 Suppose that G does not have a spanning subgraph satisfying (3.87) and (3.88). Then by Theorem 3.40, there exists a pair (S, T) of subsets $S \subseteq A$ and $T \subseteq B$ such that

$$\delta^*(T, S) = b|T| + \sum_{x \in S} \deg_{G-T}(x) - a|S|$$
$$= b|T| + e_G(S, B - T) - a|S| < 0. \tag{3.89}$$

Choose such a pair (S, T) so that $S \cup (B - T)$ is minimal. Then it is easy to see that $S \neq \emptyset$ and $T \neq B$.

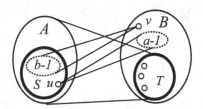

Fig. 3.19. A bipartite graph G of Theorem 3.45.

Claim 1. *For every $u \in S$, it follows that $\deg_{G-T}(u) \leq a - 1$, in particular, $e_G(S, B - T) \leq (a - 1)|S|$ (Fig. 3.19).*

Let $S' = S - \{u\}$. Then

$$1 \le \delta^*(T, S') - \delta^*(T, S)$$
$$= b|T| + \sum_{x \in S'} \deg_{G-T}(x) - a|S'| - \left(b|T| + \sum_{x \in S} \deg_{G-T}(x) - a|S| \right)$$
$$= -\deg_{G-T}(u) + a$$

Hence $\deg_{G-T}(u) \le a - 1$. The second inequality follows immediately from this.

Claim 2. *For every $v \in B - T$, it follows that $e_G(S, v) \le b - 1$ (Fig. 3.19).*

Let $T' = T \cup \{v\}$. Then

$$1 \le \delta^*(T', S) - \delta^*(T, S)$$
$$= b|T'| + \sum_{x \in S} \deg_G(x) - e_G(T', S) - a|S|$$
$$- \left(b|T| + \sum_{x \in S} \deg_G(x) - e_G(T, S) - a|S| \right)$$
$$= b - e_G(v, S)$$

Hence $e_G(v, S) \le b - 1$.

Claim 3. $|N_G(S)| < \lambda|S|$.

Since G is a bipartite simple graph, we have $|N_G(S)| \le |T| + e_G(S, B - T)$, and by (3.89) we obtain

$$|N_G(S)| \le |T| + e_G(S, B - T)$$
$$< \frac{a}{b}|S| + \left(1 - \frac{1}{b} \right) e_G(S, B - T)$$
$$\le \frac{a}{b}|S| + \left(1 - \frac{1}{b} \right)(a - 1)|S| \qquad \text{(by Claim 1)}$$
$$= \left(a - 1 + \frac{1}{b} \right)|S| = \lambda|S|. \tag{3.90}$$

Thus Claim 3 holds.

By Claim 3 and the condition (3.86), we have $N_G(S) = B$. Note that we have shown that $B - T \ne \emptyset$. Let

$$h = \min\{e_G(S, v) : v \in B - T\}.$$

By Claim 2 and the fact that $N_G(S) = B$, we have

$$1 \le h \le b - 1.$$

By the definition of h and since $N_G(S) = B$, we have

$$|B| = |N_G(S)| \le |T| + \frac{1}{h} e_G(S, B - T)$$

$$< \frac{a}{b}|S| + \left(\frac{1}{h} - \frac{1}{b}\right) e_G(S, B - T) \quad \text{(by (3.89))}$$

$$\le \frac{a}{b}|S| + \left(\frac{1}{h} - \frac{1}{b}\right)(a - 1)|S| \quad \text{(by Claim 1)}$$

$$= \left(\frac{1}{b} + \frac{a - 1}{h}\right)|S|.$$

Hence

$$|B| < \left(\frac{1}{b} + \frac{a - 1}{h}\right)|S|. \tag{3.91}$$

Let $v_0 \in B - T$ be a vertex such that $e_G(S, v_0) = h$, and let $S_0 = S - N_G(v_0)$. Since $N_G(S_0) \subseteq B - \{v_0\}$, we have by (3.86) that

$$|S_0| = |S| - h \le \frac{|B|}{\lambda} - 1. \tag{3.92}$$

By (3.92) and (3.91), we obtain

$$\lambda(|S| - h + 1) \le |B| < \left(\frac{1}{b} + \frac{a - 1}{h}\right)|S|,$$

$$\left(\lambda - \frac{1}{b} - \frac{a - 1}{h}\right)|S| = (a - 1)\left(1 - \frac{1}{h}\right)|S| < \lambda(h - 1),$$

$$|S| < \frac{\lambda h}{a - 1} = \left(a - 1 + \frac{1}{b}\right)\frac{h}{a - 1}$$

$$\le \left(b - 1 + \frac{b - 1}{b(a - 1)}\right) < b. \quad \text{(by } 1 \le h \le b - 1\text{)}$$

Hence $|S| < b$. Choose $u_0 \in S$ so that

$$\rho = e_G(u_0, B - T) = \min\{e_G(u, B - T) : u \in S\}.$$

Then by Claim 1, $\rho \le a - 1$ and thus

$$|N_G(u_0)| \le |T| + e_G(u_0, B - T)$$

$$< \frac{a}{b}|S| - \frac{1}{b}e_G(S, B - T) + e_G(u_0, B - T) \quad \text{(by (3.89))}$$

$$\le \left(\frac{a}{b} - \frac{\rho}{b}\right)|S| + \rho$$

$$\le \left(\frac{a}{b} - \frac{a - 1}{b}\right)(b - 1) + a - 1 \quad \text{(by } |S| < b \text{ and } \rho \le a - 1 \text{)}$$

$$= 1 - \frac{1}{b} + a - 1.$$

Hence $|N_G(u_0)| \le a - 1 < \lambda$ by $b \ge 2$, which implies $N_G(u_0) = B$ by (3.86). However, this contradicts $a \le |B| = |N_G(u_0)| \le a - 1$ (see Condition (i)). Consequently the theorem is proved. $\square$

Theorem 3.45 is best possible in the sense that there exist bipartite graphs that have no spanning subgraph satisfying (3.88) and (3.88), but satisfy $|N_G(S)| \geq \lceil \lambda |S| \rceil - 1$ if $|S| < \lfloor \lambda |B|/\lambda \rfloor$, and $N_G(S) = B$ otherwise.

Problems

3.1. Consider graphs $H(r, \lambda^*; \lambda)$ and $J(r, \lambda^*; \lambda)$ when r is odd and $\lambda \geq 1$ in order to obtain a graph $G(r, \lambda)$. Then show that $G = G(r, \lambda)$ has no k-regular factor if either k is even and $k > r(1 - 1/\lambda^*)$, or k is odd and $k < r/\lambda^*$.

3.2. Show that (5) of Theorem 3.10 holds.

3.3. Prove that if $tough(G) \geq t > 0$, then $\delta(G) \geq 2t$.

4

(g, f)-Factors and $[a, b]$-Factors

In this chapter we investigate (g, f)-factors and $[a, b]$-factors, which are natural generalizations of f-factors and regular factors, respectively. We first give a criterion for a graph to have a (g, f)-factor, which is called the (g, f)-factor theorem. Then we obtain some sufficient conditions for graphs to have (g, f)-factors and $[a, b]$-factors.

4.1 The (g, f)-factor theorem

For a general graph G and two functions $g, f : V(G) \to \mathbb{Z} = \{\ldots, -1, 0, 1, 2, \ldots\}$ with $g(x) \leq f(x)$ for all $x \in V(G)$, a (g, f)-**factor** is a spanning subgraph F of G that satisfies

$$g(x) \leq \deg_F(x) \leq f(x) \qquad \text{for all} \quad x \in V(G).$$

It is allowed that g and f satisfy $g(x) < 0$ for some vertices x and $\deg_G(y) < f(y)$ for some vertices y. For two integers a and b such that $0 \leq a \leq b$, a spanning subgraph H of G is called an $[a, b]$-**factor** if

$$a \leq \deg_H(x) \leq b \qquad \text{for all} \quad x \in V(G).$$

Of course, an $[a, b]$-factor is simply a (g, f)-factor with $g(x) = a$ and $f(x) = b$ for all vertices x.

A necessary and sufficient condition for a graph to have a (g, f)-factor was obtained by Lovász [176] in 1970, and is called the (g, f)-factor theorem. The original proof is difficult and long, however, in 1981 Tutte [231] found an elegant, short proof to this theorem, which is presented here.

Theorem 4.1 (The (g, f)-Factor Theorem, Lovász [175]). *Let G be a general graph and let $g, f : V(G) \to \mathbb{Z}$ be functions such that $g(x) \leq f(x)$ for all $x \in V(G)$. Then G has a (g, f)-factor if and only if for all disjoint subsets S and T of $V(G)$,*

J. Akiyama and M. Kano, *Factors and Factorizations of Graphs*,
Lecture Notes in Mathematics 2031, DOI 10.1007/978-3-642-21919-1_4,
© Springer-Verlag Berlin Heidelberg 2011

$$\gamma(S, T) = \sum_{x \in S} f(x) + \sum_{x \in T} (\deg_G(x) - g(x)) - e_G(S, T) - q^*(S, T) \geq 0, \quad (4.1)$$

where $q^*(S, T)$ denotes the number of components C of $G - (S \cup T)$ such that $g(x) = f(x)$ for all $x \in V(C)$ and

$$\sum_{x \in V(C)} f(x) + e_G(C, T) \equiv 1 \pmod 2. \qquad (4.2)$$

The function $\gamma(S, T)$ is defined by (4.1).

For convenience, we call a component C of $G - (S \cup T)$ satisfying $g(x) = f(x)$ for all $x \in V(C)$ and (4.2) a (g, f)-**odd component** of $G - (S \cup T)$. It is immediate that

$$\sum_{x \in T} \deg_G(x) - e_G(S, T) = \sum_{x \in T} \deg_{G-S}(x),$$

and thus (4.1) can be expressed as

$$\gamma(S, T) = \sum_{x \in S} f(x) + \sum_{x \in T} (\deg_{G-S}(x) - g(x)) - q^*(S, T) \geq 0. \qquad (4.3)$$

Notice that inequality (4.1) includes the following condition, which we will prove later:

$$g(x) \leq \deg_G(x) \quad \text{and} \quad 0 \leq f(x) \qquad \text{for every } x \in V(G).$$

Moreover, as we mentioned before, the (g, f)-factor theorem holds even if

$$g(x) < 0 \quad \text{and} \quad \deg_G(y) < f(y) \qquad \text{for some } x, y \in V(G).$$

This relaxation occasionally plays a useful role, though the following natural condition is assumed in the original papers [175] and [231]: $0 \leq g(x) \leq f(x) \leq \deg_G(x)$ for all $x \in V(G)$.

Proof of the necessity of the (g, f)-Factor Theorem.   Suppose that G has a (g, f)-factor F. We shall show that $\gamma(S, T) \geq 0$. It is clear that we may assume that G is connected.

Consider first $\gamma(\emptyset, \emptyset)$. By the definition of a (g, f)-odd component, if $g(v) < f(v)$ for some vertex v of G, then $q^*(\emptyset, \emptyset) = 0$. If $g(x) = f(x)$ for all $x \in V(G)$, then

$$\sum_{x \in V(G)} f(x) = \sum_{x \in V(G)} \deg_F(x) = 2\|F\|,$$

and so G itself is not a (g, f)-odd component of G, which implies $q^*(\emptyset, \emptyset) = 0$. Thus we have $\gamma(\emptyset, \emptyset) = -q^*(\emptyset, \emptyset) = 0$.

Let S and T be two disjoint subsets of $V(G)$ such that $S \cup T \neq \emptyset$. Let $G - F$ denote the spanning subgraph $G - E(F)$, and let $C_1, C_2, \ldots, C_m$ be the (g, f)-odd components of $G - (S \cup T)$, where $m = q^*(S, T)$. Since

$$\deg_G(x) - g(x) \geq \deg_G(x) - \deg_F(x) = \deg_{G-F}(x),$$

we have

$$\sum_{x \in T}(\deg_G(x) - g(x)) \geq \sum_{x \in T} \deg_{G-F}(x)$$

$$\geq e_{G-F}(T, S) + \sum_{i=1}^{m} e_{G-F}(T, C_i).$$

Furthermore,

$$\sum_{x \in S} f(x) \geq \sum_{x \in S} \deg_F(x) \geq e_F(S, T) + \sum_{i=1}^{m} e_F(S, C_i).$$

By the previous inequalities, we have

$$\gamma(S, T) \geq e_F(S, T) + \sum_{i=1}^{m} e_F(S, C_i)$$

$$+ e_{G-F}(T, S) + \sum_{i=1}^{m} e_{G-F}(T, C_i) - e_G(S, T) - m$$

$$= \sum_{i=1}^{m}\left(e_F(S, C_i) + e_{G-F}(T, C_i) - 1\right).$$

Therefore, in order to prove (4.1), it is sufficient to show that for every $C = C_i$,

$$e_F(S, C) + e_{G-F}(T, C) - 1 \geq 0. \tag{4.4}$$

If $e_{G-F}(T, C) \geq 1$, then (4.4) holds, and so we may assume that $e_{G-F}(T, C) = 0$. This implies

$$e_G(T, C) = e_F(T, C).$$

Since $g(x) = f(x) = \deg_F(x)$ for every $x \in V(C)$, we obtain

$$\sum_{x \in V(C)} f(x) + e_G(C, T) = \sum_{x \in V(C)} \deg_F(x) + e_F(C, T)$$

$$= 2\|\langle V(C)\rangle_F\| + e_F(C, S \cup T) + e_F(C, T)$$

$$\equiv e_F(C, S) \pmod 2.$$

By (4.2) and by the above equation, we have $e_F(C, S) \geq 1$, which implies (4.4) holds. Consequently the necessity is proved. $\square$

Our proof of sufficiency is based on Tutte [231].

Proof of the sufficiency of the (g, f)-factor theorem. It is easy to see that we may assume that G is connected. For each vertex v of G, it follows from (4.1) that

$$\gamma(\emptyset, \{v\}) = \deg_G(v) - g(v) - q^*(\emptyset, \{v\}) \geq 0, \qquad \text{and}$$
$$\gamma(\{v\}, \emptyset) = f(v) - q^*(\{v\}, \emptyset) \geq 0,$$

Thus

$$g(v) \leq \deg_G(v) \quad \text{and} \quad 0 \leq f(v) \quad \text{for every} \quad v \in V(G). \qquad (4.5)$$

If $g(x) = f(x)$ for all $x \in V(G)$, then a (g, f)-factor is an f-factor, and $\delta(S, T) = \gamma(S, T)$. So the (g, f)-factor theorem holds by the f-factor theorem 3.2. Therefore, we may assume that G has at least one vertex u such that $g(u) < f(u)$.

We now construct a new connected graph G^* from G as follows (see Fig. 4.1): Add a new vertex w to G, and join w to every vertex x of G by $f(x) - g(x)$ multiple edges, and add M loops to w, where M is a sufficiently large integer expressed as

$$M = \sum_{x \in V(G)} f(x) + 2N. \qquad (4.6)$$

Note that N can be defined as $\sum_{x \in V(G)} |g(x)| + |G|$ (see Problem 4.1). We next define the function $f^* : V(G^*) \to \{0, 1, 2, 3, \ldots\}$ as

$$f^*(x) = \begin{cases} f(x) & \text{if} \quad x \in V(G) \\ M & \text{if} \quad x = w. \end{cases}$$

We first show that G has a (g, f)-factor if and only if G^* has an f^*-factor. Assume that G^* has an f^*-factor F^*. Then $F = F^* - \{w\}$ becomes a (g, f)-factor of G since for every vertex $x \in V(G)$, we have

$$\deg_F(x) = \deg_{F^*}(x) - e_{F^*}(x, w)$$
$$\geq \deg_{F^*}(x) - e_{G^*}(x, w)$$
$$= f(x) - (f(x) - g(x)) = g(x), \qquad \text{and}$$
$$\deg_F(x) \leq \deg_{F^*}(x) = f(x).$$

Conversely, assume that G has a (g, f)-factor F. For every vertex x of G, add $f(x) - \deg_F(x)$ multiple edges joining x and w to F, and add $N + \left(\sum_{x \in V(G)} \deg_F(x)\right)/2$ loops incident with w. Then the degree of each vertex x of G becomes $f(x)$, and that of w is

$$\sum_{x \in V(G)} \left(f(x) - \deg_F(x)\right) + 2N + \sum_{x \in V(G)} \deg_F(x) = M.$$

Hence the resulting spanning subgraph is an f^*-factor of G^*.

We now prove that G^* has an f^*-factor by making use of the f-factor theorem (Theorem 3.2). Since G^* is connected and

$$\sum_{x \in V(G^*)} f^*(x) = \sum_{x \in V(G)} f(x) + M \equiv 0 \pmod 2, \qquad \text{(by 4.6)}$$

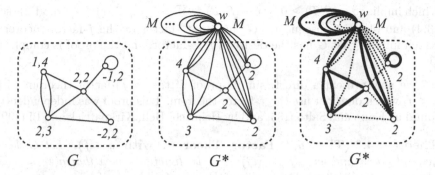

Fig. 4.1. Connected general graphs G and G^*, and an f^*-factor of G^*; numbers denote $g(x)$ and $f(x)$.

we have $\delta_{G^*}(\emptyset, \emptyset) = -q_{G^*}(\emptyset, \emptyset) = 0$.

Let S and T be two disjoint subsets of $V(G^*)$ such that $S \cup T \neq \emptyset$. If $w \in S$, then $T \subseteq V(G)$. Since M is sufficiently large, we have

$$\delta_{G^*}(S, T) = \sum_{x \in S \setminus \{w\}} f(x) + M + \sum_{x \in T} (\deg_{G^*}(x) - f(x)) \geq 0.$$

If $w \in T$, then $S \subseteq V(G)$ and $\deg_{G^*}(w) - f^*(w) \geq 2M - M = M$, and thus

$$\delta_{G^*}(S, T) \geq \sum_{x \in S} f(x) + \sum_{x \in T \setminus \{w\}} (\deg_{G^*}(x) - f(x)) + M$$

$$- e_{G^*}(S, T) - q_{G^*}(S, T) \geq 0.$$

Therefore we may assume that $w \notin S \cup T$. Then

$$\delta_{G^*}(S, T) = \sum_{x \in S} f^*(x) + \sum_{x \in T} (\deg_{G^*}(x) - f^*(x)) - e_{G^*}(S, T) - q_{G^*}(S, T)$$

$$= \sum_{x \in S} f(x) + \sum_{x \in T} (\deg_G(x) - g(x)) - e_G(S, T) - q_{G^*}(S, T).$$

Let C be an f^*-odd component of $G^* - (S \cup T)$ not containing w. Then since $e_{G^*}(C, w) = 0$, we have $g(x) = f(x)$ for all $x \in V(C)$. Moreover,

$$\sum_{x \in V(C)} f^*(x) + e_{G^*}(C, T) = \sum_{x \in V(C)} f(x) + e_G(C, T) \equiv 1 \pmod 2.$$

Hence C is a (g, f)-odd component of $G - (S \cup T)$. Therefore $q_{G^*}(S, T) \leq q^*(S, T) + 1$, where $+1$ corresponds to a component of $G - (S \cup T)$ containing w, which might be an f^*-odd component. Consequently

$$\delta_{G^*}(S, T) \geq \sum_{x \in S} f(x) + \sum_{x \in T} (\deg_G(x) - g(x)) - e_G(S, T) - q^*(S, T) - 1$$

$$= \gamma(S, T) - 1 \geq -1,$$

which implies $\delta_{G^*}(S,T) \geq 0$ by $\delta_{G^*}(S,T) \equiv \sum_{x \in V(G^*)} f^*(x) \equiv 0 \pmod 2$ (see (3.4) and (4.6)). Consequently G^* has an f^*-factor by the f-factor theorem, which implies that G has the desired (g,f)-factor, and the proof is complete. $\square$

We now present a direct, simple proof of the (g,f)-factor theorem with $g < f$, which does not use the f-factor theorem. This proof was independently found by Kano and Saito [133] and by Heinrich, Hell, Kirkpatrick and Liu [99].

Theorem 4.2 (The (g,f)-Factor Theorem with $g < f$). *Let G be a general graph and let $g, f : V(G) \to \mathbb{Z}$ be functions such that $g(x) < f(x)$ for all $x \in V(G)$. Then G has a (g,f)-factor if and only if*

$$\gamma^*(S,T) = \sum_{x \in S} f(x) + \sum_{x \in T} \big(\deg_G(x) - g(x) \big) - e_G(S,T) \geq 0 \qquad (4.7)$$

for all disjoint subsets S and T of $V(G)$.

Proof. Since the necessity can be proved in the same way as the (g,f)-factor theorem, we prove only the sufficiency. Suppose that a general graph G satisfies (4.7) but has no (g,f)-factor. Note that for every $x \in V(G)$, $\gamma^*(\emptyset, \{x\}) \geq 0$ and $\gamma^*(\{x\}, \emptyset) \geq 0$. Hence we have

$$g(x) \leq \deg_G(x) \qquad \text{and} \qquad 0 \leq f(x).$$

We define a $(0,f)$-factor of G as a spanning subgraph K satisfying

$$0 \leq \deg_K(x) \leq f(x) \qquad \text{for all} \quad x \in V(G).$$

Then a spanning subgraph with no edge is a $(0,f)$-factor of G. Choose a $(0,f)$-factor H among all $(0,f)$-factors of G so that

$$\rho(H) = \sum_{x \in U} (g(x) - \deg_H(x)) \qquad \text{is minimum,}$$
$$\text{where} \quad U = \{x \in V(G) : \deg_H(x) < g(x)\}.$$

Note that if $U = \emptyset$, then H is a (g,f)-factor and so we may assume $U \neq \emptyset$. An H-alternating trail is a trail whose edges are alternately in H and not in H. Take one vertex u from U, and define

$$OV = \{x \in V(G) : \text{There exists an } H\text{-alternating trail of odd}$$
$$\text{length connecting } u \text{ to } x \},$$
$$EV = \{x \in V(G) : \text{There exists an } H\text{-alternating trail of even}$$
$$\text{length connecting } u \text{ to } x\} \cup \{u\}.$$

Then the following four statements hold (Fig. 4.2):

(1) If $(ux_1y_1x_2y_2 \cdots x_r(\text{or } y_r))$, $x_i, y_i \in V(G)$, is an H-alternating trail,
     then $\deg_H(x_i) = f(x_i)$ and $\deg_H(y_i) \leq g(y_i)$ for every i.
(2) $OV \cap EV = \emptyset$.
(3) $\deg_{G-OV}(y) = \deg_{H-OV}(y)$ for all $y \in EV$.
(4) $e_H(OV, EV) = \sum_{x \in OV} \deg_H(x)$.

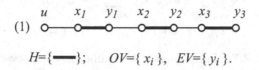

$$H=\{\text{———}\}; \quad OV=\{\,x_i\,\}, \quad EV=\{\,y_i\,\}.$$

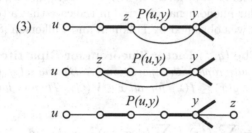

Fig. 4.2. H-alternating trails to illustrate the proofs of (1) and (3), where $\deg_H(u) < g(u)$.

(1) If $\deg_H(x_1) < f(x_1)$, then $H + ux_1$ is a $(0, f)$-factor and $\rho(H + ux_1) < \rho(H)$ (see Fig. 4.2 (1)). This contradicts the choice of H, and thus $\deg_H(x_1) = f(x_1)$. If $\deg_H(y_1) > g(y_1)$, then we get a contradiction by considering $H + ux_1 - x_1y_1$. Thus $\deg_H(y_1) \leq g(y_1)$. If $\deg_H(x_2) < f(x_2)$, then we derive a contradiction by considering $H + ux_1 - x_1y_1 + y_1x_2$. Hence $\deg_H(x_2) = f(x_2)$. We can prove (1) by repeating this argument.

(2) If $v \in EV \cap OV$, then $\deg_H(v) = f(v)$ as $v \in OV$ and $\deg_H(v) \leq g(v)$ as $v \in EV$. However, this is impossible since $g(v) < f(v)$. Hence (2) holds.

(3) We show that if $y \in EV$ and $yz \in E(G) - E(H)$, then $z \in OV$, which implies the desired property

$$\deg_{G-OV}(y) = \deg_{H-OV}(y).$$

Let $y \in EV$, $yz \in E(G) - E(H)$, and let $P(u, y)$ denote a shortest H-alternating trail connecting u and y (see Fig. 4.2 (3)). Then $P(u, y)$ terminates at y and does not pass through y on the way. If $z \in P(u, y) \cap EV$, then we can find an H-alternating trail $P(u, z) + zy$ of odd length connecting u and y, which contradicts $y \in EV$ and (2). If $z \in P(u, y) \cap OV$, then of course $z \in OV$. Thus we may assume that $z \notin P(u, y)$. Again $P(u, y) + yz$ is an H-alternating trail of odd length connecting u and z, and so $z \in OV$. Therefore (3) is proved.

(4) We can similarly show that if $x \in OV$ and $xz \in E(H)$, then $z \in EV$, which implies (4).

By (1), (2), (3), (4) and by $g(u) > \deg_H(u)$, we obtain

$$\gamma^*(OV, EV) = \sum_{x \in OV} f(x) + \sum_{x \in EV} (\deg_{G-OV}(x) - g(x))$$

$$< \sum_{x \in OV} \deg_H(x) + \sum_{x \in EV} \left(\deg_{H-OV}(x) - \deg_H(x) \right)$$

$$= e_H(OV, EV) - e_H(EV, OV) = 0.$$

This contradicts (4.7). Consequently the sufficiency is proved.   □

When we consider a (g,f)-factor in a bipartite graph, the criterion becomes simpler. The similar phenomena occurred in the case that a (g,f)-factor with $g < f$. This result was obtained by Folkman and Fulkerson [86].

Theorem 4.3 (The (g,f)-Factor Theorem for Bipartite Graphs). *Let G be a bipartite multigraph with bipartition (A, B) and let $g, f : V(G) \to \mathbb{Z}$ be functions such that $g(x) \leq f(x)$ for all $x \in V(G)$. Then G has a (g,f)-factor if and only if*

$$\gamma^*(X, Y) = \sum_{x \in X} f(x) + \sum_{x \in Y} (\deg_G(x) - g(x)) - e_G(X, Y) \geq 0, \quad (4.8)$$

and

$$\gamma^*(Y, X) = \sum_{x \in Y} f(x) + \sum_{x \in X} (\deg_G(x) - g(x)) - e_G(Y, X) \geq 0 \quad (4.9)$$

for all subsets $X \subseteq A$ and $Y \subset B$.

Proof. Since the necessity can be proved in the same way as the (g,f)-factor theorem, we prove only the sufficiency. Suppose that a bipartite multigraph G satisfies (4.8) and (4.9). We first show that for any disjoint subsets S and T of $V(G)$, we have

$$\gamma^*(S, T) = \sum_{x \in S} f(x) + \sum_{x \in T} (\deg_G(x) - g(x)) - e_G(S, T) \geq 0. \quad (4.10)$$

Let $S_A = S \cap A$, $S_B = S \cap B$, $T_A = T \cap A$ and $T_B = T \cap B$. Then

$$\gamma^*(S, T) = \sum_{x \in S} f(x) + \sum_{x \in T} (\deg_G(x) - g(x)) - e_G(S, T)$$

$$= \sum_{x \in S_A} f(x) + \sum_{x \in T_B} (\deg_G(x) - g(x)) - e_G(S_A, T_B)$$

$$+ \sum_{x \in S_B} f(x) + \sum_{x \in T_A} (\deg_G(x) - g(x)) - e_G(S_B, T_A)$$

$$= \gamma^*(S_A, T_B) + \gamma^*(S_B, T_A).$$

Hence, by (4.8) and (4.9), we have $\gamma^*(S, T) \geq 0$.

We now apply the same argument as in the proof of Theorem 4.2, since the statements (2), (3) and (4) in the proof hold even if $g(v) = f(v)$ for some vertices v of G. This is because if $u \in A$, then OA and EV are subsets of A and B, respectively. Therefore, we can prove the theorem in exactly the same way as Theorem 4.2. □

We next give a criterion for a graph to have a special (g, f)-factor that satisfies $0 \leq g(x) \leq 1$ for all vertices x. This criterion, obtained by Las Vergnas [165], is a natural extension of the 1-factor theorem.

Theorem 4.4 (The (g, f)-Factor Theorem with $g \leq 1$, Las Vergnas [165]**).** *Let G be a simple graph and let $g, f : V(G) \to \mathbb{Z}^+ = \{0, 1, 2, \cdots\}$ be functions such that $0 \leq g(x) \leq 1$ and $g(x) \leq f(x)$ for all $x \in V(G)$. Then G has a (g, f)-factor if and only if for every subset $S \subset V(G)$, we have*

$$odd(g; G - S) \leq \sum_{x \in S} f(x), \qquad (4.11)$$

where $odd(g; G - S)$ denotes the number of components C of $G - S$ such that either (i) $C = \{x\}$ and $g(x) = 1$ or (ii) $|C|$ is odd, $|C| \geq 3$ and $g(x) = f(x) = 1$ for all $x \in V(C)$.

Las Vergnas gave two proofs for Theorem 4.4. In one proof, he considered a graph G that satisfies the following conditions: (a) (4.11) holds but G has no (g, f)-factor, and (b) for any new edge e not in G, $G + e$ has a (g, f)-factor. That is, G is a counterexample to the theorem with maximal edge set. Thus G possesses some nice properties, and we can derive a contradiction. In the second proof, he used the (g, f)-factor theorem. Here we present another proof using a proof technique that we have seen before.

Proof of Theorem 4.4. Let C be a component of $G - S$ satisfying condition (i) or (ii) in the theorem. It is convenient to refer to C as a g-**odd component**, and to call any other component of $G - S$ a g-**even component**.

We first prove the necessity. Assume that G has a (g, f)-factor F. Then it is obvious that (4.11) holds for $S = \emptyset$. Let S be a non-empty vertex subset of G, and let C be a g-odd component of $G - S$. Then we can easily show that $e_F(S, C) \geq 1$, that is, there exists at least one edge of F joining C to S. Hence

$$odd(g; G - S) \leq e_F(S, V(G) - S) \leq \sum_{x \in S} \deg_F(x) \leq \sum_{x \in S} f(x).$$

We next prove sufficiency by induction on the size $\|G\|$. We may assume that G is connected, since otherwise we can apply the inductive hypothesis to each component in order to get a (g, f)-factor of G. If G has a vertex v with

$g(v) = f(v) = 0$, then $G - v$ satisfies condition (4.9), since for every subset $S \subset V(G) - \{v\}$, we have

$$odd(g; (G - v) - S) = odd(g; G - (S \cup \{v\})) \leq \sum_{x \in S \cup \{v\}} f(x) = \sum_{x \in S} f(x).$$

Hence by the inductive hypothesis, $G - v$ has a (g, f)-factor, which is the desired (g, f)-factor of G since $g(v) = f(v) = 0$. Therefore, we may assume that G has no vertex v with $g(v) = f(v) = 0$, which implies that $f(x) \geq 1$ for very vertex x.

Claim 1. *There exists $\emptyset \neq X \subset V(G)$ for which $odd(g; G - X) \geq \sum_{x \in X} f(x) - 1$.*

Assume that $odd(g; G - S) \leq \sum_{x \in S} f(x) - 2$ for all $\emptyset \neq S \subset V(G)$. Then for any chosen edge e of G, and for any subset $S \subset V(G)$, we have

$$odd(g; G - e - S) \leq odd(g; G - S) + 2 \leq \sum_{x \in S} f(x).$$

Thus by the inductive hypothesis, $G - e$ has a (g, f)-factor, which is the desired (g, f)-factor of G. Hence we may assume that Claim 1 holds.

Let us define an integer β by

$$\beta = \min_{\emptyset \neq X \subseteq V(G)} \left\{ \sum_{x \in X} f(x) - odd(g; G - X) \right\}.$$

By condition (4.11) and Claim 1, we have $\beta \in \{0, 1\}$. Choose a maximal vertex subset S of G subject to

$$\sum_{x \in S} f(x) - odd(g; G - S) = \beta. \tag{4.12}$$

Then $S \neq \emptyset$, and

$$\sum_{x \in X} f(x) - odd(g; G - X) > \beta, \qquad \text{i.e.,}$$

$$odd(g; G - X) \leq \sum_{x \in X} f(x) - \beta - 1 \quad \text{for all} \quad S \subset X \subseteq V(G). \tag{4.13}$$

Claim 2. *For every g-odd component C of $G - S$, and any vertex v of C, $C - v$ has a 1-factor, which is a (g, f)-factor of $C - v$.*

We may assume $|C| \geq 3$. Let $T \subset V(C) - \{v\}$. It is clear that every odd component of $(C - v) - T$ is a g-odd component of $G - (S \cup T \cup \{v\})$ since $g(x) = f(x) = 1$ for $x \in V(C)$. Since $S \subset S \cup \{v\} \cup T$, we have by (4.13)

$$odd(g; G - (S \cup \{v\} \cup T)) = odd(g; G - S) - 1 + odd(C - v - T)$$
$$\leq \sum_{x \in S \cup \{v\} \cup T} f(x) - \beta - 1 = \sum_{x \in S} f(x) - \beta + |T|.$$

Hence $odd(C - v - T) \leq |T| + 1$, which implies $odd(C - v - T) \leq |T|$ since $odd(C - v - T) \equiv |T| \pmod 2$. Therefore $C - v$ has a 1-factor by the 1-factor theorem.

Claim 3. *Every g-even component D of $G - S$ has a (g, f)-factor. Moreover, if $g(v) < f(v)$ for some vertex v of D, then for every edge uw of G joining $u \in S$ to $w \in V(D)$, $D + uw$ has a (g, f)-factor, where $g(u) = f(u) = 1$.*

Let $\emptyset \neq T \subset V(D)$. Then by (4.13), we have

$$odd(g; G - S) + odd(g; D - T)$$
$$= odd(g; G - (S \cup T)) \leq \sum_{x \in S \cup T} f(x) - \beta - 1.$$

By (4.12) we have

$$odd(g; D - T) \leq \sum_{x \in T} f(x) - 1, \tag{4.14}$$

and so D has a (g, f)-factor by induction. Moreover, by the above inequality (4.14), we have

$$odd(g; (D + uw) - T) \leq odd(g; D - T) + 1 \leq \sum_{x \in T} f(x).$$

It is clear that

$$odd(g; (D + uw) - (T \cup \{u\})) = odd(g; D - T) \leq \sum_{x \in T \cup \{u\}} f(x) - 2.$$

Furthermore, the existence of a vertex v with $g(v) < f(v)$ guarantees $odd(g; D + uw) = 0$. Hence $D + uw$ has a (g, f)-factor by induction when $D + uw \neq G$.

Note that if $G = D + uw$, then $S = \{u\}$, $f(u) = 1$, $\beta = 1$ and $odd(g; G - S) = 0$. In such a case D has a (g, f')-factor with $f'(x) = f(x)$, $x \in V(D) - \{w\}$, and $f'(w) = f(w) - 1$ by (4.14). Hence by adding an edge uw to this factor, we can get the desired (g, f)-factor of $G = D + uw$.

Let $\{C_1, C_2, \ldots, C_m\}$ be the set of g-odd components of $G - S$, where $m = odd(g; G - S)$. We construct a bipartite graph B with bipartition $(S, \{C_1, C_2, \ldots, C_m\})$ as follows: $x \in S$ and C_i are joined by an edge of B if and only if x and C_i are joined by at least one edge of G (see Fig. 4.3). Then B has the following properties.

Claim 4. *If $\beta = 0$, then B has a factor F such that (i) $\deg_F(x) = f(x)$ for all $x \in S$ and (ii) $\deg_F(C_i) = 1$ for all $1 \leq i \leq m$. If $\beta = 1$, then for any*

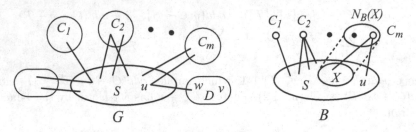

Fig. 4.3. The graph G with S, and the bipartite graph B.

given vertex $u \in S$, B has a factor F such that (iii) $\deg_F(x) = f(x)$ for all $x \in S - u$, $\deg_F(u) = f(u) - 1$, and (iv) $\deg_F(C_i) = 1$ for all $1 \le i \le m$.

We shall prove only the case $\beta = 1$ since the the case $\beta = 0$ is similarly proved. Let $\emptyset \ne X \subset S$. Assume that $|N_B(X)| < \sum_{x \in X} f(x)$. Then

$$odd(g; G - (S - X)) \ge odd(g; G - S) - |N_B(X)|$$
$$> \sum_{x \in S} f(x) - \beta - \sum_{x \in X} f(x) = \sum_{x \in S - X} f(x) - \beta.$$

This contradicts the definition of β. Hence $|N_B(X)| \ge \sum_{x \in X} f(x)$. Moreover,

$$|N_B(S)| = m = odd(g; G - S) = \sum_{x \in S} f(x) - 1.$$

Consequently, by Theorem 2.10 we have that B has a factor F such that (i) $\deg_F(x) = f(x)$ for every $x \in S - u$, $\deg_F(u) = f(u) - 1$, and (ii) $\deg_F(C_i) = 1$ for every $1 \le i \le m$.

Claim 5. *If $G - S$ has a g-even component D such that $g(v) < f(v)$ for some vertex v of D, then G has a (g, f)-factor.*

We may assume that $\beta = 1$ since if $\beta = 0$, then we can similarly prove the claim. Suppose that $G - S$ has a g-even component D that contains a vertex v with $g(v) < f(v)$. Let uw be an edge of G joining $u \in S$ to $w \in D$. Let $\{C_1, \ldots, C_m\}$ be the set of g-odd components of $G - S$. Then by Claim 4, the bipartite graph B with bipartition $(S, \{C_1, C_2, \ldots, C_m\})$ has a factor F such that $\deg_F(x) = f(x)$ for all $x \in S - u$, $\deg_F(u) = f(u) - 1$ and $\deg_F(C_i) = 1$ for all $1 \le i \le m$. For every edge xC_i of F, we can choose an edge of G joining x to a vertex $v_i \in V(C_i)$ in order to obtain a subgraph H of G by collecting these edges. By Claims 3 and 4 it follows that $D + uw$, the other g-even components, and $C_i - v_i$ all have (g, f)-factors. By combining these factors and H, we can obtain the desired (g, f)-factor of G (see Fig. 4.4).

By Claim 5, we may assume that no g-even component D of $G - S$ satisfies the condition of Claim 5. Thus every g-even component D satisfies $g(x) = f(x) = 1$ for all $x \in V(D)$, and has a 1-factor by Claim 3.

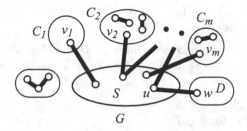

Fig. 4.4. A (g, f)-factor, denoted by bold edges, of G with $\beta = 1$.

If $\beta = 0$, then we can get the desired (g, f)-factor of G in the same way as in the proof of Claim 5, that is, we obtain the desired (g, f)-factor of G by considering the factor F given in Claim 4 with $\beta = 0$, and by the fact of Claim 2.

Thus we may assume that $\beta = 1$. If $g(u) < f(u)$ for some vertex $u \in S$, then by considering the (g, f)-factor F in Claim 4 with $\beta = 1$, and by Claim 2, we can get the desired (g, f)-factor of G. So we may assume $g(x) = f(x)$ for all $x \in S$. In this case, if $G - S$ has an isolated vertex v, then v and S must be joined by at least two edges since $\beta = 1$. If an isolated vertex v_1 of $G - S$ satisfies $f(v_1) \geq 2$, then we can find the desired (g, f)-factor of G. If every isolated vertex v of $G - S$ satisfies $f(v) = 1$, then $g(x) = f(x) = 1$ for all $x \in V(G)$, and thus

$$|G| \equiv |S| + m \equiv \beta \equiv 1 \pmod 2.$$

Hence G itself is a g-odd component, which contradicts (4.9) with $S = \emptyset$. Consequently the proof is complete. $\square$

We now consider factors combining f-factors and (g, f)-factors. Let $g, f : V(G) \to \mathbb{Z}^+$ be functions such that $g(x) < f(x)$ for all $x \in V(G)$. Then we say that G has **all (g, f)-factors** if for every function $h : V(G) \to \mathbb{Z}^+$ such that $g(x) \leq h(x) \leq f(x)$ for all $x \in V(G)$ and $\sum_{x \in V(G)} h(x) \equiv 0 \pmod 2$, G has an h-factor. A necessary and sufficient condition for a graph to have all (g, f)-factors is given in the next theorem.

Theorem 4.5 (Niessen [196]). *Let G be a connected general graph and let $g, f : V(G) \to \mathbb{Z}^+$ be functions such that $g(x) \leq f(x)$ for all $x \in V(G)$ and $g(w) < f(w)$ for at least one vertex w. Then G has all (g, f)-factors if and only if for all disjoint subsets S and T of $V(G)$,*

$$\sum_{x \in S} g(x) + \sum_{x \in T} (\deg_G(x) - f(x)) - e_G(S, T) - \hat{q}(S, T) \geq -1, \qquad (4.15)$$

where $\hat{q}(S, T)$ denotes the number of components C of $G - (S \cup T)$ such that either (i) $g(u) < f(u)$ for some vertex $u \in V(C)$ or (ii) $g(x) = f(x)$ for all $x \in V(C)$ and

$$\sum_{x\in V(C)} f(x) + e_G(C,T) \equiv 1 \pmod 2. \tag{4.16}$$

Proof. We first prove the necessity. Suppose that G has all (g,f)-factors. Let S and T be disjoint subsets of $V(G)$. We define a function $h: V(G) \to \mathbb{Z}^+$ as follows:

(1) $h(x) = g(x)$ for $x \in S$,
(2) $h(x) = f(x)$ for $x \in T$,
(3) for every component C of $G - (S \cup T)$ having a vertex u with $g(u) < f(u)$, define $h(x) = f(x)$ for all $x \in V(C) - \{u\}$, and define $h(u) = f(u)$ or $f(u)-1$ so that $\sum_{x\in V(C)} h(x) + e_G(C,T) \equiv 1 \pmod 2$; and
(4) for every component C of $G - (S \cup T)$ with $g(x) = f(x)$ for all $x \in V(C)$, define $h(x) = f(x)$ for all $x \in V(C)$.

If $\sum_{x\in V(G)} h(x)$ is even, then G has an h-factor, and so by the f-factor theorem, we have

$$0 \le \delta(h;S,T)$$
$$= \sum_{x\in S} h(x) + \sum_{x\in T}(\deg_G(x) - h(x)) - e_G(S,T) - q(h;S,T)$$
$$= \sum_{x\in S} g(x) + \sum_{x\in T}(\deg_G(x) - f(x)) - e_G(S,T) - \hat{q}(S,T).$$

If $\sum_{x\in V(G)} h(x)$ is odd, then since G has a vertex w with $g(w) < f(w)$, we change $h(w)$ by one so that $g(w) \le h(w) \le f(w)$ and $\sum_{x\in V(G)} h(x)$ is even. If $w \in S \cup T$ then $q(h;S,T) = \hat{q}(S,T)$ and $h(w) = f(w) - 1$ or $h(w) = g(w)+1$; otherwise $q(h;S,T) \ge \hat{q}(S,T) - 1$. Since G has an h-factor, by the f-factor theorem, we have

$$0 \le \delta(h;S,T)$$
$$= \sum_{x\in S} h(x) + \sum_{x\in T}(\deg_G(x) - h(x)) - e_G(S,T) - q(h;S,T)$$
$$\le \sum_{x\in S} g(x) + \sum_{x\in T}(\deg_G(x) - f(x)) - e_G(S,T) - \hat{q}(S,T) + 1.$$

Therefore (4.15) holds.

We next prove the sufficiency. Let $h: V(G) \to \mathbb{Z}^+$ be a function such that $g(x) \le h(x) \le f(x)$ for all $x \in V(G)$ and $\sum_{x\in V(G)} h(x) \equiv 0 \pmod 2$. Let S and T be disjoint subsets of $V(G)$. Then by the f-factor theorem, it suffices to show that

$$\delta(h;S,T) = \sum_{x\in S} h(x) + \sum_{x\in T}(\deg_G(x) - h(x))$$
$$-e_G(S,T) - q(h;S,T) \ge 0, \tag{4.17}$$

where $q(h; S, T)$ denotes the number of components C of $G - (S \cup T)$ satisfying

$$\sum_{x \in V(C)} h(x) + e_G(C, T) \equiv 1 \pmod 2.$$

Since every component C of $G - (S \cup T)$ counted in $q(h; S, T)$ satisfies condition (i) or (ii) in the theorem, we have $q(h; S, T) \leq \hat{q}(S, T)$. Hence by the inequality $g(x) \leq h(x) \leq f(x)$, we have

$$\delta(h; S, T) = \sum_{x \in S} h(x) + \sum_{x \in T} (\deg_G(x) - h(x)) - e_G(S, T) - q(S, T)$$

$$\geq \sum_{x \in S} g(x) + \sum_{x \in T} (\deg_G(x) - f(x)) - e_G(S, T) - \hat{q}(S, T)$$

$$\geq -1.$$

The following congruence holds by the f-factor theorem:

$$\delta(h; S, T) \equiv \sum_{x \in V(G)} h(x) \equiv 0 \pmod 2.$$

Hence $\delta(h; S, T) \geq 0$. Consequently G has an h-factor. □

4.2 Graphs having the odd-cycle property

Before going to the next stage of (g, f)-factors and $[a, b]$-factors, we investigate factor theory in a special class of graphs: those which possess the odd-cycle property. Many results on factors in bipartite graphs can be extended to those in graphs from this class. A graph is said to have the **odd-cycle property** if any two odd cycles either have a vertex in common or are joined by an edge. Of course, every bipartite graph has the old-cycle property, and so the class of graphs having the odd-cycle property includes bipartite graphs. For example, the graph G shown in Fig. 4.5 has the odd-cycle property because each of its odd cycles passes through u, v or w, and these three vertices are adjacent.

The following theorem is similar to Hall's marriage theorem in bipartite graphs.

Theorem 4.6 (Berge [27]). *Let G be a connected simple graph of even order possessing the odd-cycle property. Then G has a 1-factor if and only if*

$$|N_G(S)| \geq |S| \qquad \text{for all} \quad S \subset V(G). \tag{4.18}$$

Moreover, the above theorem is equivalent to the following theorem, which is similar to Tutte's 1-factor theorem but counts only isolated vertices.

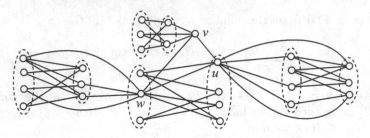

Fig. 4.5. A graph possessing the odd-cycle property.

Theorem 4.7 ([27]). *Let G be a connected simple graph of even order possessing the odd-cycle property. Then G has a 1-factor if and only if*

$$iso(G - S) \leq |S| \qquad for\ all \quad S \subset V(G). \qquad (4.19)$$

Proof of Theorems 4.6 and 4.7. We first show that if G has a 1-factor, then (4.18) holds. Suppose that G has a 1-factor F. Then for any subset $\emptyset \neq S \subset V(G)$, we have $|N_G(S)| \geq |N_F(S)| = |S|$.

We next show that (4.18) implies (4.19). Assume (4.18) holds. Then for every subset $S \subset V(G)$, we have

$$|S| \geq |N_G(Iso(G - S))| \geq |Iso(G - S)| = iso(G - S).$$

Hence (4.19) holds.

Finally, we show that (4.19) implies the existence of a 1-factor in G. Suppose that G satisfies (4.19) but has no 1-factor. Then by the 1-factor theorem, there exists a subset $S \subset V(G)$ such that $odd(G - S) \geq |S| + 2$. Choose such a maximal subset S. Then by the proof of Theorem 2.28, every component of $G - S$ is factor-critical. Since every factor-critical component of order at least three contains an odd cycle (see Problem 4.4) and G has the odd-cycle property, $G - S$ has at most one such component and the other components are isolated vertices. Hence

$$iso(G - S) \geq odd(G - S) - 1 \geq |S| + 1.$$

This contradicts (4.19). Therefore the proof is complete. □

We now give a criterion of f-factors and (g, f)-factors in graphs with the odd-cycle property. The proof presented here was given by Mahmoodian [184].

Theorem 4.8 (Folkman, Hoffman and McAndrew [90]). *Let G be a connected multigraph possessing the odd-cycle property, and let $f : V(G) \to \mathbb{Z}^+$ be a function. Then G has an f-factor if and only if $\sum_{x \in V(G)} f(x)$ is even and for all disjoint subsets S and T of $V(G)$,*

$$\delta^*(S, T) = \sum_{x \in S} f(x) + \sum_{x \in T} (\deg_G(x) - f(x)) - e_G(S, T) \geq 0. \qquad (4.20)$$

Proof. Assume first G has an f-factor F. Then for any disjoint subsets S and T of $V(G)$, it follows from the f-factor theorem that

$$\delta^*(S,T) = \delta(S,T) + q(S,T) \geq \delta(S,T) \geq 0.$$

Hence (4.20) holds.

We next prove the sufficiency by the f-factor theorem. In order to do so, it suffices to show that $\delta(S,T) \geq 0$ for any disjoint subsets S and T of $V(G)$. Since $\sum_{x \in V(G)} f(x)$ is even, we have $\delta(\emptyset, \emptyset) = 0$, and so we may assume $S \cup T \neq \emptyset$.

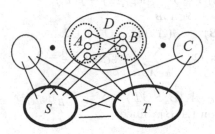

Fig. 4.6. A bipartite f-odd component D of $G - (S \cup T)$ with bipartition (A, B), and another f-odd component C.

Let C be an f-odd component of $G - (S \cup T)$. Then C satisfies

$$\sum_{x \in V(C)} f(x) + e_G(C,T) \equiv 1 \pmod 2.$$

If an f-odd component D of $G - (S \cup T)$ is a bipartite graph with bipartition (A, B), then

$$q(S \cup A, T \cup B) - q(S \cup B, T \cup A) = q(S,T) - 1. \qquad \text{(see Fig. 4.6)}$$

Thus we obtain

$$\delta(S \cup A, T \cup B) + \delta(S \cup B, T \cup A) - 2\delta(S, T)$$

$$= \sum_{x \in S \cup A} f(x) + \sum_{x \in T \cup B} (\deg_G(x) - f(x))$$

$$-e_G(S \cup A, T \cup B) - q(S \cup A, T \cup B)$$

$$+ \sum_{x \in S \cup B} f(x) + \sum_{x \in T \cup A} (\deg_G(x) - f(x))$$

$$-e_G(S \cup B, T \cup A) - q(S \cup B, T \cup A)$$

$$-2\sum_{x \in S} f(x) - 2\sum_{x \in T} (\deg_G(x) - f(x)) + 2e_G(S, T) + 2q(S, T)$$

$$= \sum_{x \in A \cup B} f(x) + \sum_{x \in A \cup B} (\deg_G(x) - f(x))$$

$$-2e_G(A, B) - e_G(A \cup B, S \cup T) + 2$$

$$= \sum_{x \in A \cup B} \deg_G(x) - \sum_{x \in A \cup B} \deg_G(x) + 2 = 2,$$

where $2e_G(A, B) + e_G(A \cup B, S \cup T) = \sum_{x \in A \cup B} \deg_G(x)$ since D is a bipartite graph with bipartition (A, B). Since

$$\delta(S, T) \equiv \delta(S \cup A, T \cup B) \equiv \delta(S \cup B, T \cup A) \pmod 2,$$

it follows from the above equation that

$$\delta(S, T) \geq \delta(S \cup A, T \cup B) \quad \text{or} \quad \delta(S \cup B, T \cup A).$$

Namely, by setting $(S', T') = (S \cup A, T \cup B)$ or $(S \cup B, T \cup A)$, we have

$$\delta(S, T) \geq \delta(S', T').$$

By repeating the same procedure for every bipartite f-odd component of $G - (S \cup B)$ one by one, we finally obtain

$$\delta(S, T) \geq \delta(S^*, T^*),$$

where $G - (S^* \cup T^*)$ has at most one f-odd component, since $G - (S \cup T)$ has at most one non-bipartite component. Therefore, since $\delta^*(S^*, T^*) \geq 0$ and $q(S^*, T^*) \leq 1$, we obtain

$$\delta(S, T) \geq \delta(S^*, T^*) = \delta^*(S^*, T^*) - q(S^*, T^*) \geq -1.$$

This implies $\delta(S, T) \geq 0$ since $\delta(S, T) \equiv \sum_{x \in V(G)} f(x) \equiv 0 \pmod 2$. Consequently, the proof is complete by the f-factor theorem. $\square$

The above proof of the f-factor theorem for a graph having the odd-cycle property uses the f-factor theorem. On the other hand, we gave an elementary direct proof to the f-factor theorem for a bipartite graph without using the f-factor theorem. Therefore, it might be possible to give an elementary proof of Theorem 4.8. Actually, Folkman, Hoffman, and McAndrew gave another proof of it using flow theory, without any appeal to the f-factor theorem.

The situation changes when we consider a (g, f)-factor in a graph with the odd-cycle property, that is, a criterion for such a graph to have a (g, f)-factor is not obtained by removing the term $q^*(S, T)$ from $\gamma(S, T)$. In fact, a necessary and sufficient condition for such a graph to have a (g, f)-factor is given in the next theorem. Notice that the assumption of the existence of a vertex u with $g(u) < f(u)$ in the following theorem is natural since otherwise a (g, f)-factor becomes an f-factor.

Theorem 4.9 (Chen and Wang [50] (1993)). *Let G be a connected multigraph possessing the odd-cycle property. Let $g, f : V(G) \to \mathbb{Z}$ be functions such that $g(x) \leq f(x)$ for all $x \in V(G)$ and $g(u) < f(u)$ for some vertex u. Then G has a (g, f)-factor if for all disjoint subsets S and T of $V(G)$ with $S \cup T \neq \emptyset$, we have*

$$\gamma^*(S, T) = \sum_{x \in S} f(x) + \sum_{x \in T} (\deg_G(x) - g(x)) - e_G(S, T) \geq \epsilon_0, \quad (4.21)$$

where $\epsilon_0 = 1$ if G has an odd cycle with $g(x) = f(x)$ for all its vertices; otherwise $\epsilon_0 = 0$.

The constant ϵ_0 cannot be removed, that is, there are graphs that have no (g, f)-factors but have the odd-cycle property and satisfy $\gamma^*(S, T) \geq 0$ for all disjoint subsets S and T of $V(G)$ with $S \cup T \neq \emptyset$.

Proof. Our proof is similar to that of Theorem 4.8 and uses the (g, f)-factor theorem instead of the f-factor theorem.

Let S and T be two disjoint subsets of $V(G)$ such that $S \cup T \neq \emptyset$. In order to prove the theorem, it suffices to show that $\gamma(S, T) \geq 0$ and $\gamma(\emptyset, \emptyset) = 0$, where $\gamma(\emptyset, \emptyset) = 0$ holds because of the existence of a vertex u with $g(u) < f(u)$.

Let C be a (g, f)-odd component of $G - (S \cup T)$. Then C satisfies

$$g(x) = f(x) \quad \text{for all} \quad x \in V(C), \quad \text{and}$$
$$\sum_{x \in V(C)} f(x) + e_G(C, T) \equiv 1 \pmod 2.$$

Assume that a (g, f)-odd component D of $G - (S \cup T)$ is a bipartite graph with bipartition (A, B). Then

$$q^*(S \cup A, T \cup B) = q^*(S \cup B, T \cup A) = q^*(S, T) - 1,$$

and thus we obtain

$$\gamma(S \cup A, T \cup B) + \gamma(S \cup B, T \cup A) - 2\gamma(S, T)$$

$$= \sum_{x \in S \cup A} f(x) + \sum_{x \in T \cup B} (\deg_G(x) - g(x))$$

$$\quad - e_G(S \cup A, T \cup B) - q^*(S \cup A, T \cup B)$$

$$+ \sum_{x \in S \cup B} f(x) + \sum_{x \in T \cup A} (\deg_G(x) - g(x))$$

$$\quad - e_G(S \cup B, T \cup A) - q^*(S \cup B, T \cup A)$$

$$- 2\sum_{x \in S} f(x) - 2\sum_{x \in T} (\deg_G(x) - g(x)) + 2e_G(S, T) + 2q^*(S, T)$$

$$= \sum_{x \in A \cup B} f(x) + \sum_{x \in A \cup B} (\deg_G(x) - g(x))$$

$$\quad - 2e_G(A, B) - e_G(A \cup B, S \cup T) + 2$$

$$= \sum_{x \in A \cup B} \deg_G(x) - \sum_{x \in A \cup B} \deg_G(x) + 2 \quad \text{(by } g(x) = f(x) \text{ for } x \in A \cup B)$$

$$= 2.$$

On the other hand, it follows that

$$\gamma(S \cup A, T \cup B) - \gamma(S, T)$$

$$= \sum_{x \in A} f(x) + \sum_{x \in B} (\deg_G(x) - g(x)) - e_G(S, B) - e_G(A, T) - e_G(A, B) + 1$$

$$\equiv \sum_{x \in A \cup B} f(x) + \sum_{x \in B} \deg_G(x) \quad \text{(by } g(x) = f(x) \text{ for } x \in A \cup B)$$

$$\quad - e_G(S, B) - e_G(A, T) - e_G(A, B) + \sum_{x \in V(D)} f(x) + e_G(D, T) \quad (\text{mod } 2)$$

$$\equiv 2 \sum_{x \in V(D)} f(x) + e_G(B, A) + e_G(B, S) + e_G(B, T)$$

$$\quad - e_G(S, B) - e_G(A, T) - e_G(A, B) + e_G(A \cup B, T) \quad (\text{mod } 2)$$

$$\equiv 2 \sum_{x \in V(D)} f(x) + 2e_G(B, T) \equiv 0 \quad (\text{mod } 2).$$

Hence

$$\gamma(S, T) \equiv \gamma(S \cup A, T \cup B) \equiv \gamma(S \cup B, T \cup A) \quad (\text{mod } 2).$$

Therefore we obtain

$$\gamma(S, T) \ge \gamma(S \cup A, T \cup B) \quad \text{or} \quad \gamma(S \cup B, T \cup A).$$

Namely, by setting $(S', T') = (S \cup A, T \cup B)$ or $(S \cup B, T \cup A)$, we have

$$\gamma(S, T) \ge \gamma(S', T').$$

By repeating the above procedure for every bipartite (g, f)-odd component of $G - (S \cup B)$, we obtain

$$\gamma(S, T) \geq \gamma(S^*, T^*),$$

where $G - (S^* \cup T^*)$ has at most one (g, f)-odd component since $G - (S \cup T)$ has at most one non-bipartite component. Define $\epsilon = 1$ if $G - (S \cup T)$ has exactly one non-bipartite component and $\epsilon = 0$ otherwise. If $\epsilon = 1$ then $\epsilon_0 = 1$ by the definition of (g, f)-odd components, and therefore it follows from $\gamma^*(S^*, T^*) \geq \epsilon_0$ that

$$\gamma(S, T) \geq \gamma(S^*, T^*) = \gamma^*(S^*, T^*) - q^*(S^*, T^*) = \gamma^*(S^*, T^*) - \epsilon$$
$$\geq \epsilon_0 - \epsilon = 0.$$

This implies $\gamma(S, T) \geq 0$. Consequently, the proof is complete by the (g, f)-factor theorem. □

4.3 $[a, b]$-factors and (g, f)-factors

In this section we shall present several sufficient conditions for a graph to have an $[a, b]$-factor or a (g, f)-factor. Recall that for two integers $0 \leq a \leq b$, a spanning subgraph F of a graph G is called an $[a, b]$-**factor** if

$$a \leq \deg_F(x) \leq b \qquad \text{for all} \quad x \in V(G).$$

Similarly, a graph G is called an $[a, b]$-**graph** if $a \leq \deg_G(x) \leq b$ for all vertices x of G.

We begin with a criterion for a graph to have an $[a, b]$-factor, which is easily obtained from the (g, f)-factor theorem with $g < f$ simply by setting $g(x) = a$ and $f(x) = b$ for all vertices x. Notice that if $a = b$, then an $[a, b]$-factor becomes a regular a-factor, which were dealt with in Chapter 3, and so in this section we always assume $a < b$. If $a = 0$, then every graph has an $[a, b]$-factor that contains no edges, and so we may restrict ourselves to the case of $a \geq 1$.

Theorem 4.10 (The $[a, b]$-Factor Theorem, Lovász). *Let a and b be integers such that $1 \leq a < b$. Then a general graph G has an $[a, b]$-factor if and only if for all two disjoint subsets S and T of $V(G)$, we have*

$$\gamma^*(S, T) = b|S| + \sum_{x \in T} \deg_G(x) - a|T| - e_G(S, T) \geq 0. \qquad (4.22)$$

It is obvious that $\gamma^*(\emptyset, \emptyset) = 0$ and so we should check only the case $S \cup T \neq \emptyset$. Note that (4.22) is equivalent to

$$\gamma^*(S, T) = b|S| + \sum_{x \in T} \deg_{G-S}(x) - a|T| \geq 0. \tag{4.23}$$

Conditions (4.22) and (4.23) can be replaced by the following inequality (4.24), which is sometimes useful. For any integer $j \geq 0$, let $n_j(G)$ denote the number of vertices of G with degree j.

Theorem 4.11 (Heinrich, Hell, Kirkpatrick and Liu, [99]). *Let G be a general graph, and a and b be integers such that $1 \leq a < b$. Then G has an $[a, b]$-factor if and only if*

$$\sum_{0 \leq j < a} (a - j) \cdot n_j(G - S) \leq b|S| \qquad \text{for all} \quad S \subset V(G). \tag{4.24}$$

Proof. It suffices to show that (4.24) and (4.23) are equivalent. Let S and T be two disjoint subsets of $V(G)$. Let $U = \{x \in V(G) - S : \deg_{G-S}(x) < a\}$. Then

$$a|T| - \sum_{x \in T} \deg_{G-S}(x)$$

$$= \sum_{x \in T} (a - \deg_{G-S}(x))$$

$$\leq \sum_{x \in U} (a - \deg_{G-S}(x))$$

$$= \sum_{0 \leq j < a} (a - j) \, n_j(G - S).$$

Hence (4.24) implies (4.23). Moreover, for any $S \subset V(G)$, by letting $T = U$, (4.23) implies (4.24). $\square$

To show the existence of an $[a, b]$-factor, we may apply the next lemma instead of Theorem 4.10.

Lemma 4.12. *Let G be a general graph, and a and b be integers such that $1 \leq a < b$. Suppose that G does not have an $[a, b]$-factor. Then there exist disjoint subsets S and T of $V(G)$ such that $T \neq \emptyset$,*

$$\gamma^*(S, T) = b|S| + \sum_{x \in T} \deg_{G-S}(x) - a|T| < 0, \qquad \text{and}$$

$$\deg_{G-S}(x) \leq a - 1 \qquad \text{for all} \quad x \in T.$$

Proof. Choose a pair S and T so that T is minimal subject to $\gamma^*(S, T) < 0$. The existence of such a pair is guaranteed by the non-existence of an $[a, b]$-factor in G. It is trivial that $T \neq \emptyset$. Let $v \in T$. Then by the choice of (S, T), we have $\gamma^*(S, T - v) \geq 0$, and thus

$$1 \leq \gamma^*(S, T - v) - \gamma^*(S, T)$$
$$= b|S| + \sum_{x \in T-v} \deg_{G-S}(x) - a|T - v|$$
$$- \left(b|S| + \sum_{x \in T} \deg_{G-S}(x) - a|T| \right)$$
$$= -\deg_{G-S}(v) + a.$$

Hence $\deg_{G-S}(v) \leq a - 1$. □

By setting $a = 1$ and $b = n$ in Theorem 4.11, we can obtain Theorem 4.13 below, since
$$n_0(G - S) = iso(G - S).$$
This theorem is also obtained from Las Vergnas's (g, f)-factor theorem with $g \leq 1$ (Theorem 4.4) because if $g(x) = 1$ and $f(x) = n \geq 2$ for all $x \in V(G)$, then $odd(g; G - S)$ is equal to $iso(G - S)$ (see (4.9)). We shall consider a $[1, n]$-factor in Chapter 5 from the point of view of component factors.

Theorem 4.13 (Las Vergnas [165]). *Let $n \geq 2$ be an integer. Then a general graph G has a $[1, n]$-factor if and only if*

$$iso(G - S) \leq n|S| \qquad \text{for all} \quad S \subset V(G). \tag{4.25}$$

Furthermore, (4.25) is equivalent to

$$|N_G(S)| \geq \frac{|S|}{n} \qquad \text{for all independent} \quad S \subset V(G). \tag{4.26}$$

The equivalence of (4.25) and (4.26) are left to the reader (Problem 4.3). We now give a simple sufficient condition for a graph to have a (g, f)-factor, which includes some results on $[a, b]$-factors as its corollaries.

Theorem 4.14 (Kano and Saito [132]). *Let G be a connected general graph and let $g, f : V(G) \to \mathbb{Z}$ be functions such that $g(x) < f(x)$ for all $x \in V(G)$. If there exists a real number $0 \leq \theta \leq 1$ such that*

$$g(x) \leq \theta \deg_G(x) \leq f(x) \qquad \text{for all} \quad x \in V(G), \tag{4.27}$$

then G has a (g, f)-factor.

Proof. We use the (g, f)-factor theorem with $g < f$ (Theorem 4.2). Let S and T be disjoint subsets of $V(G)$ such that $S \cup T \neq \emptyset$. Then

$$\gamma^*(S, T) = \sum_{x \in S} f(x) + \sum_{x \in T}(\deg_G(x) - g(x)) - e_G(S, T)$$
$$\geq \theta \sum_{x \in S} \deg_G(x) + \sum_{x \in T}(1 - \theta) \deg_G(x) - e_G(S, T)$$
$$\geq \theta e_G(S, T) + (1 - \theta)e_G(S, T) - e_G(S, T) = 0.$$

Therefore G has a (g, f)-factor. □

Theorem 4.15 (Kano and Saito [132] (1983)). *Let m, n, a and b be integers such that $1 \le m \le n$ and $1 \le a < b \le n$. If*

$$\frac{a}{b} \le \frac{m}{n},$$

then every general $[m, n]$-graph has an $[a, b]$-factor.

Proof. Let G be an $[m, n]$-graph. Define $g(x) = a$ and $f(x) = b$ for all $x \in V(G)$, and $\theta = b/n$. Then $\theta \le 1$ and

$$g(x) = a \le \theta m \le \theta \deg_G(x) \le \theta n = b = f(x).$$

Hence by Theorem 4.14, G has an $[a, b]$-factor.   □

We now show that the condition in Theorem 4.15 is sharp, that is, we can prove that if $a/b > m/n$, then the complete bipartite graph $K(m, n)$, which is an $[m, n]$-graph, does not have an $[a, b]$-factor. Let $G = K(m, n)$, and let S and T be the partite sets of $K(m, n)$ such that $|S| = m$ and $|T| = n$. Then

$$b|S| + \sum_{x \in T} \deg_G(x) - a|T| - e_G(S, T)$$
$$= bm + mn - an - mn = bm - an < 0. \qquad \text{(by } a/b > m/n)$$

Hence by Theorem 4.10, G has no $[a, b]$-factor.

As was shown in Chapter 3, some r-regular simple graphs have no k-regular factors, but the next theorem, which was conjectured by Erdős and proved by Tutte, shows that every r-regular graph has a $[k, k + 1]$-factor. Such a factor is sometimes called a **semi-regular factor**.

Theorem 4.16 (Tutte [230] (1978)). *Let k and r be integers such that $1 \le k < r$. Then every r-regular general graph has a $[k, k + 1]$-factor.*

The next theorem is a generalization of the previous theorem. Thomassen gave an elementary proof for it by using Hall's Marriage Theorem. Note that Theorems 4.16 and 4.17 follow from Theorem 4.15, since $k/(k+1) \le r/r$ and $k/(k+1) \le r/(r+1)$. We shall later show that if $1 \le k \le 2r/3$, then every r-regular simple graph has a $[k, k + 1]$-factor, each of whose components is regular.

Theorem 4.17 (Thomassen [221] (1981)). *Let k and r be integers such that $1 \le k < r$. Then every general $[r, r + 1]$-graph has a $[k, k + 1]$-g,factor.*

Theorem 4.18 (Heinrich, Hell, Kirkpatrick and Liu [99] (1990)). *Let a, b, m, n and k be integers such that $1 \le a < b$, $1 \le m \le n$ and $1 \le k$. Then a general $[m, n]$-graph has k edge-disjoint $[a, b]$-factors if*

$$(k - 1)(b^2 - a^2) \le mb - na \qquad and \qquad a < b \le n - (k - 1)a. \qquad (4.28)$$

In particular, if $(k - 1)(a + b) \le r$ and $a < b \le r - (k - 1)a$, then every r-regular general graph has k edge disjoint $[a, b]$-factors.

Proof. Let G be a general $[m, n]$-graph. We prove the theorem by induction on k. If $k = 1$, then $0 \leq mb - na$ by (4.28), which implies $a/b \leq m/n$, and so G has an $[a, b]$-factor by Theorem 4.15. Assume $k \geq 2$. Then $0 \leq mb - na$, and so G has an $[a, b]$-factor F by Theorem 4.15. It is clear that $G - E(F)$ is an $[m - b, n - a]$-graph. Then

$$(k - 2)(b^2 - a^2) \leq (m - b)b - (n - a)a$$
$$\Leftrightarrow (k - 1)(b^2 - a^2) \leq mb - na,$$

and

$$b \leq (n - a) - (k - 2)a \quad \Leftrightarrow \quad b \leq n - (k - 1)a.$$

Hence by applying the inductive hypothesis to $G - F$ and $k - 1$, we can obtain $k - 1$ edge disjoint $[a, b]$-factors of $G - F$, and thus G has k edge disjoint $[a, b]$-factors. □

The following theorem was independently obtained by Heinrich, Hell, Kirkpatrick and Liu [99], and by Egawa and Kano ([70], [234] p. 121).

Theorem 4.19. *Let G be a connected general graph and let $g, f : V(G) \to \mathbb{Z}$ be functions such that $g(x) \leq \deg_G(x)$, $0 \leq f(x)$ and $g(x) < f(x)$ for all $x \in V(G)$. If for every pair of adjacent vertices x and y of G, we have*

$$\frac{g(x)}{\deg_G(x)} \leq \frac{f(y)}{\deg_G(y)}, \tag{4.29}$$

then G has a (g, f)-factor.

Proof. We use the (g, f)-factor theorem with $g < f$ (Theorem 4.2), since $g(x) < f(x)$ for all $x \in V(G)$. Let S and T be disjoint subsets of $V(G)$ such that $S \cup T \neq \emptyset$. Then we have

$$\gamma^*(S, T) = \sum_{s \in S} f(s) + \sum_{t \in T} (\deg_G(t) - g(t)) - e_G(S, T)$$

$$= \sum_{s \in S} \deg_G(s) \frac{f(s)}{\deg_G(s)} + \sum_{t \in T} \deg_G(t) \left(1 - \frac{g(t)}{\deg_G(t)} \right) - e_G(S, T).$$

For each vertex $s \in S$, add the weight $f(s)/\deg_G(s)$ to all the edges incident with s; for each vertex $t \in T$, add the weight $1 - g(t)/\deg_G(t)$ to all edges incident with t; and finally add the weight -1 to all the edges between S and T (see Fig. 4.7). Since $f(s)/\deg_G(s) \geq 0$ and $1 - g(t)/\deg_G(t) \geq 0$, it follows from the above expression for $\gamma^*(S, T)$ that

$$\gamma^*(S, T) = \sum_{s \in S} \deg_G(s) \frac{f(s)}{\deg_G(s)} + \sum_{t \in T} \deg_G(t) \left(1 - \frac{g(t)}{\deg_G(t)} \right) - e_G(S, T)$$

$\geq$ The total sum of weights of edges between S and T

$$= \sum_{st \in E(G), \ s \in S, \ t \in T} \left(\frac{f(s)}{\deg_G(s)} + 1 - \frac{g(t)}{\deg_G(t)} - 1 \right)$$

$\geq 0. \qquad$ by (4.29)

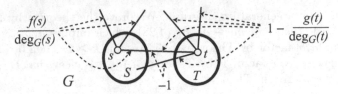

Fig. 4.7. The weights of edges of G with $S \cup T$.

Hence $\gamma^*(S,T) \geq 0$ and thus G has a (g,f)-factor. $\square$

A general graph G is said to be **locally s-almost regular** if

$$|\deg_G(x) - \deg_G(y)| \leq s \qquad \text{for all adjacent vertices } x \text{ and } y.$$

In particular, a locally 0-almost regular connected graph is a regular graph. This notion was introduced by Joentgen and Volkmann [108], and they proved the following theorem.

Theorem 4.20 (Joentgen and Volkmann [108]). *Let k, s and t be integers such that $1 \leq k, t$ and $0 \leq s$, and let G be a locally s-almost regular general graph. If*

$$k \leq \delta(G) \qquad and \qquad \frac{s}{\delta(G)} \leq \frac{t}{k}, \tag{4.30}$$

then G has a $[k, k+t]$-factor. In particular, G has a $[\delta(G), \delta(G)+s]$-factor.

Proof. We define $g(v) = k$ and $f(v) = k+t$ for all $v \in V(G)$. Let x and y be two adjacent vertices of G. Since $\delta(G) \leq \deg_G(x)$, $sk \leq \delta(G)t$ and $\deg_G(y) \leq \deg_G(x) + s$, we have

$$\frac{g(x)}{\deg_G(x)} \leq \frac{k}{\delta(G)} \leq \frac{k+t}{\delta(G)+s} \leq \frac{f(y)}{\deg_G(y)}.$$

Therefore by Theorem 4.19, G has a (g,f)-factor, which is a $[k, k+t]$-factor.
$\square$

The following theorem strengthens the results of Theorem 4.19 and results in several strong corollaries.

Theorem 4.21 (Egawa and Kano [70] (1996)). *Let G be a connected general graph and let $g, f : V(G) \to \mathbb{Z}$ be functions such that $g(x) \leq \deg_G(x)$, $0 \leq f(x)$ and $g(x) \leq f(x)$ for all vertices x of G. If the following three conditions hold, then G has a (g,f)-factor.*
(i) G has at least one vertex u with $g(u) < f(u)$.
(ii) For any two adjacent vertices x and y of G, we have

$$\frac{g(x)}{\deg_G(x)} \leq \frac{f(y)}{\deg_G(y)}. \tag{4.31}$$

(iii) For every non-empty proper subset X of $V(G)$ such that $\langle X \rangle_G$ is connected and $g(x) = f(x)$ for all $x \in X$, we have

$$\sum_{v \in V(G)-X} \left(e_G(v, X) \min \left\{ \frac{f(v)}{\deg_G(v)}, 1 - \frac{g(v)}{\deg_G(v)} \right\} \right) \geq 1. \qquad (4.32)$$

Proof. We shall show that $\gamma(S, T) \geq 0$ as is required in the (g, f)-factor theorem. Let S and T be two disjoint subsets of $V(G)$. By condition (i), we have $\gamma(\emptyset, \emptyset) = 0$, and so we may assume $S \cup T \neq \emptyset$. Let $C_1, C_2, \ldots, C_m$ be the (g, f)-odd components of $G - (S \cup T)$, where $m = q^*(S, T)$. Then we have

$$\gamma(S, T) = \sum_{s \in S} f(s) + \sum_{t \in T}(\deg_G(t) - g(t)) - e_G(S, T) - q^*(S, T)$$

$$= \sum_{s \in S} \deg_G(s) \frac{f(s)}{\deg_G(s)} + \sum_{t \in T} \deg_G(t) \left(1 - \frac{g(t)}{\deg_G(t)} \right)$$

$$- e_G(S, T) - m.$$

For each vertex $s \in S$, add the weight $f(s)/\deg_G(s)$ to all the edges incident with s; for each vertex $t \in T$, add the weight $1 - g(t)/\deg_G(t)$ to all edges incident with t; and add the weight -1 to every edge joining S and T (see Fig. 4.8).

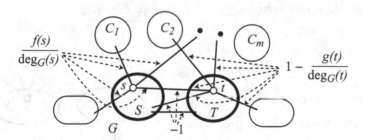

Fig. 4.8. The graph G with $S \cup T$, and the components $C_1, \ldots, C_m$, where $m = q^*(S, T)$.

Then $f(s)/\deg_G(s) \geq 0$, $1 - g(t)/\deg_G(t) \geq 0$, and $V(C_i) = X$ satisfies (4.32) and

$$\sum_{v \in V(G)-V(C_i)} e_G(v, C_i) = \sum_{v \in S \cup T} e_G(v, C_i).$$

Hence we have

$$\gamma(S,T) = \sum_{s\in S}\deg_G(s)\frac{f(s)}{\deg_G(s)} + \sum_{t\in T}\deg_G(t)\left(1-\frac{g(t)}{\deg_G(t)}\right)$$
$$-e_G(S,T) - m$$

$\geq$ (The total sum of weights of edges connecting S and T)

+ (The total sum of weights of edges connecting $S\cup T$

and $V(C_1)\cup\cdots\cup V(C_m)$)

$- m$

$$\geq \sum_{st\in E(G),\ s\in S,\ t\in T}\left(\frac{f(s)}{\deg_G(s)}+1-\frac{g(t)}{\deg_G(t)}-1\right)$$

$$+\sum_{i=1}^{m}\left(\sum_{v\in S\cup T}e_G(v,C_i)\min\left\{\frac{f(v)}{\deg_G(v)},1-\frac{g(v)}{\deg_G(v)}\right\}-1\right)$$

$\geq 0.$ (by (4.32))

Consequently, G has a (g,f)-factor.   □

Theorem 4.22 (Egawa and Kano [70]). *Let G be an r-regular simple graph and let $g,f : V(G) \to \mathbb{Z}$ be functions such that $g(x) < r$, $0 < f(x)$ and $g(x) \leq f(x)$ for all $x \in V(G)$. If the following two conditions hold, then G has a (g,f)-factor.*
(i) $g(x) \leq f(y)$ for all two adjacent vertices x and y; and
(ii) for every vertex u with $g(u) = f(u)$, there exists at least one vertex w that is adjacent to u and satisfies $g(w) < f(w)$ (see Fig. 4.9).

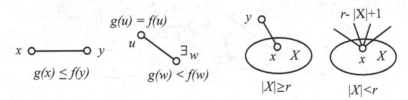

Fig. 4.9. The conditions in Theorem 4.21, and X in the proof.

Proof. We shall show that the three conditions of Theorem 4.21 hold. Since (i) and (ii) of Theorem 4.21 follows immediately, it suffices to show that condition (iii) holds. Let $\emptyset \neq X \subset V(G)$ such that $\langle X\rangle_G$ is connected and $g(x) = f(x)$ for all $x \in X$. By condition (ii), for every vertex $x \in X$ (see Fig. 4.9), there exists an edge xy such that $y \in V(G) - X$ and $g(y) < f(y)$. Hence if $|X| \geq r$, then

$$e_G(X,V(G) - X) \geq |X| \geq r.$$

If $|X| < r$, then since G is a simple graph, $x \in X$ is incident with at least $r - |X| + 1$ edges joining x to $V(G) - X$ (see Fig. 4.9), and thus we have

$$e_G(X, V(G) - X) \geq (r - |X| + 1)|X| \geq r.$$

Therefore, by

$$\frac{1}{r} \leq \frac{f(v)}{\deg_G(v)} \quad \text{and} \quad \frac{g(v)}{\deg_G(v)} \leq \frac{r-1}{r},$$

we obtain

$$\sum_{v \in V(G)-X} e_G(v, X) \min\left\{\frac{f(v)}{\deg_G(v)}, 1 - \frac{g(v)}{\deg_G(v)}\right\}$$

$$\geq e_G(X, V(G) - X) \min\left\{\frac{1}{r}, 1 - \frac{r-1}{r}\right\} \geq r \cdot \frac{1}{r} = 1.$$

Consequently G has a (g, f)-factor by Theorem 4.21. □

The following theorem is an improvement of Tutte's Theorem 4.16, which says that every r-regular graph has a $[k, k+1]$-factor. Namely, the following theorem guarantees the existence of a $[k, k+1]$-factor that has many vertices of degree k (or $k+1$) and a small number of vertices of degree $k+1$ (or k).

Theorem 4.23 (Egawa and Kano [70]). *Let k and r be integers such that $0 \leq k < r$, and G be an r-regular simple graph. Let W be a maximal independent vertex subset of G. Then G has a $[k, k+1]$-factor F such that $\deg_F(x) = k$ for all $x \in V(G) - W$, as well as a $[k, k+1]$-factor H such that $\deg_H(x) = k+1$ for all $x \in V(G) - W$.*

Proof. This theorem is an easy consequence of Theorem 4.22. For any vertex $u \in V(G) - W$, there exists at least one vertex w in W adjacent to u since otherwise $W \cup \{u\}$ becomes an independent set, which contradicts the maximality of W. Define g and f as

$$g(x) = f(x) = k \qquad \text{for all} \quad x \in V(G) - W; \quad \text{and}$$
$$g(y) = k \quad \text{and} \quad f(y) = k+1 \quad \text{for all} \quad y \in W.$$

Then g and f satisfy the conditions of Theorem 4.22, which implies that G has the desired $[k, k+1]$-factor F. We can similarly prove the existence of a $[k, k+1]$-factor H. □

We again consider $[a, b]$-graphs instead of regular graphs.

Theorem 4.24 (Kano and Saito [132]). *Let l, k, r, s, t, u be integers such that $0 \leq l \leq k \leq r$, $l + u \leq k + t \leq r + s$, $l \neq r$, $l + u \neq r + s$, $0 \leq s$, $0 \leq u$ and $1 \leq t$. Let G be a general $[r, r+s]$-graph and H be an $[l, l+u]$-factor of G. If*

$$(l - k)s + (k - r)u + (r - l)t \geq 0, \tag{4.33}$$

then G has a $[k, k+t]$-factor that contains the given factor H as a subgraph.

By setting $s = u = t$ in Theorem 4.24, we obtain the next theorem.

Theorem 4.25 ([132]). *Let l, k, r, t be integers such that $0 \le \ell < k < r$ and $1 \le t$. Then for any given $[\ell, \ell + t]$-factor H of a general $[r, r + t]$-graph G, G has a $[k, k + t]$-factor that contains H as a subgraph.*

Proof of Theorem 4.24. Let G be a general $[r, r+s]$-graph, H be an $[\ell, \ell+u]$-factor of G, and $K = G - E(H)$. Define two functions $g, f : V(K) \to \mathbb{Z}^+$ by

$$g(x) = k - \deg_H(x) \quad \text{and} \quad f(x) = k + t - \deg_H(x) \qquad \text{for all} \quad x \in V(G).$$

If K has a (g, f)-factor F, then $F \cup H$ is the desired $[k, k + t]$-factor. Let

$$\theta = \frac{k - \ell}{r - \ell} \quad \text{and} \quad \lambda = \frac{k + t - \ell - u}{r + s - \ell - u}.$$

Then $\theta \le \lambda$ by (4.33). We shall show that K, g and f satisfy condition (4.27) in Theorem 4.14. It is clear that $0 \le \theta \le 1$ and $g(x) < f(x)$. Since

$$k = \theta r + (1 - \theta)\ell \le \theta \deg_G(x) + (1 - \theta) \deg_H(x)$$

and

$$\deg_K(x) = \deg_G(x) - \deg_H(x),$$

we have

$$\begin{aligned} g(x) = k - \deg_H(x) &\le \theta \deg_G(x) + (1 - \theta) \deg_H(x) - \deg_H(x) \\ &= \theta(\deg_G(x) - \deg_H(x)) = \theta \deg_K(x). \end{aligned}$$

Similarly, since

$$k + t = \lambda(r + s) + (1 - \lambda)(\ell + u) \ge \lambda \deg_G(x) + (1 - \lambda) \deg_H(x),$$

we have

$$\begin{aligned} f(x) &= k + t - \deg_H(x) \\ &\ge \lambda \deg_G(x) + (1 - \lambda) \deg_H(x) - \deg_H(x) \\ &= \lambda(\deg_G(x) - \deg_H(x)) = \lambda \deg_K(x) \\ &\ge \theta \deg_K(x). \qquad \text{(by } \theta \le \lambda) \end{aligned}$$

Hence $g(x) \le \theta \deg_K(x) \le f(x)$. Consequently (4.27) holds, and the theorem is proved. $\square$

It was shown in Theorem 3.8 that for odd integers a, b, k such that $1 \le a < k < b$, if a graph G has both an a-factor and a b-factor, then G has a k-factor. The following theorem gives a similar result for $[a, b]$-factors. Note that condition (4.34) is equivalent to the condition that the point (m, n) in the plane lies above the straight line passing through the points (a, b) and (c, d). This will be explained after the proof.

Theorem 4.26 ([117]). *Let a, b, c, d, m, n, be integers such that $0 \le a < m < c$ and $a \le b$, $c \le d$ and $m+1 \le n$. Suppose that a general graph G has both an $[a,b]$-factor and a $[c,d]$-factor. If*

$$\frac{n-b}{m-a} \ge \frac{n-d}{m-c}, \tag{4.34}$$

then G has an $[m,n]$-factor.

Proof. Let S and T be disjoint subsets of $V(G)$ such that $S \cup T \ne \emptyset$. By the (g,f)-factor theorem with $g < f$ (Theorem 4.2), it suffices to show that $\gamma^*(S,T) \ge 0$. Assume that

$$\gamma^*(S,T) = n|S| + \sum_{x \in T} \deg_{G-S}(x) - m|T| < 0$$

for some $S, T \subset V(G)$, $S \cap T = \emptyset$. Since G has both an $[a,b]$-factor and a $[c,d]$-factor, and since $a \le b$ and $c \le d$, it follows from the (g,f)-factor theorem that

$$\gamma(S,T) = b|S| + \sum_{x \in T} \deg_{G-S}(x) - a|T| - q^*(S,T) \ge 0,$$
$$\gamma(S,T) = d|S| + \sum_{x \in T} \deg_{G-S}(x) - c|T| - q^*(S,T) \ge 0.$$

Hence

$$\gamma_1^*(S,T) = b|S| + \sum_{x \in T} \deg_{G-S}(x) - a|T| \ge 0,$$
$$\gamma_2^*(S,T) = d|S| + \sum_{x \in T} \deg_{G-S}(x) - c|T| \ge 0.$$

Therefore we obtain

$$0 > \gamma^*(S,T) - \gamma_1^*(S,T) \ge (n-b)|S| - (m-a)|T| \tag{4.35}$$
$$0 > \gamma^*(S,T) - \gamma_2^*(S,T) \ge (n-d)|S| - (m-c)|T|. \tag{4.36}$$

If $S = \emptyset$, then $0 > -(m-c)|T|$ by (4.36), which contradicts $m < c$. Thus $S \ne \emptyset$. By (4.35) and (4.36), we obtain

$$\frac{n-b}{m-a} < \frac{|T|}{|S|} \quad \text{and} \quad \frac{|T|}{|S|} < \frac{n-d}{m-c}.$$

Thus

$$\frac{n-b}{m-a} < \frac{n-d}{m-c}.$$

This contradicts condition (4.34). Consequently G has an $[m,n]$-factor. $\quad\square$

It is clear that the point (m, n) lies above the line passing through (a, b) and (c, d) if and only if (c, d) lies below the line passing through (a, b) and (m, n). The line passing through (a, b) and (m, n) is

$$y = \frac{n - b}{m - a}(x - m) + n.$$

Hence (c, d) lies below this line if and only if

$$d \leq \frac{n - b}{m - a}(c - m) + n.$$

This is equivalent to (4.34).

We now turn our attention to other types of sufficient conditions for a graph to have an $[a, b]$-factor. When we consider a general graph with a given binding number or given toughness, we may restrict ourselves to a simple graph because a general graph and its underlying simple graph have the same binding number and toughness.

Theorem 4.27 (Kano [119]). *Let a and b be integers such that $2 \leq a < b$, and G be a simple graph with $|G| \geq 6a + b$. Put $\lambda = 1 + (a - 1)/b$. Suppose*

$$|N_G(X)| \geq \lambda|X| \qquad if \quad |X| < \left\lfloor \frac{|G|}{\lambda} \right\rfloor; \quad and \qquad (4.37)$$

$$|N_G(X)| = V(G) \qquad otherwise. \qquad (4.38)$$

Then G has an $[a, b]$-factor.

We first note that a similar condition for a graph to have a $[1, n]$-factor with $n \geq 2$ was given in Theorem 4.13, which says that if $|N_G(X)| \geq |X|/n$ for all independent $X \subseteq V(G)$, then G has a $[1, n]$-factor. From this theorem, it follows that if $bind(G) \geq 1/n$, then G has a $[1, n]$-factor. On the other hand, if a connected simple graph G with order at least $4k - 6$ satisfies

$$bind(G) > \frac{(2k - 1)(|G| - 1)}{k(|G| - 2) + 3} = 2 - \frac{1}{k} + \epsilon, \quad \epsilon = \frac{2 - 7/k + 3/k^2}{|G| - 2 + 3/k},$$

then G has a k-regular factor (see Theorem 3.17). If we substitute $a = b = k$ into λ in Theorem 4.27, then $\lambda = 2 - 1/k$, which is almost the same as the lower bound of the above $bind(G)$, though Theorem 4.27 holds only in the case $a < b$. We give one more remark about Theorem 4.27. Conditions (4.37) and (4.38) cannot be replaced by the condition that $|N_G(X)| \geq \lambda|X|$ or $N_G(X) = V(G)$ for all $X \subset V(G)$.

The proof of Theorem 4.27 is fairly long and the proof technique used is similar to that of the next theorem, and so we shall prove only the latter.

Theorem 4.28 (Chen [46]). *Let a and b be integers such that $2 \leq a < b$, and G be a simple graph with $|G| \geq b + 3a$. Put $\lambda = 1 + (a-1)/b$. If*

$$bind(G) \geq \lambda \quad and \quad \delta(G) \geq 1 + \frac{(\lambda - 1)|G|}{\lambda}, \tag{4.39}$$

then G has an $[a, b]$-factor.

In the proof of the above theorem, we need Lemma 3.19, which says that if $bind(G) \geq \lambda$, then

$$|N_G(X)| \geq \frac{(\lambda - 1)|G| + |X|}{\lambda} \tag{4.40}$$

for every independent vertex subset X of G.

Proof of Theorem 4.28. Suppose that G does not have an $[a, b]$-factor. Then by Lemma 4.12, there exist disjoint subsets S and T of $V(G)$ such that $T \neq \emptyset$, $\deg_{G-S}(x) \leq a - 1$ for all $x \in T$, and

$$b|S| + \sum_{x \in T} \deg_{G-S}(x) - a|T| < 0. \tag{4.41}$$

Define

$$h = \min\{\deg_{G-S}(x) : x \in T\}.$$

Then $0 \leq h \leq a - 1$. We consider the following two cases.

Case 1. $h = 0$.

Let $I = Iso(G - S) \cap T = \{x \in T : \deg_{G-S}(x) = 0\}$. Then I is an independent vertex subset of G and $I \neq \emptyset$. By (4.40), we have

$$|N_G(I)| \geq \frac{(\lambda - 1)|G| + |I|}{\lambda}. \tag{4.42}$$

On the other hand, by (4.41) we have

$$
\begin{aligned}
0 &> b|S| + \sum_{x \in T} \deg_{G-S}(x) - a|T| \\
&\geq b|S| + |T - I| - a|T| \\
&= b|S| - a|I| + (1 - a)|T - I| \\
&\geq b|S| - a|I| + (1 - a)(|G| - |S| - |I|) \\
&= (a + b - 1)|S| - |I| - (a - 1)|G|.
\end{aligned}
$$

Thus

$$|N_G(I)| \leq |S| < \frac{|I| + (a-1)|G|}{a + b - 1} = \frac{|I| + (\lambda - 1)|G|}{\lambda}.$$

This contradicts (4.42).

Case 2. $1 \le h \le a - 1$.

Since

$$0 > b|S| + \sum_{x \in T} \deg_{G-S} - a|T| \ge b|S| + (h - a)|T|$$
$$\ge b|S| + (h - a)(|G| - |S|) = (a + b - h)|S| - (a - h)|G|,$$

we have

$$|S| < \frac{(a - h)|G|}{a + b - h}.$$

By considering a vertex $v \in T$ with $\deg_{G-S}(v) = h$, we have

$$\delta(G) \le \deg_G(v) \le h + |S| < h + \frac{(a - h)|G|}{a + b - h}.$$

If $h = 1$, then we get a contradiction by the assumption that

$$\delta(G) \ge 1 + \frac{(\lambda - 1)|G|}{\lambda} = 1 + \frac{(a - 1)|G|}{a + b - 1}.$$

Therefore we may assume $2 \le h \le a - 1$. Let

$$f(h) = h + \frac{(a - h)|G|}{a + b - h} = h + |G| - \frac{b|G|}{a + b - h}.$$

Then $f(h)$ takes its maximum value at $h = 2$ by noting that its derivative

$$f'(h) = 1 - \frac{b|G|}{(a + b - h)^2}$$
$$\le 1 - \frac{b(b + 3a)}{(a + b - 2)^2} \le 0$$

since $b + 3a \le |G|$ and $2 \le h$. Hence

$$1 + \frac{(a - 1)|G|}{a + b - 1} \le \delta(G) < f(h) \le f(2) = 2 + \frac{(a - 2)|G|}{a + b - 2}.$$

This implies $|G| < (a + b - 1)(a + b - 2)/b < b + 3a$, which contradicts $|G| \ge b + 3a$. Consequently the theorem is proved. $\square$

Theorem 4.29 (Katerinis [141]). *Let G be connected simple graph, and a and b be integers such that $1 \le a < b$. If*

$$tough(G) \ge a + \frac{a}{b} - 1, \tag{4.43}$$

then G has an $[a, b]$-factor.

Proof of Theorem 4.29 in the case $a = 1$. Let S be a non-empty subset of $V(G)$. By the definition of toughness, $tough(G) \geq 1/b$. Therefore, $\omega(G - S) \geq 2$ implies that

$$iso(G - S) \leq \omega(G - S) \leq \frac{|S|}{tough(G)} \leq b|S|.$$

If $iso(G - S) = 1$, then obviously $iso(G - S) = 1 \leq b|S|$. Hence by Theorem 4.13, G has a $[1,b]$-factor. Therefore the theorem holds when $a = 1$. $\square$

In order to prove the above theorem in the case $a \geq 2$, we need the following lemma. Recall that a vertex subset D of a graph G **covers** G if $N_G[D] = N_G(D) \cup D = V(G)$, that is, for any vertex v of $V(G) - D$, at least one vertex of D is adjacent to v. Note that the minimum cardinality of covering vertex subsets of G is called the **vertex covering number** of G, though we will not use it here.

Lemma 4.30 (Katerinis [141]). *Let $a \geq 2$ be an integer, G be a simple graph, and $V(G) = T_1 \cup T_2 \cup \cdots \cup T_{a-1}$ be a partition such that $\deg_G(x) \leq j$ for all $x \in T_j$ and every $1 \leq j \leq a - 1$, where some T_j's may be empty sets. Then there exist a covering vertex subset D and an independent vertex subset I of G such that $V(G) = D \cup I$ and*

$$\sum_{j=1}^{a-1}(a - j)|D \cap T_j| \leq \sum_{j=1}^{a-1} j(a - j)|I \cap T_j|. \tag{4.44}$$

Proof. We prove the lemma by induction on the order $|G|$. If $|G| = 1$, then $D = I = V(G)$ satisfies (4.44). Hence we assume $|G| \geq 2$. Let $m = \min\{j : T_j \neq \emptyset\}$. Choose a vertex $y \in T_m$, and let

$$G' = G - (\{y\} \cup N_G(y)) \qquad \text{(see Fig. 4.10)}.$$

Then by the induction hypothesis, there exist a covering vertex subset D' and an independent vertex subset I' of G' such that

$$\sum_{j=1}^{a-1}(a - j)|D' \cap T_j| \leq \sum_{j=1}^{a-1} j(a - j)|I' \cap T_j|. \tag{4.45}$$

Let $I = I' \cup \{y\}$ and $D = D' \cup N_G(y)$. Then

$$\sum_{j=1}^{a-1} j(a - j)|I \cap T_j|$$

$$= \sum_{j=1}^{a-1} j(a - j)|I' \cap T_j| + m(a - m) \qquad \text{(by } y \in T_m\text{)}$$

$$\geq \sum_{j=1}^{a-1}(a - j)|D' \cap T_j| + m(a - m). \qquad \text{(by (4.45))}$$

Since $\deg_G(y) \le m$ and $m = \min\{j : T_j \ne \emptyset\}$, we have

$$\sum_{j=1}^{a-1}(a-j)|N_G(y) \cap T_j| \le (a-m)|N_G(y)| \le (a-m)m,$$

and so

$$\sum_{j=1}^{a-1}(a-j)|D \cap T_j| \le \sum_{j=1}^{a-1}(a-j)|D' \cap T_j| + (a-m)m.$$

Therefore

$$\sum_{j=1}^{a-1}(a-j)|D \cap T_j| \le \sum_{j=1}^{a-1}j(a-j)|I \cap T_j|.$$

□

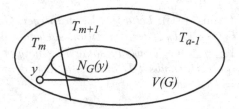

Fig. 4.10. The vertex y and $N_G(y)$ of G.

Lemma 4.31. *Let G be a connected simple graph that is not complete. Then $2 \cdot tough(G) \le \delta(G)$, in particular, if $tough(G) \ge 1$, then $tough(G)+1 \le \delta(G)$.*

Proof. Let v be a vertex of G with $\deg_G(v) = \delta(G)$. Since G is not a complete graph, we have $N_G(v) \ne V(G) - v$. Then $\omega(G - N_G(v)) \ge 2$, and so

$$2 \le \omega(G - N_G(v)) \le \frac{|N_G(v)|}{tough(G)} = \frac{\delta(G)}{tough(G)}.$$

Hence $2tough(G) \le \delta(G)$. The latter part is immediate by this inequality.   □

Proof of Theorem 4.29 in the case $a \ge 2$. By Lemma 4.31, a complete graph trivially has the desired $[a,b]$-factor since $\delta(G) \ge a+1$. Hence we may assume that G is not a complete graph. Let $\emptyset \ne S \subset V(G)$. By Theorem 4.11, it suffices to show that

$$\sum_{j=0}^{a-1}(a-j)n_j(G-S) \le b|S|, \tag{4.46}$$

where $n_j(G - S)$ denotes the number of vertices in $G - S$ with degree j. Let

$$T = \{x : x \in V(G) - S \quad \text{and} \quad 1 \leq \deg_{G-S}(x) \leq a - 1\},$$
$$T_j = \{x : x \in T \quad \text{and} \quad \deg_{G-S}(x) = j\} \quad (1 \leq j \leq a - 1), \quad \text{and}$$
$$H = \langle T \rangle_G.$$

Then $T_1 \cup T_2 \cup \cdots \cup T_{a-1}$ is a partition of $V(H)$ such that $\deg_H(x) \leq j$ for all $x \in T_j$ $(1 \leq j \leq a - 1)$. Note that we may assume $T \neq \emptyset$ since otherwise (4.46) holds by noting that $a \cdot n_0(G - S) \leq a \leq b|S|$ if $n_0(G - S) \leq 1$, and

$$a \cdot n_0(G - S) \leq a\omega(G - S) \leq \frac{a|S|}{tough(G)} \leq b|S| \quad \text{if} \quad n_0(G - S) \geq 2.$$

By Lemma 4.30, there exist a covering vertex subset D and an independent vertex subset I of H such that $V(H) = D \cup I$ and

$$\sum_{j=1}^{a-1}(a - j)|D \cap T_j| \leq \sum_{j=1}^{a-1} j(a - j)|I \cap T_j|. \tag{4.47}$$

It is obvious that $I \neq \emptyset$, and it follows that

$$\sum_{j=0}^{a-1}(a - j)n_j(G - S) = a \cdot n_0(G - S) + \sum_{j=1}^{a-1}(a - j)|T_j|$$

$$= a \cdot n_0(G - S) + \sum_{j=1}^{a-1}(a - j)(|I \cap T_j| + |D \cap T_j|)$$

$$\leq a \cdot n_0(G - S) + \sum_{j=1}^{a-1}(a - j)|I \cap T_j|$$

$$\quad + \sum_{j=1}^{a-1} j(a - j)|I \cap T_j| \qquad \text{(by (4.47))}$$

$$= a \cdot n_0(G - S) + \sum_{j=1}^{a-1}(a - j)(1 + j)|I \cap T_j|. \tag{4.48}$$

Let $X = S \cup N_{G-S}(I)$ (see Fig. 4.11). Then

$$|X| \leq |S| + \sum_{x \in I} \deg_{G-S}(x) = |S| + \sum_{j=1}^{a-1} j|I \cap T_j|, \tag{4.49}$$

and

$$\omega(G - X) \geq |I \cup Iso(G - S)| = \sum_{j=1}^{a-1}|I \cap T_j| + n_0(G - S). \tag{4.50}$$

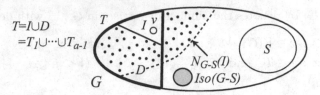

Fig. 4.11. $V(H) = T = I \cup D$, $I \cap D = \emptyset$ and $X = S \cup N_{G-s}(I)$.

If $\omega(G - X) \geq 2$, then we have

$$|X| \geq tough(G)\omega(G - X).$$

If $\omega(G - X) = 1$, then for a vertex $v \in I$, we have by Lemma 4.31 that

$$|X| \geq |S| + |N_{G-s}(v)| \geq \deg_G(v) \geq \delta(G) \geq tough(G)\omega(G - X).$$

Therefore in any case, by (4.49), (4.50) and since $|X| \geq tough(G)\omega(G - X)$, we obtain

$$|S| + \sum_{j=1}^{a-1} j|I \cap T_j| \geq |X| \geq \left(a + \frac{a}{b} - 1\right)\left(\sum_{j=1}^{a-1}|I \cap T_j| + n_0(G - S)\right).$$

Thus

$$b|S| \geq (ab + a - b)n_0(G - S) + \sum_{j=1}^{a-1}(ab + a - b - bj)|I \cap T_j|.$$

Hence in order to prove (4.46), by the above inequality and (4.48), it suffices to show that

$$a \cdot n_0(G - S) + \sum_{j=1}^{a-1}(a - j)(1 + j)|I \cap T_j|$$

$$\leq (ab + a - b)n_0(G - S) + \sum_{j=1}^{a-1}(ab + a - b - bj)|I \cap T_j|. \qquad (4.51)$$

It is clear that $a \leq ab + a - b$ as $2 \leq a < b$, and the desired inequality

$$(a - j)(1 + j) \leq ab + a - b - bj$$

follows from

$$(ab + a - b - bj) - (a - j)(1 + j) = (a - j - 1)(b - j) \geq 0$$

as $1 \leq j \leq a - 1$. Consequently (4.51) holds, and the proof is complete. □

Theorem 4.32 (Li and Cai [166]). *Let G be a connected simple graph, and a and b be integers such that $1 \le a < b$. Assume that $\delta(G) \ge a$, $|G| \ge 3a + b$ and*

$$\max\{\deg_G(x), \deg_G(y)\} \ge \frac{a|G|}{a+b} \tag{4.52}$$

for any two non-adjacent vertices x and y of G. Then G has an $[a, b]$-factor.

Proof. Suppose that G satisfies the conditions of the theorem, but has no $[a, b]$-factors. By Lemma 4.12, there exist disjoint subsets S and T of $V(G)$ such that $T \ne \emptyset$,

$$b|S| + \sum_{x \in T} \deg_{G-S}(x) - a|T| < 0, \quad \text{and} \tag{4.53}$$

$$0 \le \deg_{G-S}(x) \le a - 1 \quad \text{for all} \quad x \in T. \tag{4.54}$$

Define

$$h_1 = \min\{\deg_{G-S}(x) : x \in T\},$$

and choose a vertex $v_1 \in T$ such that $\deg_{G-S}(v_1) = h_1$. Then, if $T \ne N_T[v_1]$, where $N_T[v_1] = (N_G(v_1) \cap T) \cup \{v_1\}$, define

$$h_2 = \min\{\deg_{G-S}(x) : x \in T - N_T[v_1]\},$$

and choose a vertex $v_2 \in T - N_T[v_1]$ satisfying $\deg_{G-S}(v_2) = h_2$. Then

$$h_1 \le h_2 \le a - 1 \quad \text{and} \quad \deg_G(v_i) \le |S| + h_i \quad \text{for} \quad i = 1, 2.$$

If $T \ne N_T[v_1]$, then since v_1 and v_2 are not adjacent in G, we have

$$|S| + h_2 \ge \max\{\deg_G(v_1), \deg_G(v_2)\} \ge \frac{a|G|}{a+b}. \tag{4.55}$$

We consider the following two cases.

Case 1. $T = N_T[v_1]$.

In this case, we obtain

$$|T| = |N_T[v_1]| \le \deg_{G-S}(v_1) + 1 = h_1 + 1 \le a < b.$$

Then we have

$$\begin{aligned}
0 > b|S| + \sum_{x \in T} \deg_{G-S}(x) - a|T| \\
\ge b|S| + (h_1 - a)|T| \\
\ge b(a - h_1) + (h_1 - a)|T| \quad \text{(by } a \le \delta(G) \le \deg_G(v_1) \le |S| + h_1) \\
= (a - h_1)(b - |T|) \\
\ge 1. \quad \text{(by } |T| \le h_1 + 1 \le a < b)
\end{aligned}$$

This is a contradiction.

Case 2. $T \neq N_T[v_1]$.

Let $p = |N_T[v_1]|$. Then $1 \leq p < |T|$ and $p \leq h_1 + 1$. It follows that

$$(|G| - |S| - |T|)(a - h_2) \geq 0.$$

Moreover, we have

$$
\begin{aligned}
0 > b|S| + \sum_{x \in T} \deg_{G-S}(x) - a|T| \\
\geq b|S| + h_1 p + h_2(|T| - p) - a|T| \\
= b|S| + (h_1 - h_2)p + (h_2 - a)|T| \\
\geq b|S| + (h_1 - h_2)(h_1 + 1) + (h_2 - a)|T|. \quad (\text{by } h_1 \leq h_2)
\end{aligned}
$$

By the previous two inequalities, we obtain

$$(|G| - |S| - |T|)(a - h_2) > b|S| + (h_1 - h_2)(h_1 + 1) + (h_2 - a)|T|$$
$$(a - h_2)|G| - (a + b - h_2)|S| > (h_1 - h_2)(h_1 + 1). \quad (4.56)$$

By (4.55), we have $|S| - a|G|/(a + b) \geq -h_2$. Combining this inequality and $a + b - h_2 \geq 1$, we have

$$\left(|S| - \frac{a|G|}{a+b}\right)(a + b - h_2) \geq -h_2(a + b - h_2). \quad (4.57)$$

Since $|G| \geq 3a + b$, we have

$$\frac{b|G|}{a+b} \geq \frac{b(3a + b)}{a+b} \geq a + b,$$

and hence

$$h_2 \frac{b|G|}{a+b} \geq h_2(a + b). \quad (4.58)$$

If we add the three inequalities (4.56), (4.57) and (4.58), then the left hand-side becomes 0, and therefore

$$
\begin{aligned}
0 &> (h_1 - h_2)(h_1 + 1) - h_2(a + b - h_2) + h_2(a + b). \\
&= h_1^2 - (h_2 - 1)h_1 + h_2^2 - h_2.
\end{aligned}
$$

The right hand-side of the above inequality is a function of h_1, and takes its minimum value at $h_1 = (h_2 - 1)/2$. Substituting $h_1 = (h_2 - 1)/2$ into the function, we have

$$
\begin{aligned}
-1 &\geq h_1^2 - (h_2 - 1)h_1 + h_2^2 - h_2 \\
&\geq -\frac{1}{4}(h_2 - 1)^2 + h_2^2 - h_2 \\
&= \frac{1}{4}(3h_2^2 - 2h_2 - 1) = \frac{1}{4}\left(3\left(h_2 - \frac{1}{3}\right)^2 - \frac{2}{3}\right) > -1.
\end{aligned}
$$

This is a contradiction and consequently the theorem is proved. $\square$

We now present some other results on $[a, b]$-factors without proof.

Theorem 4.33 (Chen and Liu and [49]). *Let G be a connected simple graph, and a and b be integers such that $2 \le a < b$. If*

$$tough(G) \ge a + \frac{a}{b} - 1, \tag{4.59}$$

then for any edge e of G, G has an $[a, b]$-factor including e and another $[a, b]$-factor excluding e.

We say that a graph G is $(a, b; k)$-**critical** if for every vertex subset $X \subset G$ with $|X| = k$, $G - X$ has an $[a, b]$-factor.

Theorem 4.34 (Liu and Wang [171]). *Let G be a connected simple graph, and a, b and k be integers such that $1 \le a < b$ and $1 \le k$. Then G is $(a, b; k)$-critical if and only if*

$$\sum_{j=0}^{a-1}(a - j)n_j(G - S) \le b|S| - bk$$

for all subset $S \subset V(G)$ with $|S| \ge k$.

Corollary 4.35 (Liu and Wang [171]). *Let G be a connected simple graph, and let k, a, b, m, n be integers such that $1 \le a < b \le n$ and $1 \le k$. If*

$$\frac{a}{b} \le \frac{m - k}{n},$$

then every $[m, n]$-graph is $(a, b; k)$-critical.

Theorem 4.36 (Liu and Wang [171]). *Let G be a connected simple graph, and a, b and k be integers such that $1 \le a < b$, $m \le n$ and $1 \le k$. Then G is $(a, b; k)$-critical if and only if*

$$\sum_{j=0}^{a-1}(a - j)n_j(G - S) \le b|S| - bk$$

for all subsets $S \subset V(G)$ with $|S| \ge k$.

We conclude this section with the following theorem on a semi-regular factor with regular components of regular graphs. This result gives an additional property to Theorem 4.16, which says that every r-regular graph has a $[k - 1, k]$-factor. Notice that by Petersen's 2-factorable theorem, for every integer $1 \le k < 2r$, every $2r$-regular graph has a k-regular factor or a $(k - 1)$-regular factor, which is obviously a $[k - 1, k]$-factor with regular components. We may therefore restrict ourselves to odd regular graphs for such a $[k - 1, k]$-factor.

Theorem 4.37 (Kano [117]). *Let $r \geq 3$ be an odd integer and k an integer such that $1 \leq k \leq 2r/3$. Then every r-regular multigraph has a $[k-1,k]$-factor, each of whose components is regular.*

In order to prove the above theorem, we need the following theorem.

Theorem 4.38 (Kano [115]). *Let G be an n-edge connected multigraph ($n \geq 1$), θ be a real number such that $0 \leq \theta \leq 1$, and $f : V(G) \rightarrow \{0,2,4,6,\ldots\}$ be a function. If the following two conditions hold, then G has an f-factor.*
(i) $\epsilon = \sum_{x \in V(G)} |f(x) - \theta \deg_G(x)| < 2$.
(ii) $m(1-\theta) \geq 1$, where $m \in \{n, n+1\}$ and $m \equiv 1 \pmod 2$.

Proof. Let S and T be disjoint subsets of $V(G)$. By the f-factor theorem, it suffices to show that $\delta(S,T) \geq 0$. Since G is connected and $\sum_{x \in V(G)} f(x) \equiv 0 \pmod 2$, we have $\delta(\emptyset, \emptyset) = 0$. Thus we may assume that $S \cup T \neq \emptyset$. Let $\ell = q(S,T)$, and $C_1, C_2, \ldots, C_\ell$ be the f-odd components of $G - (S \cup T)$. Then for every f-odd component C_i, we have

$$\sum_{x \in V(C_i)} f(x) + e_G(C_i, T) \equiv e_G(C_i, T) \equiv 1 \pmod 2.$$

Hence $e_G(C_i, T) \geq 1$. If $e_G(S, C_i) = 0$, then $n \leq e_G(C_i, S \cup T) = e_G(C_i, T)$, and so $e_G(C_i, T) \geq m$, where m is defined in condition (ii). It follows that

$$\delta(S,T) = \sum_{x \in S} f(x) + \sum_{x \in T} \left(\deg_G(x) - f(x) \right) - e_G(S,T) - q(S,T)$$

$$= \theta \sum_{x \in S} \deg_G(x) + (1-\theta) \sum_{x \in T} \deg_G(x) - e_G(S,T) - \ell$$

$$+ \sum_{x \in S} \left(f(x) - \theta \deg_G(x) \right) - \sum_{x \in T} \left(f(x) - \theta \deg_G(x) \right)$$

$$\geq \theta \left(e_G(S,T) + \sum_{i=1}^{\ell} e_G(S,C_i) \right)$$

$$+ (1-\theta) \left(e_G(T,S) + \sum_{i=1}^{\ell} e_G(T,C_i) \right)$$

$$- e_G(S,T) - \ell - \sum_{x \in S \cup T} |f(x) - \theta \deg_G(x)|$$

$$\geq \sum_{i=1}^{\ell} \left(\theta e_G(S,C_i) + (1-\theta)(T,C_i)) - 1 \right) - \epsilon.$$

If $e_G(S,C_i) \geq 1$, then the following holds since $e_G(T,C_i) \geq 1$.

$$\theta e_G(S,C_i) + (1-\theta)e_G(T,C_i)) - 1 \geq \theta + (1-\theta) - 1 = 0.$$

If $e_G(S, C_i) = 0$, then $e_G(T, C_i)) \geq m$ as shown above, and so by condition (ii), we have

$$\theta e_G(S, C_i) + (1 - \theta)e_G(T, C_i)) - 1 \geq (1 - \theta)m - 1 \geq 0.$$

Therefore, $\delta(S, T) \geq -\epsilon > -2$ by (i), which implies $\delta(S, T) \geq 0$ since $\delta(S, T) \equiv 0 \pmod 2$ (see (3.4)). Therefore G has an f-factor. $\square$

For a set $\mathcal{I}$ of integers, an $\mathcal{I}$-**factor** of a graph G is a spanning subgraph F of G such that

$$\deg_F(x) \in \mathcal{I} \qquad \text{for all} \quad x \in V(G).$$

Analogously, an $\mathcal{I}$-**graph** can be defined as a graph G with $\deg_F(x) \in \mathcal{I}$ for every $x \in V(G)$.

Lemma 4.39. *Let $r \geq 3$ be an odd integer, and G be a 2-edge connected r-regular multigraph. Then for every even integer h, $2 \leq h \leq 2r/3$, G has a h-regular factor. Moreover, for every odd integer h', $r/3 \leq h' \leq r$, G has an h'-regular factor. In particular, for every integer $k, 1 \leq k \leq r$, G has a $[k-1, k]$-factor, each of whose components is regular.*

Proof. This lemma follows from (5) of Theorem 3.4, and can also be proved using Theorem 4.38. Define $\theta = h/r$ and $f(x) = h$ for all $x \in V(G)$. Then $\epsilon = 0$, $m = 3$ and

$$m(1 - \theta) = 3\left(1 - \frac{h}{r}\right) \geq 1 \qquad \text{as} \quad h \leq \frac{2r}{3}.$$

Hence G has an f-factor, which is the desired h-regular factor F. It is clear that $G - F$ is a $(r-h)$-regular factor, which implies the existence of the desired h'-regular factor of G. $\square$

Lemma 4.40. *Let $r \geq 3$ be an odd integer, and G be a 2-edge connected $[r-1, r]$-multigraph having exactly one vertex w of degree $r-1$. Then (i) for every even integer h, $2 \leq h \leq 2r/3$, G has an h-regular factor; and (ii) for every odd integer h', $r/3 \leq h' \leq r$, G has an $[h'-1, h']$-factor F such that $\deg_F(w) = h' - 1$ and $\deg_F(x) = h'$ for every $x \in V(G) - \{w\}$.*

Proof. We use Theorem 4.38. We first prove (i). Define $\theta = h/r$ and a function f by $f(x) = h$ for every $x \in V(G)$. Then

$$\epsilon = \sum_{x \in V(G)} |f(x) - \theta \deg_G(x)| = |f(w) - \theta \deg_G(w)|$$

$$= h - \theta(r - 1) = \frac{h}{r} < 1.$$

Moreover, since $m = 3$ and $h/r \leq 2/3$, we obtain

$$m(1 - \theta) = 3\left(1 - \frac{h}{r}\right) \geq 3\left(1 - \frac{2}{3}\right) = 1.$$

Hence G has an f-factor, which is the desired h-regular factor.

We next prove (ii). Let F be a h-regular factor obtained in (i). Then $G - F$ is the desired factor of G. □

Lemma 4.41. *Let $r \geq 3$ be an odd integer, and G be a connected r-regular multigraph having at least two bridges. Then for every integer k, $(r/3) + 1 \leq k \leq 2r/3$, G has a $[k - 1, k]$-factor, each of whose components is regular.*

Proof. We first prove that for every odd integer k, $r/3 \leq k \leq 2r/3$, G has a $[k - 1, k]$-factor with regular components, that is, we first show that the lemma holds for an odd integer k.

Since G has at least two bridges, for any bridge $vw \in E(G)$, $G - vw$ has at most one 2-edge connected component, where $v, w \in V(G)$. Let $v_1 w_1, v_2 w_2, \ldots, v_s w_s$ be the bridges of G such that $G - v_i w_i$ has exactly one 2-edge connected component D_i, where $s \geq 2$ and $w_i \in D_i$ (see Fig. 4.12). Note that it may occur that $v_i = v_j$ for some $i \neq j$, and we can choose an end-vertex w_i of a bridge $v_i w_i$ so that w_i is contained in D_i (see Fig. 4.12). Let

$$H = G - \bigcup_{i=1}^{s}(V(D_i) - \{w_i\}). \qquad \text{(see Fig. 4.12)}$$

Then H is a $\{1, r\}$-graph whose end-vertices are $w_1, w_2, \ldots, w_s$.

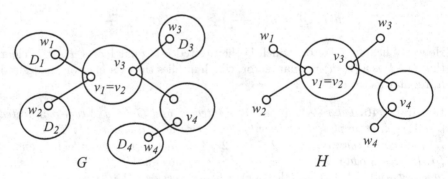

Fig. 4.12. An r-regular multigraph G having at least two bridges; and the subgraph H.

We now show that H has a $\{0, 1, k\}$-factor F such that

$$\deg_F(x) = k \qquad \text{if } x \in V(H) - \{w_1, \ldots, w_s\}; \text{ and}$$
$$\deg_F(w_i) \in \{0, 1\} \qquad \text{for every } 1 \leq i \leq s.$$

Define two functions g and f on $V(H)$ by

$$g(x) = \begin{cases} k & \text{if } x \in V(H) - \{w_1, \ldots, w_s\} \\ 0 & \text{otherwise,} \end{cases}$$

$$f(x) = \begin{cases} k & \text{if } x \in V(H) - \{w_1, \ldots, w_s\} \\ 1 & \text{otherwise.} \end{cases}$$

Then a (g, f)-factor is the desired factor F of H.

We apply the (g, f)-factor theorem. It suffices to show that $\gamma(S, T) \geq 0$ for all disjoint subsets S and T of $V(H)$. By the existence of w_1, we have $\gamma(\emptyset, \emptyset) = 0$. Suppose that $\gamma(S, T) < 0$ for some S and T. Choose such a pair (S, T) so that $|S \cup T|$ is minimum. Let $C_1, C_2, \ldots, C_m$ be the (g, f)-odd components of $H - (S \cup T)$, where $m = q^*(S, T)$. Then $g(x) = f(x) = k$ for all $x \in V(C_i)$ and

$$\sum_{x \in V(C_i)} f(x) + e_H(C_i, T)$$
$$\equiv |C_i| + e_H(C_i, T) \equiv 1 \pmod 2. \tag{4.60}$$

In particular, C_i contains no end-vertices w_j, $1 \leq j \leq s$.

Assume that S contains an end-vertex $w = w_j$. Let $v = v_j$ be the vertex adjacent to w in H. Then

$$\gamma(S - w, T) = \sum_{x \in S - w} f(x) + \sum_{x \in T} \left(\deg_H(x) - g(x) \right)$$
$$- e_H(S - w, T) - q^*(S - w, T)$$
$$= \gamma(S, T) - f(w) + e_H(w, T) + q^*(S, T) - q^*(S - w, T)$$
$$\leq \gamma(S, T) - 1 + e_H(w, T) + q^*(S, T) - q^*(S - w, T).$$

If $v \in T$, then $e_H(w, T) = 1$ and $q^*(S - w, T) = q^*(S, T)$. Thus $\gamma(S - w, T) = \gamma(S, T) < 0$, which contradicts the choice of (S, T). If $v \in S$, then $e_H(w, T) = 0$ and $q^*(S - w, T) = q^*(S, T)$, and so $\gamma(S - w, T) = \gamma(S, T) - 1 < 0$, a contradiction. If v is not contained in $S \cup T$, then $q^*(S, T) - q^*(S - w, T) \leq 1$, and so $\gamma(S - w, T) \leq \gamma(S, T) < 0$, a contradiction. Therefore S contains no w_i, $1 \leq i \leq s$.

Assume that T contains an end-vertex $w = w_j$. Then

$$\gamma(S, T - w) = \gamma(S, T) - 1 + e_H(S, w) + q^*(S, T) - q^*(S, T - w).$$

If $v = v_j \in T$, then $\gamma(S, T - w) = \gamma(S, T) - 1$, a contradiction. If $v \in S$, then $\gamma(S, T - w) = \gamma(S, T)$, a contradiction. If $v \notin S \cup T$, then $\gamma(S, T - w) \leq \gamma(S, T)$ since $q^*(S, T) - q^*(S, T - w) \leq 1$, which is again a contradiction. Hence T contains no w_i, $1 \leq i \leq s$.

We now consider a (g, f)-odd component C_i of $H - (S \cup T)$. It is clear that $e_H(C_i, S \cup T) \geq 1$. If $e_H(C_i, S \cup T) = 1$, then C_i contains at least one vertex in $\{w_1, w_2, \ldots, w_s\}$, which contradicts the fact that $g(x) = f(x)$ for all $x \in V(C_i)$. Hence

$$e_G(C_i, S \cup T) \geq 2. \tag{4.61}$$

Let $\theta = k/r$. Then $0 < \theta < 1$, and since $S \cup T$ contains no end-vertices w_j, we have

$$\gamma(S, T) = \sum_{x \in S} f(x) + \sum_{x \in T} \left(\deg_H(x) - g(x) \right) - e_H(S, T) - q^*(S, T)$$

$$= k|S| + (r - k)|T| - e_H(S, T) - m$$

$$= \theta \sum_{x \in S} \deg_H(x) + (1 - \theta) \sum_{x \in T} \deg_H(x) - e_H(S, T) - m$$

$$\geq \theta \left(e_H(S, T) + \sum_{i=1}^{m} e_H(S, C_i) \right)$$

$$+ (1 - \theta) \left(e_H(T, S) + \sum_{i=1}^{m} e_H(T, C_i) \right) - e_H(S, T) - m$$

$$\geq \sum_{i=1}^{m} \left(\theta e_H(S, C_i) + (1 - \theta) e_H(T, C_i) - 1 \right).$$

Hence it suffices to show that for every $C = C_i$,

$$\theta e_H(S, C) + (1 - \theta) e_H(T, C) - 1 \geq 0. \tag{4.62}$$

If $e_H(S, C) \geq 1$ and $e_H(T, C) \geq 1$, then (4.62) obviously holds. Since r is odd, we have

$$|C| \equiv r|C| = \sum_{x \in V(C)} \deg_H(x) = 2\|C\| + e_H(C, S \cup T) \quad (\text{mod } 2). \tag{4.63}$$

Assume $e_H(S, C) = 0$. Then by (4.63) we have $|C| \equiv e_H(C, T)$ (mod 2). But this contradicts (4.60). Next suppose $e_H(T, C) = 0$. Then by (4.63) and (4.60), we have

$$e_H(S, C) \equiv |C| \equiv 1 \quad (\text{mod } 2).$$

Thus $e_H(S, C) \geq 3$ by (4.61). Therefore

$$\theta e_H(S, C) - 1 \geq \frac{k}{r} \cdot 3 - 1 \geq 0. \quad (\text{by } k \geq r/3)$$

Consequently H has the desired factor F.

We are ready to construct the desired factor of G. Recall that k is odd and $r/3 \leq k \leq 2r/3$. If $\deg_F(w_i) = 0$, then by Lemma 4.40 we take a $(k - 1)$-regular factor $R(i)$ of D_i. If $\deg_F(w_i) = 1$, then by Lemma 4.40 we obtain a $[k-1, k]$-factor $R(i)$ of D_i such that $\deg_{R(i)}(w_i) = k - 1$ and $\deg_{R(i)}(x) = k$ for all $x \in V(D_i) - \{w_i\}$. Then the following set forms the desired $[k - 1, k]$-factor of G with regular components:

$$\bigcup_{i=1}^{s} R_i \cup F.$$

We next prove that for every even integer k, $(r/3) + 1 \leq k \leq 2r/3$, G has a $[k - 1, k]$-factor with regular components. Since $k - 1$ is an odd integer satisfying $r/3 \leq k - 1 \leq 2r/3$, by the preceding result, the graph H defined above has a $\{0, 1, k - 1\}$-factor F such that

$$\deg_F(x) = k - 1 \qquad \text{if } x \in V(H) - \{w_1, \ldots, w_s\}; \text{ and}$$
$$\deg_F(w_i) \in \{0, 1\} \qquad \text{for every } 1 \leq i \leq s.$$

If $\deg_F(w_i) = 0$, then by Lemma 4.40 we can take a k-regular factor $R(i)$ of D_i. If $\deg_F(w_i) = 1$, then by Lemma 4.39 we can take a $[k - 2, k - 1]$-factor R_i of D_i such that $\deg_{R(i)}(w_i) = k - 2$ and $\deg_{R(i)}(x) = k - 1$ for all $x \in V(D_i) - \{w_i\}$. Then $\cup_{i=1}^{s} R(i) \cup F$ is the desired $[k - 1, k]$-factor of G with regular components. Consequently the lemma is proved. $\square$

Proof of Theorem 4.37. We shall prove the theorem by induction on r. Let $r \geq 3$ be an odd integer and k be an integer such that $1 \leq k \leq 2r/3$. Let G be a connected r-regular multigraph. Since every regular graph has a 0-factor, which is a $[0, 1]$-factor with regular components, we may assume $k \geq 2$. By Lemma 4.39, we may assume that G is not 2-edge connected. Suppose that G has exactly one bridge vw. Then each component C of $G - vw$ is 2-edge connected, and so by Lemma 4.40, for an even integer $h \in \{k - 1, k\}$, C has an h-factor. Thus G itself has an h-factor, which is the desired $[k - 1, k]$-factor of G with regular components. Hence we may assume that G has at least two bridges.

We first show that if the theorem is true for $3 \leq r \leq 15$, which will be shown later, then the theorem holds for every odd integer r. Let $r \geq 17$ be an odd integer. Let $r = 2s + 1$, where $s \geq 8$. If $(r/3) + 1 \leq k \leq 2r/3$, then by Lemma 4.41, G has a $[k - 1, k]$-factor with regular components. So we may assume $2 \leq k < (r/3) + 1 = (2s + 4)/3$, which implies $k \leq (2s + 3)/3$. Let h be the greatest integer not exceeding $2r/3$. Then $(2r/3) - (2/3) = 4s/3 \leq h$. Since

$$\frac{r}{3} + 1 = \frac{2s + 4}{3} \leq h \leq \frac{2r}{3} < r,$$

by Lemma 4.41 G has a $[h - 1, h]$-factor F_1 with regular components. Since

$$k \leq \frac{2s + 3}{3} \leq \frac{2(4s - 3)}{9} \leq \frac{2(h - 1)}{3}, \qquad (\text{by } s \geq 8)$$

every odd-regular component C of F_1, where C is an $(h - 1)$- or h-regular graph, has a $[k - 1, k]$-factor with regular components by the induction hypothesis. On the other hand, every even-regular component of F_1 contains a regular k or $(k - 1)$-regular factor by Petersen's 2-factorable theorem. By combining these factors, we can obtain the desired $[k - 1, k]$-factor of G with regular components.

We finally show that the theorem holds for every odd integer $3 \leq r \leq 15$ by induction on r. We may assume that G is an r-regular multigraph having at least two bridges. We use Lemma 4.41 and the inductive hypothesis without mentioning them.

Case r = 3. Since $(r/3) + 1 \leq 2 \leq 2r/3$, G has a $[1,2]$-factor with regular components, and the theorem holds for $r = 3$.

Case r = 5. Since $(r/3) + 1 \leq 3 \leq 2r/3$, G has a $[2,3]$-factor $F(2,3)$ with regular components. Every 3-regular component of $F(2,3)$ has a $[1,2]$-factor with regular components. Therefore $F(2,3)$, and in particular G itself, has a $[1,2]$-factor with regular components. Thus the theorem holds for $r = 5$.

Case r = 7. Since $(r/3) + 1 \leq 4 \leq 2r/3$, G has a $[3,4]$-factor $F(3,4)$ with regular components. Every 4-regular component of $F(3,4)$ has a 2-factor. Therefore $F(3,4)$, and in particular G itself, has a $[2,3]$-factor $F(2,3)$ with regular components. Since $F(2,3)$ has a $[1,2]$-factor with regular components, the theorem holds.

Case r = 9. Since $(r/3)+1 \leq 4,5,6 \leq 2r/3$, G has a $[k-1,k]$-factor $F(k-1,k)$ with regular components for $k \in \{4,5,6\}$. Since $F(3,4)$ has a $[2,3]$-factor with regular components, G has such a factor. By the same argument as above, the theorem holds for $r = 9$.

Case r = 11. Since $(r/3) + 1 \leq 5,6,7 \leq 2r/3$, G has a $[k-1,k]$-factor $F(k-1,k)$ with regular components for $k \in \{5,6,7\}$. Every 7-regular component of $F(6,7)$ has a $[3,4]$-factor with regular components and every 6-regular component of $F(6,7)$ has a 4-regular factor. Therefore, $F(6,7)$, and in particular G itself, has a $[3,4]$-factor with regular components. Since $F(3,4)$ has a $[2,3]$-factor $F(2,3)$ with regular components and $F(2,3)$ has a $[1,2]$-factor with regular components, G has a $[k-1,k]$-factor with regular components for all $1 \leq k \leq 7$.

Case r = 13, 15. The proof of this case is left to the reader (Problem 4.5).
   Consequently the proof is complete.   □

The sharpness of the upper bound $k \leq 2r/3$ has not yet been established. It is known that if an odd integer r and an integer k satisfy $r+1-\sqrt{r+1} < k < r$, then there exists a simple connected r-regular graph that has no $[k-1,k]$-factor with regular components [117]. However, for other integers k, it is an open problem whether or not every r-regular graph has a $[k-1,k]$-factor with regular components.

Problems

4.1. Show that N in (4.6) can be defined as $N = \sum_{x \in V(G)} |g(x)| + |G|$. In order to show this, it suffices to prove that $\delta_{G^*}(S,T) \geq 0$ when $w \in S$ or $w \in T$.

4.2. Let G be a connected graph that is not complete. Show that if $tough(G) \geq$ 2, then $|G| \geq 6$ and $\delta(G) \geq 4$.

4.3. Let G be a simple graph and $n \geq 2$ be an integer.

(i) Prove that if $|N_G(X)| \geq |X|$ for all $X \subset V(G)$, then $iso(G - Y) \leq |Y|$ for all $Y \subset V(G)$.

(ii) Prove that the following two conditions are equivalent (see Theorem 4.13).

$$iso(G - S) \leq n|S| \qquad \text{for all} \quad S \subset V(G).$$

$$|N_G(S)| \geq \frac{|S|}{n} \qquad \text{for all independent} \quad S \subset V(G).$$

4.4. Prove that every factor-critical graph of order at least three contains an odd cycle.

4.5. Prove Case $r = 13, 15$ in the proof of Theorem 4.37

5

$[a, b]$-Factorizations

Let G be a graph, and $g, f : V(G) \to \mathbb{Z}$ be functions such that $g(x) \leq f(x)$ for all $x \in V(G)$. If the set of edges of G can be decomposed into disjoint subsets

$$E(G) = F_1 \cup F_2 \cup \cdots \cup F_n,$$

so that every F_i induces a (g, f)-factor of G, then we say that G is (g, f)-**factorable**, and the above decomposition is called a (g, f)-**factorization** of G. We often regard an edge set F of a graph as its spanning subgraph with edge set F. As a special case of (g, f)-factorization, we can define **1-factorization**, k-**regular factorization**, $[a, b]$-**factorization** and f-**factorization**. In this chapter, we mainly investigate $[a, b]$-factorizations of graphs. We begin with some basic results on factorizations of special graphs.

5.1 Factorizations of special graphs

A graph G is said to be **1-factorable** if $E(G)$ can be decomposed into disjoint 1-factors

$$E(G) = F_1 \cup F_2 \cup \cdots \cup F_n, \qquad \text{where every } F_i \text{ is a 1-factor of } G.$$

The next theorem shows the existence of 1-factorizations of complete graphs with even order. The readers who are interested in this topic should refer to a survey [188] by Mendelsohn and Rosa and the book [242] by Wallis.

Theorem 5.1. *Let $n \geq 2$ be an even integer. Then the complete graph K_n is 1-factorable.*

Proof. We may assume $n \geq 4$. Let $V(G) = \{v_0, v_1, \cdots, v_{n-2}\} \cup \{w\}$. For every integer $0 \leq i \leq n - 2$, let

$$F_i = \{v_{i-1}v_{i+1}, v_{i-2}v_{i+2}, \ldots, v_{i-(n-2)/2}v_{i+(n-2)/2}\} \cup \{v_i w\},$$

where the subscripts are expressed modulo $n - 1$ (see Fig. 5.1). Then we can obtain the desired 1-factorization $F_0 \cup F_1 \cup \cdots \cup F_{n-2}$ of K_n. $\square$

J. Akiyama and M. Kano, *Factors and Factorizations of Graphs*,
Lecture Notes in Mathematics 2031, DOI 10.1007/978-3-642-21919-1_5,
© Springer-Verlag Berlin Heidelberg 2011

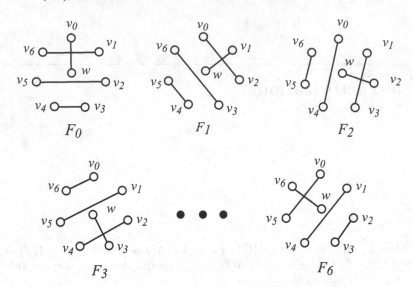

Fig. 5.1. A 1-factorization of K_8.

The following is an interesting unsolved conjecture on 1-factorization of a graph. One known result is given below, and another related result can be found in Theorem 8.40 by Egawa.

Conjecture 5.2 (Chetwynd and Hilton [52]). Let r and n be integers such that $1 \leq n \leq r$. Then every r-regular simple graph of order $2n$ is 1-factorable.

Theorem 5.3 (Chetwynd and Hilton [52], Niessen and Volkmann [198]). *If $r \geq (\sqrt{7} - 1)n \approx 1.647n$, then every r-regular simple graph of order $2n$ is 1-factorable.*

The following theorem was proved by König (see Theorem 2.2).

Theorem 5.4. *Every regular bipartite multigraph is 1-factorable.*

We rewrite the 2-factorable theorem (Theorem 3.1) as the following theorem so that we can use it more easily.

Theorem 5.5 (Petersen's 2-Factorable Theorem). *Let r and k_i be positive integers. Then every $2r$-regular general graph is 2-factorable. In particular, if $r = k_1 + k_2 + \ldots + k_m$, then every $2r$-regular general graph G can be decomposed into $F_1 \cup F_2 \cup \cdots \cup F_m$, where F_i is a $2k_i$-regular factor of G for every $1 \leq i \leq m$.*

It is known that the following theorem is equivalent to the Four Color Theorem. However, here we prove the theorem by using the Four Color Theorem (Appel, Haken and Koch [21]; Robertson, Sanders, Seymour and Thomas [213]).

Theorem 5.6. *Every 2-connected planar cubic simple graph is 1-factorable.*

Proof. Let G be a 2-connected planar cubic simple graph, which is drawn in the plane. Then the plane is partitioned into faces, sometimes called regions. Since G has no bridge, two faces having the same boundary are really distinct faces. By the Four Color Theorem, we can color all the faces with four colors so that any two faces having a boundary edge in common have different colors. We denote these four colors by $00, 01, 10, 11$. From this face-coloring, we can get an edge-coloring as follows. If an edge e of G is a boundary edge of two faces with colors x and y, then e is colored with $x+y$ (mod 2), where $x+y \in \{00, 01, 10, 11\}$ (see Fig. 5.2). We shall show that $x + y \in \{01, 10, 11\}$ and for each color $c \in \{01, 10, 11\}$, the set of edges colored with c forms a 1-factor of G.

Let v be any vertex of G. Then there exist three edges e_1, e_2, e_3 incident with v, and three faces around v, which are colored with three distinct colors, say r, s, t. Then $r + s, s + t, t + r$ are all distinct and not equal to 00. Hence e_1, e_2, e_3 are colored with distinct colors of $\{01, 10, 11\}$, that is, for every color $c \in \{01, 10, 11\}$, exactly one edge colored with c is incident with v. Therefore, each color induces a 1-factor of G. $\square$

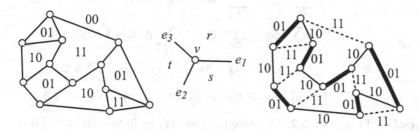

Fig. 5.2. A face-coloring of a 2-connected cubic planar simple graph and the edge-coloring induced by it.

The following theorem says that a trivial necessary condition for the existence of a (g, f)-factorization is sufficient for bipartite graphs. Thus the (g, f)-factorization problem is completely solved for bipartite graphs. Notice that an f-factorization of bipartite graphs was dealt with in Theorem 3.41.

Theorem 5.7 (Lovász [181] (p.50, Problem 7.11) (1979)). *Let $n \geq 1$ be an integer and G be a bipartite multigraph. If two functions $g, f : V(G) \to \mathbb{Z}$ satisfy*

$$g(x) \leq \frac{\deg_G(x)}{n} \leq f(x) \qquad for \ all \quad x \in V(G), \tag{5.1}$$

then G can be decomposed into n (g, f)-factors.

We define two functions $g', f' : V(G) \to \mathbb{Z}$ as

$$g'(x) = \left\lfloor \frac{\deg_G(x)}{n} \right\rfloor, \qquad f'(x) = \left\lceil \frac{\deg_G(x)}{n} \right\rceil,$$

for all $x \in V(G)$. Then g' and f' satisfy

$$g(x) \leq g'(x) \qquad \text{and} \qquad f'(x) \leq f(x)$$

for all $x \in V(G)$, and so a (g', f')-factor is a (g, f)-factor. Therefore, in order to prove Theorem 5.7, it is suffices to show that G can be decomposed into n (g', f')-factors.

Lemma 5.8. *Let G be a bipartite multigraph, and let θ be a real number such that $0 \leq \theta \leq 1$. If $g, f : V(G) \to \mathbb{Z}$ satisfy*

$$g(x) \leq \theta \deg_G(x) \leq f(x) \qquad \text{for all} \quad x \in V(G), \tag{5.2}$$

then G has a (g, f)-factor.

Proof. Let S and T be disjoint subsets of $V(G)$. Then

$$\begin{aligned}
\gamma^*(S, T) &= \sum_{x \in S} f(x) + \sum_{x \in T} \big(\deg_G(x) - g(x) \big) - e_G(S, T) \\
&\geq \theta \sum_{x \in S} \deg_G(x) + (1 - \theta) \sum_{x \in T} \deg_G(x) - e_G(S, T) \\
&\geq \theta e_G(S, T) + (1 - \theta) e_G(T, S) - e_G(S, T) = 0.
\end{aligned}$$

Hence $\gamma^*(S, T) \geq 0$, and thus G has a (g, f)-factor by Theorem 4.3. $\square$

Proof of Theorem 5.7 We prove the theorem by induction on n. For $n = 1$, the theorem holds since G itself is an h-factor and a (g, f)-factor. Hence we may assume $n \geq 2$.

As we mentioned above, it suffices to show that G is decomposed into n (g', f')-factors. It is obvious that

$$g'(x) \leq \frac{1}{n} \deg_G(x) \leq f'(x) \qquad \text{for all} \quad x \in V(G).$$

By Lemma 5.8, G has a (g', f')-factor F. We shall later show that

$$g'(x) \leq \frac{\deg_{G-F}(x)}{n - 1} \leq f'(x) \qquad \text{for all} \quad x \in V(G). \tag{5.3}$$

If (5.3) holds, then by induction $G - F$ can be decomposed into $n - 1$ (g', f')-factors, and therefore G is decomposed into n (g', f')-factors.

We now prove (5.3). Let

$$\deg_G(x) = kn + t, \qquad \text{where} \quad 0 \leq k \quad \text{and} \quad 0 \leq t < n.$$

If $t = 0$, then $g'(x) = f'(x) = k$ and so

$$(n-1)g'(x) = (n-1)k = \deg_{G-F}(x) = (n-1)f'(x).$$

If $t \geq 1$, then $g'(x) = k$ and $f'(x) = k+1$. If $\deg_F(x) = k$, then

$$(n-1)g'(x) \leq \deg_{G-F}(x) = (n-1)k + t \leq (n-1)f'(x).$$

If $\deg_F(x) = k+1$, then

$$(n-1)g'(x) \leq \deg_{G-F}(x) = (n-1)k + t - 1 \leq (n-1)f'(x).$$

Therefore (5.3) holds, and the proof is complete.   □

Theorem 5.7 can be proved by using Theorem 5.4, which is an easy consequence of the marriage theorem (Theorem 2.1). Namely, we can prove Theorem 5.7 without using (g, f)-factor theorem, and we explain this elementary and nice proof below.

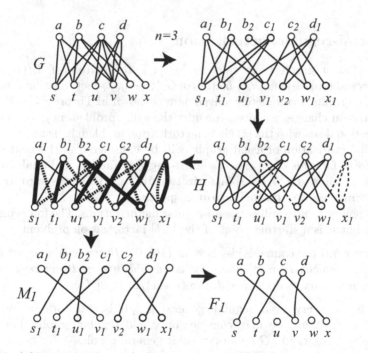

Fig. 5.3. A bipartite multigraph G, a new 3-regular bipartite multigraph H, and a (g, f)-factor F_1 of G obtained from a 1-factor M_1, where $n = 3$.

Another Proof of Theorem 5.7 (Lovász [181], Solution 7.11) The degree of any vertex v of G is expressed as

$$\deg_G(v) = kn + t, \qquad \text{where} \quad 0 \leq k \quad \text{and} \quad 0 \leq t < n,$$

where k and t depend on the vertex v. If $1 \leq t$, then we split v into $k + 1$ vertices $v_1, v_2, \ldots, v_{k+1}$ so that all vertices $v_1, v_2, \ldots, v_k$ have degree n and v_{k+1} has degree t; otherwise we split v into k vertices $v_1, v_2, \ldots, v_k$ so that all vertices v_i have degree n (Fig. 5.3). Then add some new vertices, if necessary, and some new edges between vertices with degree less than n so that the resulting graph H becomes an n-regular bipartite multigraph (Fig. 5.3).

By Theorem 5.4, H can be decomposed into n 1-factors

$$E(H) = M_1 \cup M_2 \cup \cdots \cup M_n, \qquad \text{where every } M_i \text{ is a 1-factor of } H.$$

Then for every vertex $v \in V(G)$, M_j contains at least k edges and at most $k + 1$ edges of G incident with v. Therefore $F_j = M_j \cap E(G)$ satisfies

$$k \leq \deg_{F_j}(v) \leq k + 1. \qquad \text{(see Fig. 5.3)}$$

Therefore F_j is a (g, f)-factor of G since $g(v) \leq k$ and $k + 1 \leq f(v)$. Consequently G is decomposed into n (g, f)-factors $F_1 \cup F_2 \cup \cdots \cup F_n$. $\square$

5.2 Semi-regular factorization

Recall that a graph G is said to be [a, b]-**factorable** if $E(G)$ can be decomposed into disjoint $[a, b]$-factors of G. The (g, f)-factorization problem in bipartite graphs is completely solved as was shown in Theorem 5.7. However, the situation changes when we consider the same problem in general graphs. This section deals with the $[k, k+1]$-factorization problem in usual graphs, and an $[a, b]$-factorization of usual graphs will be considered in the next section. A $[k, k + 1]$-factor is often called a **semi-regular factor**, and analogously an $[r, r+1]$-graph and a $[k, k+1]$-factorization are called a **semi-regular graph** and a **semi-regular factorization**, respectively.

We review some results on semi-regular factorizations. The following result is easy, but it is a starting point of the $[a, b]$-factorization problem.

Theorem 5.9 (Akiyama, Avis, Era [3] (1980)). *Every regular general graph is* [1, 2]-*factorable. In particular, if r is an odd integer, then every r-regular general graph can be decomposed into $(r + 1)/2$ [1, 2]-factors.*

Proof. Since every even regular general graph is 2-factorable by the 2-factorable theorem, we may consider only odd regular graphs. Let $r \geq 3$ be an odd integer, and G be an r-regular general graph.

Let G' be a copy of G. We construct a new $(r + 1)$-regular general graph H from $G \cup G'$ by joining each pair of corresponding vertices $v \in V(G)$ and $v' \in V(G')$ by a new edge (Fig. 5.4 (a)). Since $r + 1$ is even, H can be decomposed into $m = (r + 1)/2$ 2-factors $F_1, F_2, \ldots, F_m$ by the 2-factorable theorem. Then G is decomposed into m [1, 2]-factors

$$(G \cap F_1) \cup (G \cap F_2) \cup \cdots \cup (F_m \cap G). \qquad \square$$

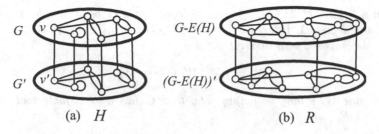

Fig. 5.4. (a) An r-regular graph G and an $(r + 1)$-regular graph H containing G as an induced subgraph; (b) An $[r, r + 1]$-graph $G - E(H)$ and an $(r + 1)$-regular graph R containing $G - E(H)$ as an induced subgraph.

From Theorem 5.9, Akiyama conjectured that for every integer $k \geq 1$, there exists an integer $\Phi(k)$ such that for every integer $r \geq \Phi(k)$, every r-regular graph is $[k, k + 1]$-factorable. This conjecture was proved by Era [81], [82] in 1985 and the sharp bound for $\Phi(k)$ was obtained by Egawa [64] in the following year. These results are given in the next theorem.

Theorem 5.10 (Era [82], [81]; Egawa [64]). *Let $k \geq 2$ be an integer. Then*
(i) if k is even and $r \geq k^2$, then every r-regular simple graph is $[k, k + 1]$-factorable;
(ii) if k is odd and $r \geq k^2 + 1$, then every r-regular simple graph is $[k, k + 1]$-factorable.
(iii) these bounds on r are sharp.

In order to prove this theorem we need some results on edge-coloring. Recall that a **proper edge-coloring** of a graph G is a coloring of the edges of G so that no two adjacent edges have the same color. The smallest number of colors by which G can be properly edge-colored is called the **chromatic index** (or **edge chromatic number**) of G and denoted by $\chi'(G)$. It is clear that for each color c of a proper edge-coloring of G, the set of edges colored with c forms a matching of G. Thus the chromatic index $\chi'(G)$ is the smallest integer k for which $E(G)$ can be decomposed into k disjoint matchings. Furthermore, a matching is simply a $[0, 1]$-factor. The famous Vizing's Theorem 1.11 says that the chromatic index of a simple graph with maximum degree Δ satisfies

$$\Delta \leq \chi'(G) \leq \Delta(G) + 1. \tag{5.4}$$

He also proved the next theorem.

Theorem 5.11 (Vizing [233]). *Let G be a simple graph with maximum degree Δ. If the set*
$$W = \{x \in V(G) : \deg_G(x) = \Delta\}$$
is independent, then $\chi'(G) = \Delta$. In particular, G can be decomposed into Δ $[0, 1]$-factors.

Lemma 5.12. *Let* $2 \leq \Delta \leq q$ *be integers, and* G *be a simple graph with maximum degree* Δ. *If the set* $W = \{x \in V(G) : \deg_G(x) = \Delta\}$ *is independent and is decomposed into disjoint subsets*

$$W = W_1 \cup W_2 \cup \cdots \cup W_q,$$

where some W_i'*s may be empty sets, then* G *has a* $[0, 1]$-*factorization* $F_1 \cup F_2 \cup \cdots \cup F_q$ *such that*

$$\text{every vertex of } W_j \text{ has degree one in } F_j \text{ for all } 1 \leq j \leq q. \qquad (5.5)$$

Proof. If $W_{\Delta+1} \cup \cdots \cup W_q = \emptyset$, then the lemma follows from Theorem 5.11 since for a $[0, 1]$-factorization $F_1 \cup \cdots \cup F_\Delta$ of G, every vertex of W_j has degree one in all F_i $(1 \leq i \leq \Delta)$. Thus we may assume that $W_{\Delta+1} \cup \cdots \cup W_q \neq \emptyset$. Let $\emptyset \neq X \subseteq W_{\Delta+1} \cup \cdots \cup W_q$. Then $N_G(X) \subseteq V(G) - W$ since W is independent. Since every vertex of $N_G(X)$ has degree at most $\Delta - 1$, we have

$$|X|\Delta = e_G(X, N_G(X)) \leq |N_G(X)|(\Delta - 1).$$

Hence $|X| \leq |N_G(X)|$. Since X is an arbitrary subset of $W_{\Delta+1} \cup \cdots \cup W_q$, by the marriage theorem (Theorem 2.1), there exists an injection

$$h : W_{\Delta+1} \cup \cdots \cup W_q \rightarrow V(G) - W \qquad \text{such that} \quad h(x) \in N_G(x).$$

So $M = \{xh(x) : x \in W_{\Delta+1} \cup \cdots \cup W_q\}$ forms a matching of G.

Let $H = G - M$. Then H satisfies the condition of Lemma 5.12 since

$$W_H = \{x \in V(H) : \deg_H(v) = \Delta\} = W_1 \cup \cdots \cup W_\Delta \subset W$$

is independent or empty. Hence H has a $[0, 1]$-factorization $F_1 \cup F_2 \cup \cdots \cup F_\Delta$ such that every j $(1 \leq j \leq \Delta)$ satisfies (5.5). Note that if W_H is empty, H can be decomposed into Δ $[0, 1]$-factors by (5.4), where some $[0, 1]$-factors may be 0-factors. For every $\Delta < i \leq q$, let us define $F_i = \{xh(x) : x \in W_i\}$. Then $F_1 \cup F_2 \cup \cdots \cup F_q$ is the desired $[0, 1]$-factorization of G. $\square$

Lemma 5.13. *Let* $r \geq 1$ *be an integer, and* G *be a connected* $2r$-*regular general graph. If* G *has even size, then* G *can be decomposed into two* r-*regular factors. If* G *has odd size, then for any vertex* u *of* G, G *can be decomposed into an* $[r - 1, r]$-*factor having exactly one vertex* u *of degree* $r - 1$ *and an* $[r, r + 1]$-*factor having exactly one vertex* u *of degree* $r + 1$.

Proof. Since every vertex of G has even degree, G has an Euler circuit C by Theorem 1.12. We traverse all the edges of G along C and alternately color the edges red and blue. Then if G is of even size, the set of red edges and that of blue edges form r-regular factors, respectively. If G is of odd size, then we start with the vertex u, and alternately color the edges red and blue. Consequently, the set of blue edges and that of red edges form the desired two factors. $\square$

Proof of Theorem 5.10 We first prove statement (i). Assume that $k \geq 2$ is an even integer. Since $r \geq k^2$, we can write

$$r = kq + s, \qquad \text{where} \quad 0 \leq s < k \leq q.$$

By Theorem 4.16, G has an $[s, s+1]$-factor. Choose such a minimal $[s, s+1]$-factor H. Then

$$W = \{v \in V(H) \;:\; \deg_H(v) = s+1\}$$

is empty or independent in H since H has no edge joining two vertices with degree $s+1$. Since $G - E(H)$ is a $[kq - 1, kq]$-graph, two copies of $G - E(H)$ can be embedded in a kq-regular graph R, which is constructed from the two copies by adding a new edge between every pair of corresponding vertices with degree $kq - 1$ (see Fig. 5.4 (b)). Since k is even, by applying the 2-factorable theorem (Theorem 5.5) to R, we obtain a k-regular factorization of R, and so we get a $[k - 1, k]$-factorization of $G - E(H)$

$$G - E(H) = A_1 \cup A_2 \cup \cdots \cup A_q, \quad \text{where all} \ A_i \ \text{are} \ [k-1, k]\text{-factors.}$$

Let

$$W_i = \{x \in V(G) \;:\; \deg_{A_i}(x) = k - 1\} \qquad \text{for} \quad 1 \leq i \leq q. \qquad (5.6)$$

Since every $v \in W$ has degree $qk - 1$ in $G - E(H)$, v has degree $k - 1$ in exactly one A_j. Hence W is decomposed into disjoint subsets

$$W = W_1 \cup W_2 \cup \cdots \cup W_q,$$

where W_i is define by (5.6). Since $s + 1 \leq k \leq q$, by Lemma 5.12, H has a $[0, 1]$-factorization $F_1 \cup F_2 \cup \cdots \cup F_q$ such that $\deg_{F_j}(x) = 1$ for all $x \in W_j$ for every $1 \leq j \leq q$. Then

$$(A_1 \cup F_1) \cup (A_2 \cup F_2) \cup \cdots \cup (A_q \cup F_q)$$

is the desired $[k, k+1]$-factorization of G.

Next we prove statement (ii). Assume that $k \geq 3$ is an odd integer. Here we consider only the case when r can be expressed as $r = k^2 + 2s + 1$, where s is an integer such that $0 \leq s \leq (k - 1)/2$. The proofs of the other cases are similar but fairly long, and so they are omitted here. Note that r can be expressed as

$$r = 2k \left(\frac{k-1}{2} - s \right) + 2(k+1)s + k + 1.$$

Since r is even, by the 2-factorable theorem (Theorem 5.5), G can be decomposed into

$$F_1 \cup \cdots \cup F_n \cup F_{n+1} \cup \cdots \cup F_{n+s} \cup F, \qquad n = \frac{k-1}{2} - s,$$

where F_i $(1 \leq i \leq n)$ is a $2k$-regular factor, F_j $(n + 1 \leq j \leq n + s)$ is a $2(k+1)$-regular factor and F is a $(k+1)$-regular factor. Since $k + 1$ is even, every F_j $(n+1 \leq j \leq n+s)$ can be decomposed into two $(k+1)$-regular factors by Lemma 5.13. Thus we shall decompose $F_1 \cup \cdots \cup F_n \cup F$ into $[k, k+1]$-factors of G.

Let $H_i(1), H_i(2), \ldots, H_i(\ell_i)$ be the components of F_i, $1 \leq i \leq n$. Then $|V(H_i(t))| \geq 2k+1$ since G is a simple graph. Set

$$I = \{(i, t) : 1 \leq i \leq n, \ 1 \leq t \leq \ell_i\}.$$

By Theorem 5.9, F has a $[1, 2]$-factor. Choose a minimal $[1, 2]$-factor Q of F. Then each component of Q is a path of order two or three. Let $Q_1, Q_2, \ldots, Q_m$ be the components of Q. For every $(i, t) \in I$, let

$$\Phi(i, t) = \{j : V(H_i(t)) \cap V(Q_j) \neq \emptyset\} \subseteq \{1, 2, \ldots, m\}.$$

Then for any $X \subseteq I$, we have

$$n \cdot \left| \bigcup_{(i,t) \in X} \Phi(i, t) \right|$$

$$\geq \#\{((i, t), j) : (i, t) \in X, \ j \in \Phi(i, t)\}$$

$$\geq |X|(2k+1)/3$$

since for each j, there are at most n (i, t)'s (i.e., $F_1, \ldots, F_n$), and for each (i, t) of X, there are at least $(2k+1)/3$ j's as $|V(H_i(t))| \geq 2k+1$. Then $\left| \bigcup_{(i,t) \in X} \Phi(i, t) \right| \geq |X|$ by $n \leq (k-1)/2 \leq (2k+1)/3$. Hence by the marriage theorem, there exists an injection

$$\phi : I \ \rightarrow \ \{1, 2, \ldots, m\} \qquad \text{such that} \quad \phi(i, t) \in \Phi(i, t).$$

Let

$$J = \{(i, t) \in I : H_i(t) \text{ has odd size}\}.$$

For every $(i, t) \in J$, choose a vertex

$$v(i, t) \in V(H_i(t)) \cap V(Q_{\phi(i,t)}).$$

By Lemma 5.13, if $H_i(t)$ has even size, then $H_i(t)$ is decomposed into two k-regular factors $A_i(t)$ and $B_i(t)$; if $H_i(t)$ has odd size, then $H_i(t)$ is decomposed into a $[k-1, k]$-factor $A_i(t)$ and a $[k, k+1]$-factor $B_i(t)$ such that only one vertex $v(i, t)$ has degree $k-1$ in $A_i(t)$ and $k+1$ in $B_i(t)$. For every $(i, t) \in J$, choose an edge $e(i, t)$ of $Q_{\phi(i,t)}$ that is incident with $v(i, t)$. Then

$$\bigcup_{t=1}^{m} A_i(t) \ \cup \ \{e(i, t) : (i, t) \in J\} \qquad \text{and} \qquad \bigcup_{t=1}^{m} B_i(t)$$

are $[k, k+1]$-factors of G. Since $F - \{e(i, t) : (i, t) \in J\}$ is a $[k, k+1]$-factor of G, $F_1 \cup \cdots \cup F_n \cup F$ is decomposed into $[k, k+1]$-factors of G. Consequently the proof of (ii) is complete.

The proof of (iii) is found in Era [81]. □

It is plausible that not only regular graphs but also some other graphs might be semi-regular factorable. Indeed, we have the following theorems, which will be proved in the next section.

Theorem 5.14 (Akiyama and Kano [8] (1985)). *Let $k \geq 2$ be an even integer, and $s \geq 0$ and $t \geq 1$ be integers. Then every $[(6k + 2)t + ks, (6k + 4)t + ks]$-multigraph is $[k, k + 1]$-factorable.*

Theorem 5.15 (Cai [38] (1991)). *Let $k \geq 1$ be an odd integer, and $s \geq 0$ and $t \geq 1$ be integers. Then every $[(6k+2)t+(k+1)s, (6k+4)t+(k+1)s]$-multigraph is $[k, k + 1]$-factorable.*

5.3 [a, b]-factorizations of graphs

We begin with the following theorem, which will play an important role in this section and contains some known results as corollaries.

Theorem 5.16 (Kano [115]). *Let G be an n-edge connected general graph $(n \geq 1)$, θ be a real number such that $0 \leq \theta \leq 1$, and $g, f : V(G) \to \mathbb{Z}$ such that $g(x) \leq f(x)$ for all $x \in V(G)$. If one of $\{(ia),(ib)\}$, (ii) and one of $\{(iiia), (iiib), (iiic), (iiid), (iiie), (iiif)\}$ hold, then G has a (g, f)-factor.*

(ia) $g(x) \leq \theta \deg_G(x) \leq f(x)$ *for all* $x \in V(G)$.
(ib) $\epsilon = \sum_{x \in V(G)} \left(\max\{0, g(x) - \theta \deg_G(x)\} + \max\{0, \theta \deg_G(x) - f(x)\} \right) < 1$.

(ii) *G has at least one vertex v such that $g(v) < f(v)$; or $g(x) = f(x)$ for all $x \in V(G)$ and $\sum_{x \in V(G)} f(x) \equiv 0 \pmod 2$.*

(iiia) $n\theta \geq 1$ *and* $n(1 - \theta) \geq 1$.
(iiib) *Both $\{\deg_G(x) | g(x) = f(x), x \in V(G)\}$ and $\{f(x) | g(x) = f(x), x \in V(G)\}$ consist of even numbers.*
(iiic) *$\{\deg_G(x) | g(x) = f(x), x \in V(G)\}$ consists of even numbers, n is odd, $(n + 1)\theta \geq 1$ and $(n + 1)(1 - \theta) \geq 1$.*
(iiid) *$\{f(x) | g(x) = f(x), x \in V(G)\}$ consists of even numbers and $m(1 - \theta) \geq 1$, where $m \in \{n, n + 1\}$ and $m \equiv 1 \pmod 2$.*
(iiie) *Both $\{\deg_G(x) | g(x) = f(x), x \in V(G)\}$ and $\{f(x) | g(x) = f(x), x \in V(G)\}$ consist of odd numbers, and $m\theta \geq 1$, where $m \in \{n, n+1\}$ and $m \equiv 1 \pmod 2$.*
(iiif) $g(x) < f(x)$ *for all* $x \in V(G)$.

Proof. Let S and T be disjoint subsets of $V(G)$. By the (g, f)-factor theorem, it suffices to show that $\gamma(S, T) \geq 0$. By (ii), it follows that $\gamma(\emptyset, \emptyset) = 0$, and so we may assume $S \cup T \neq \emptyset$. Since (ia) implies (ib) with $\epsilon = 0$, we may assume that (ib), (ii) and one of (iiia) - (iiif) hold. Let $C_1, C_2, \ldots, C_m$ be the (g, f)-odd components of $G - (S \cup T)$, where $m = q^*(S, T)$. Then

$$\gamma(S, T) = \sum_{x \in S} f(x) + \sum_{s \in T} \left(\deg_G(x) - g(x) \right) - e_G(S, T) - q^*(S, T)$$

$$\geq \theta \sum_{x \in S} \deg_G(x) + (1 - \theta) \sum_{x \in T} \deg_G(x) - e_G(S, T) - m$$

$$- \sum_{x \in S} \left(\theta \deg_G(x) - f(x) \right) - \sum_{x \in T} \left(g(x) - \theta \deg_G(x) \right)$$

$$\geq \theta \left(e_G(S, T) + \sum_{i=1}^{m} e_G(S, C_i) \right)$$

$$+ (1 - \theta) \left(e_G(T, S) + \sum_{i=1}^{m} e_G(T, C_i) \right) - e_G(S, T) - m - \epsilon$$

$$= \sum_{i=1}^{m} \left(\theta e_G(S, C_i) + (1 - \theta) e_G(T, C_i) - 1 \right) - \epsilon.$$

Since $\gamma(S, T)$ is an integer and $\epsilon < 1$, in order to prove $\gamma(S, T) \geq 0$ it suffices to show that every $C = C_i$ satisfies

$$\theta e_G(S, C) + (1 - \theta) e_G(T, C) - 1 \geq 0. \tag{5.7}$$

It is clear that if $e_G(S, C) \geq 1$ and $e_G(T, C) \geq 1$, then (5.7) follows. Hence we may assume that

$$e_G(S, C) = 0 \quad \text{or} \quad e_G(T, C) = 0. \tag{5.8}$$

Since C is a (g, f)-odd component of $G - (S \cup T)$ and G is n-edge connected, it follows that $g(x) = f(x)$ for all $x \in V(C)$,

$$\sum_{x \in V(C)} f(x) + e_G(C, T) \equiv 1 \pmod{2}, \quad \text{and} \quad e_G(S \cup T, C) \geq n. \tag{5.9}$$

Moreover, we have

$$\sum_{x \in V(C)} \deg_G(x) = 2\|C\| + e_G(C, S \cup T) \pmod{2}. \tag{5.10}$$

Assume that (iiia) holds. Then by (5.8) and (5.9), we have either $e_G(T, C) = 0$ and $e_G(S, C) \geq n$ or $e_G(S, C) = 0$ and $e_G(T, C) \geq n$, and thus

$$\theta e_G(S, C) - 1 \geq \theta n - 1 \geq 0 \quad \text{or} \quad (1 - \theta) e_G(T, C) \geq (1 - \theta) n - 1 \geq 0.$$

Hence (5.7) follows.

Assume (iiib) holds. By (5.9), $e_G(C, T) \equiv 1 \pmod{2}$. Hence $e_G(C, T) \geq 1$, and in particular we may assume $e_G(S, C) = 0$ by (5.8). So, by (5.10) we have

$e_G(C, T) \equiv 0 \pmod 2$, which contradicts $e_G(C, T) \equiv 1 \pmod 2$. Thus (5.7) follows.

Assume (iiic) holds. By (5.10), we have $e_G(C, S \cup T) \equiv 0 \pmod 2$, which implies $e_G(C, S \cup T) \geq n+1$ since n is odd. So if $e_G(C, T) = 0$, then $\theta e_G(S, C) - 1 \geq \theta(n+1) - 1 \geq 0$. If $e_G(C, S) = 0$, then $(1 - \theta)e_G(S, T) - 1 \geq (1 - \theta)(n + 1) - 1 \geq 0$. Hence (5.7) follows.

Assume (iiid) holds. By (5.9), we have $e_G(C, T) \equiv 1 \pmod 2$, which implies that we may assume $e_G(C, S) = 0$. Then $e_G(T, C) = e_G(S \cup T, C) \geq n$ and so $e_G(T, C) \geq m$, where m is defined in (iiid). Hence $(1-\theta)e_G(T, C) - 1 \geq (1 - \theta)m - 1 \geq 0$. Therefore (5.7) follows.

Assume (iiie) holds. By (5.9), $|C| + e_G(C, T) \equiv 1 \pmod 2$. By (5.10), we have $|C| \equiv e_G(C, S \cup T) \pmod 2$. Thus we have a contradiction when $e_G(S, C) = 0$. Hence $e_G(T, C) = 0$. Then $e_G(S, C) \equiv |C| \equiv 1 \pmod 2$. Therefore $\theta e_G(S, C) - 1 \geq \theta m - 1 \geq 0$, and (5.7) follows.

Assume (iiif) holds. In this case $m = q^*(S, T) = 0$ and so $\gamma(S, T) \geq -\epsilon$ from the fist calculation. Consequently the theorem is proved. □

As an easy consequence of the above theorem, we can obtain the following theorem, which is an extension of the 2-factorable theorem.

Theorem 5.17 (Kano [115] (1985)). *Let a and b be even integers such that $0 \leq a \leq b$, and let $n \geq 1$ be an integer. Then a general graph G can be decomposed into n $[a, b]$-factors if and only if G is an $[an, bn]$-graph.*

Proof. It is trivial that if a general graph G can be decomposed into n $[a, b]$-factors, then G is an $[an, bn]$-graph. So it suffices to prove that every general $[an, bn]$-graph can be decomposed into n $[a, b]$-factors. We use induction on n, and may assume $n \geq 2$.

Let G be a general $[an, bn]$-graph. Define $\theta = 1/n$ and $g, f : V(G) \to \mathbb{Z}$ as follows:

$$g(x) = f(x) = a \qquad \text{if } \deg_G(x) = an,$$
$$g(x) \leq \theta \deg_G(x) \leq f(x) \quad \text{and} \quad f(x) - g(x) = 1 \quad \text{if } an < \deg_G(x) < bn,$$
$$g(x) = f(x) = b \qquad \text{if } \deg_G(x) = bn.$$

Then g, f, θ satisfy conditions (i), (ii) and (iiib) of Theorem 5.9. Hence G has a (g, f)-factor F. For any vertex x with $an < \deg_G(x) < bn$, we have

$$a(n - 1) < (1 - \theta) \deg_G(x) < b(n - 1)$$

since $1 - \theta = (n - 1)/n$, and thus $a(n - 1) \leq \deg_{G-F}(x) \leq b(n - 1)$. Hence $G - F$ is an $[a(n-1), b(n-1)]$-graph. By the inductive hypothesis, $G - F$ can be decomposed into $n - 1$ $[a, b]$-factors, and therefore G can be decomposed into n $[a, b]$-factors. □

Here we give another application of Theorem 5.16, that is, we give another proof for the following theorem (Theorem 2.38).

Theorem 5.18. *Let $r \geq 2$ be an even integer, and G be an $(r - 1)$-edge connected r-regular multigraph of odd order. Then for every vertex v, $G - v$ has a 1-factor.*

Proof. Define $\theta = 1/r$, and $g, f : V(G) \to \mathbb{Z}$ by $g(x) = f(x) = 1$ for all $x \in V(G) - v$ and $g(v) = 0$ and $f(v) = 1$. Then conditions (ia), (ii) and (iiic) are satisfied since $n = r - 1$ is odd. Hence G has a (g, f)-factor F, which is the desired 1-factor of $G - v$ since $\deg_F(v) = 0$ by the parity of $|G|$. $\square$

We now give the main theorem in this section, which was obtained by Cai.

Theorem 5.19 (Cai [38] (1991)). *Let a, b, m, n be integers such that $1 \leq a < b$, $0 \leq m, n$ and $1 \leq m + n$. Then*
(i) If a is even, b is odd and $b \leq 2a + 1$, then every $[(6a + 2)m + an, (6b - 2)m + (b - 1)n]$-multigraph is $[a, b]$-factorable.
(ii) If a is odd, b is even and $b \leq 2a$, then every $[(6a + 2)m + (a + 1)n, (6b - 2)m + bn]$-multigraph is $[a, b]$-factorable.
(iii) If both a and b are odd and $b \leq 2a + 1$, then every $[(3a + 1)m + (a + 1)n, (3b - 1)m + (b - 1)n]$-multigraph is $[a, b]$-factorable.

The bounds in the above theorem are sharp in the following sense: consider, for example, the case when a is even and b is odd. Then there exist a $[6a + 1, 6b - 2]$-multigraph and a $[6a + 2, 6b - 1]$-multigraph that are not $[a, b]$-factorable. For other cases, the same situations hold.

We now show that the graphs G_1 and G_2 of Fig. 5.5, which are a $[6a+1, 6b-2]$-multigraph and a $[6a+2, 6b-1]$-multigraph, are not $[a, b]$-factorable, where a is even and b is odd. Assume that G_1 is $[a, b]$-factorable. Since $\deg_G(v) = 6b-2$ and $\deg_G(u) = 6a+1$, G_1 must be decomposed into six $[a, b]$-factors. We may assume that an $[a, b]$-factor F_1 contains the minimum number of edges joining u to w. Then F_1 contains at most $a - (b+1)/2$ edges joining u to w. Since the degree of u in F_1 is at least a, F_1 contains at least $(b + 1)/2$ edges joining u to v. Similarly F_1 contains at least $(b + 1)/2$ edges joining w to v. Therefore the degree of v in F_1 is at least $b + 1$, which is a contradiction. Therefore G_1 is not $[a, b]$-factorable.

Assume that G_2 is $[a, b]$-factorable, where G_2 consists of two components. Then G_2 must be decomposed into six $[a, b]$-factors. It is clear that there exist at least three $[a, b]$-factors that contain at least $(b+1)/2$ edges each joining y to z. Similarly, there exist at least two $[a, b]$-factors that contain at least $(b+1)/2$ edges each joining x to y, and another two $[a, b]$-factors that contain at least $(b + 1)/2$ edges each joining x to z. Moreover, no $[a, b]$-factor F_1 contains at least $(b + 1)/2$ edges joining x to y and at least $(b + 1)/2$ edges joining x to z, since the degree of x in F_1 must be at most b. Therefore, there exists an $[a, b]$-factor F_2 that contains at least $(b+1)/2$ edges joining y to z and at least $(b+1)/2$ edges joining either x to y or x to z. Then one of $\{y, z\}$ has degree at least $b+1$ in F_2, which is a contradiction. Therefore G_2 is not $[a, b]$-factorable.

Some consequences of Theorem 5.19 are obtained as follows. By setting $a = k$ and $b = k + 1$ in (i), we can obtain Theorem 5.14. Similarly, by setting

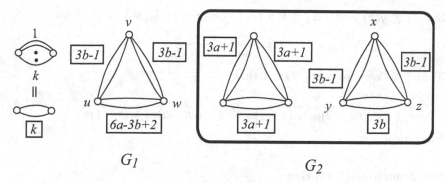

Fig. 5.5. (a) A $[6a + 1, 6b - 2]$-multigraph G_1 that is not $[a, b]$-factorable; (b) an $[6a + 2, 6b - 1]$-multigraph that is not $[a, b]$-factorable.

$a = k$ and $b = k+1$ in (ii) we obtain Theorem 5.15. Moreover, by taking $a = 1$ and $b = 2$ in (ii), we obtain the next theorem, which is the main result in [115] and whose proof yields techniques that are used in the proof of Theorem 5.19.

Theorem 5.20 ([115]). *Let $m \geq 1$ and $n \geq 0$ be integers. Then every $[8m + 2n, 10m + 2n]$-multigraph is $[1, 2]$-factorable.*

We shall prove only (i) of Theorem 5.19, and give a brief sketch of proofs of (ii) and (iii) since they can be proved in a similar way. We prove (i) of the theorem in several steps, which are given in the following lemmas.

Lemma 5.21. *Let a be an even integer and b be an odd integers such that $1 \leq a < b \leq 2a + 1$, and $m, n \geq 1$ be integers. Then every $[(6a + 2)m + an, (6b - 2)m + (b - 1)n]$-multigraph G contains an $[a, b]$-factor F such that $G - F$ is a $[(6a + 2)m + a(n - 1), (6b - 2)m + (b - 1)(n - 1)]$-graph.*

Proof. Let G be a $[(6a + 2)m + an, (6b - 2)m + (b - 1)n]$-multigraph. Define $g, f : V(G) \to \mathbb{Z}$ by

$$g(x) = \max\{a, \deg_G(x) - ((6b - 2)m + (b - 1)(n - 1))\},$$
$$f(x) = \min\{b, \deg_G(x) - ((6a + 2)m + a(n - 1))\}$$

for all $x \in V(G)$. Then

$$a \leq \deg_G(x) - ((6a + 2)m + a(n - 1)),$$
$$\deg_G(x) - ((6b - 2)m + (b - 1)(n - 1)) \leq b - 1 < b,$$
$$(6a + 2)m + a(n - 1) < (6b - 2)m + (b - 1)(n - 1).$$

Hence

$$g(x) \leq f(x) \quad \text{for all} \quad x \in V(G), \text{ and}$$
$$g(x) = f(x) \quad \text{if and only if} \quad \deg_G(x) = (6a + 2)m + an,$$
$$\text{which implies } g(x) = f(x) = a.$$

Since $a \leq g(x)$ and $f(x) \leq b$, a (g,f)-factor is an $[a,b]$-factor. Define

$$\theta = \frac{a}{(6a+2)m + an}.$$

Then it follows from $b \leq 2a+1$ that

$$\frac{b-1}{(6b-2)m + (b-1)n} \leq \theta < \frac{b}{(6b-2)m + (b-1)n}. \qquad (5.11)$$

In particular,

$$a \leq \theta \deg_G(x) < b.$$

We obtain from (5.11) that

$$(1-\theta)\deg_G(x) \leq (1-\theta)\Big((6b-2)m + (b-1)n\Big)$$
$$\leq (6b-2)m + (b-1)n - (b-1)$$
$$= (6b-2)m + (b-1)(n-1),$$

which implies $\deg_G(x) - \big((6b-2)m + (b-1)(n-1)\big) \leq \theta \deg_G(x)$, and thus

$$g(x) \leq \theta \deg_G(x).$$

Similarly,

$$(1-\theta)\deg_G(x) \geq (1-\theta)\big((6a+2)m + an\big)$$
$$= (6a+2)m + an - a = (6a+2)m + a(n-1),$$

which implies $\deg_G(x) - \big((6a+2)m + a(n-1)\big) \geq \theta \deg_G(x)$, and so

$$\theta \deg_G(x) \leq f(x).$$

Therefore θ, $n = 1$, g and f satisfy conditions (ia), (ii) and (iiib) of Theorem 5.16. Thus G has an (g,f)-factor F.

It follows from

$$A - \max\{B,C\} \leq A - B \quad \text{and} \quad A - \min\{B,C\} \geq A - B$$

that

$$\deg_G(x) - g(x) \leq \deg_G(x)$$
$$- \Big(\deg_G(x) - ((6b-2)m + (b-1)(n-1))\Big)$$
$$= (6b-2)m + (b-1)(n-1),$$

and

$$\deg_G(x) - f(x) \geq \deg_G(x)$$
$$- \Big(\deg_G(x) - ((6a+2)m + a(n-1))\Big)$$
$$= (6a+2)m + a(n-1).$$

Hence $G - F$ is a $[(6a+2)m + a(n-1), (6b-2)m + (b-1)(n-1)]$-graph, and the lemma is proved. □

Lemma 5.22. *Let a and b be integers such that* $1 \le a < b$. *Then every connected* $[6a + 2, 6b - 2]$*-multigraph can be decomposed into a* $[3a, 3b - 1]$*-factor having at most one vertex of degree 3a and a* $[3a + 1, 3b - 1]$*-factor.*

Proof. Let G be a connected $[6a + 2, 6b - 2]$-multigraph. If G has a vertex of degree $6a + 2$, then choose one such vertex w. Define $g, f : V(G) \to \mathbb{Z}$ by

$$g(x) = \max\{3a + 1, \deg_G(x) - (3b - 1)\} \quad \text{for} \quad x \in V(G) - w,$$
$$g(w) = 3a, \qquad \text{and}$$
$$f(x) = \min\{3b - 1, \deg_G(x) - (3a + 1)\} \quad \text{for all} \quad x \in V(G).$$

Then

$$\deg_G(x) - (3b - 1) \le \deg_G(x) - \frac{1}{2} \cdot \deg_G(x) = \frac{1}{2} \cdot \deg_G(x),$$
$$\deg_G(x) - (3a + 1) \ge \deg_G(x) - \frac{1}{2} \cdot \deg_G(x) = \frac{1}{2} \cdot \deg_G(x).$$

Thus

$$g(x) \le \frac{1}{2} \cdot \deg_G(x) \le f(x) \qquad \text{for all} \quad x \in V(G), \quad \text{and}$$
$$\text{if} \quad g(x) = f(x) \quad \text{then} \quad \deg_G(x) = 6a + 2 \quad \text{or} \quad 6b - 2.$$

Then $\theta = 1/2$, $n = 1$, g and f satisfy (ia), (ii) and (iiic) of Theorem 5.16, and thus G has a (g, f)-factor F. It is easy to verify that F is the desired $[3a, 3b - 1]$-factor and $G - F$ is the desired $[3a + 1, 3b - 1]$-factor. $\square$

Lemma 5.23. *Let a be an even integer and b be an odd integer such that* $2 \le a < b \le 2a + 1$. *Then every connected* $[3a, 3b - 1]$*-multigraph having at most one vertex of degree 3a can be decomposed into three* $[a, b]$*-factors.*

Since the proof of Lemma 5.23 is long, we postpone it and now prove Theorem 5.19 under the assumption that Lemma 5.23 holds.

Proof of (i) of Theorem 5.19 Let G be a $[(6a + 2)m + an, (6b - 2)m + (b - 1)n]$-multigraph. By considering each component of G, we may assume that G is connected. We use this argument below without mentioning it. If $m = 0$, then G is an $[an, (b-1)n]$-graph, and so G can be decomposed into n $[a, b-1]$-factors by Theorem 5.17. Hence we may assume $m \ge 1$. By Lemma 5.21, G can be decomposed into n $[a, b]$-factors and one $[(6a + 2)m, (6b - 2)m]$-factor. By Theorem 5.17 the $[(6a+2)m, (6b-2)m]$-factor can be decomposed into m $[(6a + 2), (6b - 2)]$-factors. By Lemmas 5.22 and 5.23, every $[(6a + 2), (6b - 2)]$-factor can be decomposed into six $[a, b]$-factors. Consequently, G is decomposed into $[a, b]$-factors, and the theorem is proved. $\square$

Proof of Lemma 5.23 Let G be a connected $[3a, 3b - 1]$-graph having at most one vertex of degree $3a$. We shall prove the lemma by induction on the order $|G|$. Since G has no loops, $|G| \geq 2$. It is easy to verify that the assertion holds if $|G| \leq 3$. Thus we may assume $|G| \geq 4$. Define two functions g and f on $V(G)$ by

$$g(x) = \max\{a, \deg_G(x) - (2b - 1)\},$$
$$f(x) = \min\{b, \deg_G(x) - 2a\}$$

for all $x \in V(G)$. However, if G has no vertex x with $3a < \deg_G(x) < 3b - 1$, then choose one vertex w of degree $3b - 1$ and modify g and f as follows:

$$g(x) = \max\{a, \deg_G(x) - (2b - 1)\} \quad \text{for} \quad x \in V(G) - w,$$
$$g(w) = b - 1, \qquad \text{and}$$
$$f(x) = \min\{b, \deg_G(x) - 2a\} \qquad \text{for all} \quad x \in V(G).$$

Then

$$g(x) \leq f(x) \qquad \text{for all} \quad x \in V(G), \text{ and}$$
$$\text{if} \quad g(x) = f(x), \quad \text{then either} \quad \deg_G(x) = 3a \quad \text{or}$$
$$\deg_G(x) = 3b - 1 \quad \text{and} \quad x \neq w.$$

We consider the following two cases.

Case 1. G has a (g, f)-factor F.

Let $H = G - F$. Then

$$\deg_G(x) - g(x) \leq \deg_G(x) - \big(\deg_G(x) - (2b - 1)\big) = 2b - 1, \quad (x \neq w)$$
$$\deg_G(w) - g(w) = (3b - 1) - (b - 1) = 2b,$$
$$\deg_G(x) - f(x) \geq \deg_G(x) - \big(\deg_G(x) - 2a\big) = 2a.$$

Hence H is a $[2a, 2b]$-graph with at most one vertex of degree $2b$. Furthermore, if H has a vertex of degree $2b$, G has no vertices x with $3a < \deg_G(x) < 3b-1$. In particular, G has at least two vertices of degree $3b - 1$, and thus H has a vertex u of degree $2b-1$, for which the following g' and f' satisfy $g'(u) < f'(u)$.

Define g', f' on $V(H)$ by

$$g'(x) = \max\{a, \deg_H(x) - b\},$$
$$f'(x) = \min\{b, \deg_H(x) - a\}$$

for all $x \in V(H)$. Then

$$g'(x) \leq \frac{1}{2} \cdot \deg_H(x) \leq f'(x) \qquad \text{for all} \quad x \in V(H), \text{ and}$$
$$\text{if} \quad g'(x) = f'(x), \quad \text{then} \quad \deg_G(x) = 2a \text{ or } 2b.$$

Then $\theta = 1/2$, $n = 1$, g', f' satisfies (ia), (ii) and (iiic) of Theorem 5.16, and thus H has a (g', f')-factor F'. It is clear that $F \cup F' \cup (H - F')$ is the required $[a, b]$-factorization of G.

Case 2. *G has no (g, f)-factor.*

By the (g, f)-factor theorem, there exist two disjoint subsets S and T of $V(G)$ such that

$$\gamma(S, T) = \sum_{x \in S} f(x) + \sum_{x \in T} (\deg_G(x) - g(x))$$

$$-e_G(S, T) - q^*(S, T) \leq -1. \tag{5.12}$$

Let $C_1, C_2, \ldots, C_m$ be the set of (g, f)-odd components of $G - (S \cup T)$, where $m = q^*(S, T)$. Then every $C = C_i$ satisfies $g(x) = f(x)$ for all $x \in V(C)$ and $\sum_{x \in V(C)} f(x) + e_G(C, S \cup T) \equiv 1 \pmod 2$. Therefore, every vertex of C has even degree in G, and thus

$$e_G(C, S \cup T) \geq 2 \quad \text{and} \quad e_G(C, S \cup T) \equiv 0 \pmod 2.$$

Let

$$\beta = \#\{C_i : e_G(C_i, S \cup T) = 2,\ 1 \leq i \leq m\}.$$

Then

$$\sum_{x \in S \cup T} \deg_G(x) \geq \sum_{i=1}^m e_G(S \cup T, C_i) + 2e_G(S, T)$$

$$> 4(m - \beta) + 2\beta + 2e_G(S, T),$$

and thus

$$\sum_{x \in S \cup T} \deg_G(x) + 2\beta \geq 4m + 2e_G(S, T). \tag{5.13}$$

For any $0 \leq \theta \leq 1/2$, we have

$$\gamma(S, T) = \sum_{x \in S} f(x) + \sum_{x \in T} (\deg_G(x) - g(x)) - e_G(S, T) - m$$

$$= \sum_{x \in S} f(x) + \sum_{x \in T} (\deg_G(x) - g(x)) - \theta e_G(S, T)$$

$$- \frac{1}{4}\left(4m + 2e_G(S, T)\right) - \left(\frac{1}{2} - \theta\right) e_G(S, T)$$

$$\geq \sum_{x \in S} f(x) + \sum_{x \in T} (\deg_G(x) - g(x)) - \theta \sum_{x \in S} \deg_G(x)$$

$$- \frac{1}{4}\left(\sum_{x \in S \cup T} \deg_G(x) + 2\beta\right) - \left(\frac{1}{2} - \theta\right) \sum_{x \in T} \deg_G(x) \quad \text{(by (5.13))}$$

$$= \sum_{x \in S} \left(f(x) - (\frac{1}{4} + \theta) \deg_G(x)\right)$$

$$+ \sum_{x \in T} \left((\frac{1}{4} + \theta) \deg_G(x) - g(x)\right) - \frac{\beta}{2}. \tag{5.14}$$

In particular, by taking $\theta = (b+1)/4(3b-1)$, we have $1/4 + \theta = b/(3b-1)$. For every vertex $x \in S$, it follows that

$$f(x) - \left(\frac{1}{4} + \theta\right) \deg_G(x) = f(x) - \frac{b}{3b-1} \deg_G(x)$$

$$= \frac{1}{3b-1}\left((3b-1)f(x) - b\deg_G(x)\right)$$

$$= \frac{1}{3b-1} \min\{b((3b-1) - \deg_G(x)), \ (2b-1)\deg_G(x) - 2a(3b-1)\}$$

$$\geq \begin{cases} -a/(3b-1) & \text{if } \deg_G(x) = 3a \\ 0 & \text{otherwise.} \end{cases}$$

Similarly, for every $x \in T$, it follows that

$$\left(\frac{1}{4} + \theta\right) \deg_G(x) - g(x)$$

$$\geq \frac{1}{3b-1} \min\{b\deg_G(x) - a(3b-1), \ (2b-1)((3b-1) - \deg_G(x))\}$$

$$\geq 0.$$

Let $S(3a) = \{x \in S | \deg_G(x) = 3a\}$. Then $|S(3a)| \leq 1$ since G has at most one vertex of degree $3a$. It follows that

$$\sum_{x \in S} \left(f(x) - \left(\frac{1}{4} + \theta\right) \deg_G(x)\right)$$

$$\geq -\frac{a}{3b-1} \cdot |S(3a)| \geq -\frac{1}{3} \cdot |S(3a)|. \qquad (5.15)$$

By combining (5.12), (5.14) and (5.15), we have

$$\frac{|S(3a)|}{3} + \frac{\beta}{2} \geq 1.$$

From the above inequality and $|S(3a)| \leq 1$, it follows that $\beta \geq 2$, and thus the next Claim holds.

Claim $G - (S \cup T)$ *has a* (g, f)-*odd component, say* C_t, *such that*

$$e_G(C_t, S \cup T) = 2 \quad \text{and} \quad \deg_G(x) = 3b - 1 \quad \text{for all} \quad x \in V(C_t).$$

Since G has no loops, $|C_t| \geq 2$. Furthermore, since

$$\sum_{x \in V(C_t)} f(x) + e_G(C_t, S \cup T) = b|C_t| + 2 \equiv |C_t| \equiv 1 \pmod 2,$$

we have $|C_t| \geq 3$. Let $\{u_1 w_1, u_2 w_2\}$ be the edges joining C_t and $S \cup T$, where $u_1, u_2 \in S \cup T$ and $w_1, w_2 \in V(C_t)$. Let K be the graph obtained from C_t by adding a new edge $w_1 w_2$, which might be a loop of K. Let

$$b = 2s + 1, \qquad \text{where} \quad s \geq 1.$$

Then $3b - 1 = 6s + 2$ and K is a $(6s + 2)$-regular graph, so by the 2-factorable theorem, K can be decomposed into $3s + 1$ 2-factors $R_0, R_1, R_2, \cdots, R_{3s}$, where we may assume $w_1 w_2 \in R_1$. Let

$$Q_1 = \bigcup_{i=1}^{s} R_i, \quad Q_2 = \bigcup_{i=s+1}^{2s} R_i, \quad Q_3 = \bigcup_{i=2s+1}^{3s} R_i.$$

Then every Q_i is a $2s$-factor of K. We can assign the edges of R_0 one by one to Q_1, Q_2, Q_3 so that

$$\deg_{Q_1}(w_1) = 2s + 1, \qquad \deg_{Q_2}(w_1) = 2s, \quad \text{and}$$
$$\deg_{Q_i}(x) \leq 2s + 1 \qquad \text{for all} \quad x \in V(C_t), \quad i \in \{1, 2, 3\}. \qquad (5.16)$$

To see this, first assign an edge of R_0 incident with w_1 to Q_1, and in general, for any unassigned edge xy of R_0, there exists at most one Q_i with $\deg_{Q_i}(x) = 2s + 1$ and at most one Q_j with $\deg_{Q_j}(y) = 2s + 1$, and so we can assign xy to Q_k, where $k \notin \{i, j\}$. Since $\deg_{Q_1}(w_1) = 2s + 1$, we may assume $\deg_{Q_2}(w_1) = 2s$ (and $\deg_{Q_3}(w_1) = 2s + 1$).

We consider two subcases according to whether $u_1 = u_2$ or not.

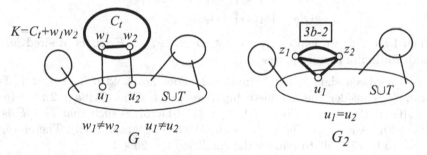

Fig. 5.6. A graph G with a component C_t, where it may occur $w_1 = w_2$; and a $[3a, 3b - 1]$-graph G_2, where $u_1 = u_2$.

Subcase 2.1 $u_1 \neq u_2$.

By adding a new edge $u_1 u_2$ to $G - V(C_t)$, we obtain a $[3a, 3b-1]$-multigraph G_1, which contains at most one vertex of degree $3a$ (Fig. 5.6). By the induction hypothesis, G_1 has an $[a, b]$-factorization $A_1 \cup A_2 \cup A_3$. We may assume $u_1 u_2 \in A_1$. Let

$$F_1 = (A_1 - u_1 u_2) \cup (Q_1 - w_1 w_2) \cup \{u_1 w_1, u_2 w_2\},$$
$$F_2 = A_2 \cup Q_2,$$
$$F_3 = A_3 \cup Q_3.$$

Then $F_1 \cup F_2 \cup F_3$ is the desired $[a, b]$-factorization of G. Note the above arguments hold when $w_1 = w_2$ without changing anything.

Subcase 2.1 $u_1 = u_2$.

In this case, we obtain a new graph G_2 from $G - V(C_t)$ by adding two new vertices z_1 and z_2, by joining them by $3b - 2$ multiple edges, and by adding two new edges $u_1 z_1$ and $u_1 z_2$. Then G_2 is a $[3a, 3b - 1]$-multigraph having at most one vertex of degree $3a$. Moreover $|G_2| < |G|$ as $|C_t| \geq 3$. Hence by the induction hypothesis, G_2 can be decomposed into three $[a, b]$-factors A_1, A_2, A_3.

If $u_1 z_1, u_1 z_2 \in A_1$, then let

$$F_1 = (A_1 - \{z_1, z_2\}) \cup (Q_1 - w_1 w_2) \cup \{u_1 w_1, u_1 w_2\},$$
$$F_2 = (A_2 - \{z_1, z_2\}) \cup Q_2,$$
$$F_3 = (A_3 - \{z_1, z_2\}) \cup Q_3,$$

where $A_i - \{z_1, z_2\}$ denotes the graph obtained from A_i by removing z_1 and z_2 together with all edges incident with them.

If $u_1 z_1 \in A_1$ and $u_1 z_2 \in A_2$, then define

$$F_1 = (A_1 - \{z_1, z_2\}) \cup (Q_1 - w_1 w_2) \cup \{u_1 w_2\},$$
$$F_2 = (A_2 - \{z_1, z_2\}) \cup Q_2 \cup \{u_1 w_1\},$$
$$F_3 = (A_3 - \{z_1, z_2\}) \cup Q_3.$$

Then by (5.16) we can easily show that $F_1 \cup F_2 \cup F_3$ is the desired $[a, b]$-factorization of G in each case.   □

It has been shown that Lemma 5.21 does not hold if $2a + 1 < b$. For example, consider the complete bipartite graph $G = K((6b + 2)m + (b - 1)n, (6a + 2)m + an)$. Then G has no $[a, b]$-factor F such that $G - F$ is a $[(6a + 2)m + a(n - 1), (6b + 2)m + (b - 1)(n - 1)]$-graph ([38]). Therefore, it seems to be difficult to remove the condition $b \leq 2a + 1$.

Statement (ii) of Theorem 5.19 can be proved by showing the following three lemmas. We give only a brief sketch of the proof for each lemma.

Lemma 5.24. *Let a be an odd integer and b be an even integer such that $1 \leq a < b \leq 2a$, and let $m, n \geq 1$ be integers. Then every $[(6a + 2)m + (a + 1)n, (6b - 2)m + bn]$-multigraph G contains an $[a, b]$-factor F such that $G - F$ is a $[(6a + 2)m + (a + 1)(n - 1), (6b - 2)m + b(n - 1)]$-graph.*

Proof. Let G be a $[(6a + 2)m + (a + 1)n, (6b - 2)m + bn]$-multigraph. Define $g, f : V(G) \to \mathbb{Z}$ by

$$g(x) = \max\{a, \deg_G(x) - ((6b - 2)m + b(n - 1))\}$$
$$f(x) = \min\{b, \deg_G(x) - ((6a + 2)m + (a + 1)(n - 1))\}.$$

Then G has a (g, f)-factor F, which is an $[a, b]$-factor, and $G - F$ is a $[(6a + 2)m + (a + 1)(n - 1), (6b - 2)m + b(n - 1)]$-graph.   □

Lemma 5.25. *Let a and b be integers such that* $1 \le a < b \le 2a$. *Then every connected* $[6a+2, 6b-2]$-*multigraph can be decomposed into a* $[3a+1, 3b]$-*factor having at most one vertex of degree* $3b$ *and a* $[3a+1, 3b-1]$-*factor.*

Proof. Let G be a connected $[6a + 2, 6b - 2]$-multigraph. If G has a vertex of degree $6b - 2$, then choose one such vertex w. Define $g, f : V(G) \to \mathbb{Z}$ by

$$g(x) = \max\{3a + 1, \deg_G(x) - (3b - 1)\} \quad \text{for all} \quad x \in V(G),$$
$$f(x) = \min\{3b - 1, \deg_G(x) - (3a + 1)\} \quad \text{if} \quad x \in V(G) - w,$$
$$f(w) = 3b.$$

Then G has a (g, f)-factor F, which is a $[3a+1, 3b]$-factor having at most one vertex of degree $3b$, and thus we obtain the desired factorization $F \cup (G - F)$. $\square$

Lemma 5.26. *Let a be an odd integer and b be an even integer such that* $1 \le a < b \le 2a$. *Then every connected* $[3a + 1, 3b]$-*multigraph having at most one vertex of degree* $3b$ *can be decomposed into three* $[a, b]$-*factors.*

Proof. Let G be a connected $[3a+1, 3b]$-multigraph having at most one vertex of degree $3b$. We shall prove the lemma by induction on the order $|G|$. We may assume $|G| \ge 4$. Define two functions g and f on $V(G)$ as

$$g(x) = \max\{a, \deg_G(x) - 2b\},$$
$$f(x) = \min\{b, \deg_G(x) - (2a + 1)\}.$$

However, if G has no vertex x with $3a + 1 < \deg_G(x) < 3b$, then choose a vertex w of degree $3a + 1$ and modify g and f as follows:

$$g(x) = \max\{a, \deg_G(x) - 2b\} \quad \text{for all} \quad x \in V(G),$$
$$f(x) = \min\{b, \deg_G(x) - (2a + 1)\} \quad \text{if} \quad x \in V(G) - w,$$
$$f(w) = a + 1.$$

Then we consider two cases according to whether G has a (g, f)-factor or not, and prove the lemma. $\square$

Statement (iii) of Theorem 5.19 can be proved by showing the following three lemmas. We give only a brief sketch of the proof of each lemma.

Lemma 5.27. *Let a and b be integers such that* $1 \le a < b \le 2a + 1$, *and* $m, n \ge 1$ *be integers. Then every* $[(3a+1)m + (a+1)n, (3b-1)m + (b-1)n]$-*multigraph* G *contains an* $[a, b]$-*factor* F *such that* $G - F$ *is a* $[(3a + 1)m + (a+1)(n-1), (3b-1)m + (b-1)(n-1)]$-*graph.*

Proof. Let G be a $[(3a + 1)m + (a + 1)n, (3b - 1)m + (b - 1)n]$-multigraph. Define $g, f : V(G) \to \mathbb{Z}$ by

$$g(x) = \max\{a, \deg_G(x) - ((3b - 1)m + (b - 1)(n - 1))\}$$
$$f(x) = \min\{b, \deg_G(x) - ((3a + 1)m + (a + 1)(n - 1))\}.$$

Then G has a (g, f)-factor, which is the required $[a, b]$-factor. □

Lemma 5.28. *Let both a and b be odd integers such that $1 \leq a < b \leq 2a + 1$. Then every connected $[3a + 1, 3b - 1]$-multigraph can be decomposed into three $[a, b]$-factors.*

Proof. Let G be a connected $[3a + 1, 3b - 1]$-multigraph. We shall prove the lemma by induction on the order $|G|$. We may assume $|G| \geq 4$. Define two functions g and f on $V(G)$ as

$$g(x) = \max\{a, \deg_G(x) - 2b\},$$
$$f(x) = \min\{b, \deg_G(x) - (2a + 1)\}.$$

We may assume that G has a vertex of degree greater than $3a + 1$ since otherwise the lemma follows from Lemma 5.26. Then we consider two cases according to whether G has a (g, f)-factor or not, and prove the lemma. □

We now turn our attention to (g, f)-factorizations. For two functions $h, k : V(G) \to \mathbb{Z}$ with $h(x) \leq k(x)$ for all $x \in V(G)$ we say that a graph G is an (h, k)-**graph** if

$$h(x) \leq \deg_G(x) \leq k(x) \qquad \text{for all } x \in V(G).$$

Moreover, for integers s and t, $sh + t$ denotes the function defined by

$$(sh + t)(x) = sh(x) + t \qquad \text{for all } x \in V(G).$$

Some results on $[a, b]$-factorization are generalized to those on (g, f)-factorization as follows.

Theorem 5.29 (Liu [170])**.** *Let G be a $((g + 1)n - 1, (f - 1)n + 1)$-multigraph, where $g, f : V(G) \to \mathbb{Z}$ such that $1 \leq g(x) < f(x)$ for all $x \in V(G)$, and let $n \geq 1$ be an integer. Then G is (g, f)-factorable.*

Proof. Let G be a $((g + 1)n - 1, (f - 1)n + 1)$-multigraph. We shall prove the theorem by induction on n. If $n = 1$, then G itself is a (g, f)-factor. So we may assume $n \geq 2$. Define $h, k : V(G) \to \mathbb{Z}$ by

$$h(x) = \max\{g(x), \deg_G(x) - ((f(x) - 1)(n - 1) + 1)\},$$
$$k(x) = \min\{f(x), \deg_G(x) - ((g(x) + 1)(n - 1) - 1)\}.$$

Then

$$g(x) < \deg_G(x) - \Big((g(x) + 1)(n - 1) - 1\Big),$$
$$\deg_G(x) - \Big((f(x) - 1)(n - 1) + 1\Big) < f(x),$$
$$g(x) < f(x).$$

Hence

$$h(x) < k(x) \qquad \text{for all} \quad x \in V(G).$$

Since $g(x) \le h(x)$ and $k(x) \le f(x)$, a (h, k)-factor is a (g, f)-factor. Define

$$\theta = \frac{1}{n}.$$

Then $1 - \theta = (n - 1)/n$ and so

$$(1 - \theta) \deg_G(x) \le (1 - \theta)\Big((f(x) - 1)n + 1\Big)$$
$$= (f(x) - 1)(n - 1) + 1 - \frac{1}{n},$$

which implies $\deg_G(x) - \big((f(x) - 1)(n - 1) + 1\big) < \theta \deg_G(x)$, and since $\theta \deg_G(x) \ge (g(x) + 1) - 1/n > g(x)$, we have

$$h(x) < \theta \deg_G(x).$$

Similarly,

$$(1 - \theta) \deg_G(x) \ge (1 - \theta)\Big((g(x) + 1)n - 1\Big)$$
$$= (g(x) + 1)(n - 1) - 1 + \frac{1}{n},$$

which implies $\deg_G(x) - \big((g(x) + 1)(n - 1) - 1\big) > \theta \deg_G(x)$, and so

$$\theta \deg_G(x) < k(x).$$

Therefore θ, $n = 1$, h and k satisfy conditions (ia), (ii) and (iiif) of Theorem 5.16. Thus G has an (h, k)-factor F. It is easy to see that $G - F$ is a $((g + 1)(n - 1) - 1, (f - 1)(n - 1) + 1)$-graph since

$$\deg_G(x) - h(x) \le (f(x) - 1)(n - 1) + 1,$$
$$\deg_G(x) - k(x) \ge (g(x) + 1)(n - 1) - 1.$$

Consequently the theorem is proved. □

The following Theorem 5.30 can also be proved by using Theorem 5.29 and a proof technique similar to the one used in the proof of Theorem 5.19. Namely, it is shown that every $\big(2t((g+1)n-1)+(g+1)s, 2t((f-1)n+1)+(f-1)s\big)$-multigraph G has a (g, f)-factor F such that $G-F$ is a $\big(2t((g+1)n-1)+(g+1)(s-1), 2t((f-1)n+1)+(f-1)(s-1)\big)$-multigraph; a $\big(2t((g+1)n-1), 2t((f-1)n+1)\big)$-multigraph can be decomposed into t $\big(2((g+1)n-1)\big), \big(2((f-1)n+1)\big)$-graphs; and every $\big(2((g+1)n-1)\big), \big(2((f-1)n+1)\big)$-multigraph is (g, f)-factorable.

Theorem 5.30 (Yan [252]). *Let $n \geq 2$, $s \geq 0$, and $t \geq 0$ be integers. Then every $\big(2t((g+1)n-1)+(g+1)s, 2t((f-1)n+1)+(f-1)s\big)$-multigraph G is (g, f)-factorable, where $g, f : V(G) \to \mathbb{Z}$ such that $1 \leq g(x) \leq f(x)$ for all $x \in V(G)$.*

Parity Factors

In this chapter we consider (g, f)-parity factors. In particular, we consider odd factors and even factors, in which every vertex has odd degree or even degree, respectively. Then we consider partial parity factors. These factors are special cases of $\mathcal{H}$-factors. Finally, we give some remarks on $\mathcal{H}$-factors.

6.1 Parity (g, f)-factors and $(1, f)$-odd factors

Let G be a general graph, and g, $f : V(G) \to \mathbb{Z}$ be functions such that

$$g(x) \leq f(x) \qquad \text{and} \qquad g(x) \equiv f(x) \pmod 2$$

for all $x \in V(G)$. Then a spanning subgraph F of G is called a **parity (g, f)-factor** if

$$g(x) \leq \deg_F(x) \leq f(x) \qquad \text{and} \qquad \deg_F(x) \equiv f(x) \pmod 2 \qquad (6.1)$$

for all $x \in V(G)$. Notice that we allow $g(x) < 0$ and $\deg_G(y) < f(y)$ for some vertices x and y of G, and this relaxation on g and f makes the proofs of some theorems shorter and simpler.

Theorem 6.1 (The Parity (g, f)-Factor Theorem, Lovász [176] (1972)). *Let G be a general graph and let $g, f : V(G) \to \mathbb{Z}$ be functions such that $g(x) \leq f(x)$ and $g(x) \equiv f(x) \pmod 2$ for all $x \in V(G)$. Then G has a parity (g, f)-factor if and only if for all disjoint subsets S and T of $V(G)$, it follows that*

$$\eta(S, T) = \sum_{x \in S} f(x) + \sum_{x \in T} (\deg_G(x) - g(x)) - e_G(S, T) - q(S, T) \geq 0, \quad (6.2)$$

where $q(S, T)$ denotes the number of components C of $G - (S \cup T)$ such that

$$\sum_{x \in V(C)} f(x) + e_G(C, T) \equiv 1 \pmod 2. \qquad (6.3)$$

J. Akiyama and M. Kano, *Factors and Factorizations of Graphs*,
Lecture Notes in Mathematics 2031, DOI 10.1007/978-3-642-21919-1_6,
© Springer-Verlag Berlin Heidelberg 2011

Proof. We first construct a new graph G^* from G by adding $(f(x) - g(x))/2$ loops to every vertex $x \in V(G)$, and we regard f as a function $f : V(G^*) \to \mathbb{Z}$ (Fig. 6.1).

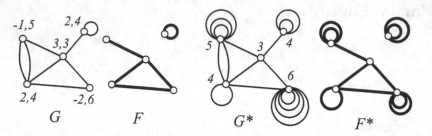

Fig. 6.1. A graph G, its parity (g, f)-factor F, the new graph G^*, and its f-factor F^*; Numbers denote $g(x), f(x)$ in G and $f(x)$ in G^*.

We can easily see that G has a parity (g, f)-factor if and only if G^* has an f-factor: we can get an f-factor of G^* from a parity (g, f)-factor F of G by adding $(f(x) - \deg_F(x))/2$ loops to every vertex x of G. Conversely, we can obtain a parity (g, f)-factor of G from an f-factor F^* of G^* by deleting all the new added loops from F^* (see Fig. 6.1). Therefore, we can apply the f-factor theorem (Theorem 3.2) to the new graph G^*. Let $S, T \subset V(G^*) = V(G)$, $S \cap T = \emptyset$. Since

$$\deg_{G^*}(x) = \deg_G(x) + f(x) - g(x),$$
$$e_{G^*}(S, T) = e_G(S, T), \qquad q_{G^*}(S, T) = q_G(S, T),$$

we have

$$\delta_{G^*}(S, T) = \sum_{x \in S} f(x) + \sum_{x \in T}(\deg_{G^*}(x) - f(x)) - e_{G^*}(S, T) - q_{G^*}(S, T)$$
$$= \sum_{x \in S} f(x) + \sum_{x \in T}(\deg_G(x) - g(x)) - e_G(S, T) - q_G(S, T)$$
$$= \eta(S, T).$$

Consequently, the theorem follows from the (g, f)-factor theorem.   □

We will mainly consider $(1, f)$-odd factors, but before moving on to this topic we first deal with even factors. A spanning subgraph F of a graph G is called an **even factor** of G if every vertex has a positive even degree in F. From the (g, f)-parity factor theorem, we can easily obtain the next theorem.

Note that by using a function $f : V(G) \to \{2, 4, 6, \cdots\}$, we can consider a $(2, f)$-even factor. However a criterion for the existence of this factor does not become simpler as the following theorem demonstrates.

Theorem 6.2. *A general graph G has an even factor if and only if*

$$\sum_{x \in X} (\deg_G(x) - 2) - q(G; X) \geq 0 \qquad \text{for all} \quad X \subset V(G), \qquad (6.4)$$

where $q(G; X)$ denotes the number of components C of $G - X$ such that $e_G(C, X) \equiv 1 \pmod 2$.

Proof. Let N be a sufficiently large even integer, and define $g(x) = 2$ and $f(x) = N$ for all $x \in V(G)$. Then G has an even factor if and only if G has a parity (g, f)-factor.

Let S and T be two disjoint subsets of $V(G)$. If S is a nonempty set, then

$$\eta(S, T) = \sum_{x \in S} f(x) + \sum_{x \in T} (\deg_G(x) - 2) - e_G(S, T) - q(S, T)$$

$$\geq N + \sum_{x \in T} (\deg_G(x) - 2) - e_G(S, T) - q(S, T) \geq 0.$$

Hence we may assume that $S = \emptyset$. Then

$$\eta(\emptyset, T) = \sum_{x \in T} (\deg_G(x) - 2) - q(\emptyset, T) = \sum_{x \in T} (\deg_G(x) - 2) - q(G; T).$$

Therefore the theorem is proved by the parity (g, f)-factor theorem. □

Theorem 6.3 (Problem 42 in Sec. 7 of [181]). *Every 2-edge connected multigraph with $\delta(G) \geq 3$ has an even factor.*

Proof. Let X be a nonempty vertex set of G, and $D_1, D_2, \ldots, D_m$ be the components of $G - X$ such that $e_G(D_i, X)$ is odd, where $m = q(G; X)$. Since G is 2-edge connected, $e_G(D_i, X) \geq 3$. Hence

$$3m \leq e_G(D_1 \cup \cdots \cup D_m, X) \leq \sum_{x \in X} \deg_G(x).$$

Thus

$$\sum_{x \in X} (\deg_G(x) - 2) - m$$

$$\geq \sum_{x \in X} \deg_G(x) - 2|X| - \frac{1}{3} \sum_{x \in X} \deg_G(x)$$

$$\geq \frac{2}{3} \sum_{x \in X} \deg_G(x) - 2|X| \geq \frac{2}{3} \cdot 3|X| - 2|X| = 0.$$

Therefore G has an even factor by Theorem 6.2. □

The next theorem gives a result on connected $\{2, 4\}$-factors in some graphs, which is of course much more difficult than usual $\{2, 4\}$-factors.

Theorem 6.4 (Broersma, Kriesell and Ryjáček [36]). *Every 4-connected claw-free simple graph has a connected $\{2,4\}$-factor, each of whose vertices has degree 2 or 4.*

We now consider a special parity factor called an odd factor. For an odd integer-valued function $f : V(G) \rightarrow \{1,3,5,\ldots\}$, a spanning subgraph F of G is called a $(1,f)$-**odd factor** if

$$\deg_F(x) \in \{1,3,\ldots,f(x)\} \qquad \text{for all} \quad x \in V(G).$$

Of course, if $f(x) = 1$ for all vertices x, then a $(1,f)$-odd factor is nothing but a 1-factor. For a constant odd integer $n \geq 1$, if $f(x) = n$ for all $x \in V(G)$, then a $(1,f)$-odd factor is called a $[1,n]$-**odd factor**. Hence a $[1,n]$-odd factor F satisfies

$$\deg_F(x) \in \{1,3,\ldots,n\} \qquad \text{for all} \quad x \in V(F).$$

It is interesting that a criterion for a graph to have a $(1,f)$-odd factor is much simpler than that for a parity (g,f)-factor. Since a $(1,f)$-odd factor is a special parity (g,f)-factor, we can prove the following theorem by making use of the parity (g,f)-factor theorem. Also, since a $(1,f)$-odd factor is an extension of a 1-factor, we can expect that the theorem can also be proved in a similar manner to the proof of the 1-factor theorem. Here we give these two proofs.

Theorem 6.5 (($1,f$)-**Odd Factor Theorem**, Cui and Kano [60]). *Let G be a general graph and let $f : V(G) \rightarrow \{1,3,5,\ldots\}$ be a function. Then G has a $(1,f)$-odd factor if and only if*

$$odd(G - S) \leq \sum_{x \in S} f(x) \qquad \text{for all} \quad S \subseteq V(G). \qquad (6.5)$$

Proof of the necessity. Suppose that G has a $(1,f)$-odd factor F. Since every component of F has even order, G has no odd components, and so (6.5) holds for $S = \emptyset$. Let $\emptyset \neq S \subset V(G)$, and C be any odd component of $G - S$. Then there exists an edge of F joining C to S. Hence

$$odd(G - S) \leq e_F(G - S, S) \leq \sum_{x \in S} \deg_F(x) \leq \sum_{x \in S} f(x).$$

Therefore (6.5) holds.

Proof of the sufficiency (1). Suppose that a general graph G satisfies (6.5). Let N be a sufficiently large odd integer. For example, N is an odd integer greater than $|G| + \|G\|$. We define $g : V(G) \rightarrow \mathbb{Z}$ by $g(x) = -N$ for all $x \in V(G)$. Then a $(1,f)$-odd factor and a parity (g,f)-factor are the same. Thus it suffices to show that G has a parity (g,f)-factor.

Let S and T be two disjoint subsets of $V(G)$. If $T \neq \emptyset$, then $\deg_G(x) - g(x) = \deg_G(x) + N$ for every $x \in T$. Since N is sufficiently large, we obtain

$$\eta(S, T) = \sum_{x \in S} f(x) + \sum_{x \in T} (\deg_G(x) - g(x)) - e_G(S, T) - h(S, T)$$
$$\geq N - e_G(S, T) - h(S, T) \geq 0.$$

Hence we may assume $T = \emptyset$. It is immediate that a component C of $G - S$ satisfying (6.3) has odd order since

$$\sum_{x \in V(C)} f(x) \equiv |C| \equiv 1 \pmod 2.$$

Hence $q(S, \emptyset) = odd(G - S)$, and therefore it follows from (6.2) and (6.5) that

$$\eta(S, \emptyset) = \sum_{x \in S} f(x) - q(S, \emptyset) = \sum_{x \in S} f(x) - odd(G - S) \geq 0.$$

Consequently, by the parity (g, f)-factor theorem, G has a parity (g, f)-factor, which is the desired $(1, f)$-odd factor of G. $\square$

In order to give another proof of the $(1, f)$-odd factor theorem, we need the next theorem, which gives a criterion for a tree to have a $(1, f)$-odd factor.

Theorem 6.6 ([60]). *Let T be a tree of even order and $f : V(G) \to \{1, 3, 5, \ldots\}$. Then T has a $(1, f)$-odd factor if and only if*

$$odd(T - x) \leq f(x) \qquad for\ all \qquad x \in V(T). \tag{6.6}$$

Proof. We shall prove only the sufficiency since the necessity can be easily shown as in the proof of necessity of Theorem 6.5. Let F be a subgraph of T induced by the edge subset

$$A = \{e \in E(T) : odd(T - e) = 2\}. \qquad \text{(see Fig. 6.2)}$$

Note that for any edge e of T, $T - e$ consists of two components, which are simultaneously odd or even. Here the set A consists of those edges e for which $T - e$ consists of two odd components. For any vertex v of T, $T - v$ has odd order since T is of even order, and so $T - v$ has at least one odd component, say C. Then the edge e of T joining C to v is contained in A, which implies that F is a spanning subgraph of T. Obviously, an edge of T joining v to an even component of $G - v$ is not contained in A. Hence the number of edges in A incident with v is equal to the number of odd components of $G - v$. Furthermore, since $T - v$ has odd order, the number of odd components of $T - v$ must be odd. Therefore

$$\deg_F(v) = odd(T - v) \qquad \text{and} \qquad odd(T - v) \equiv 1 \pmod 2.$$

Consequently, by (6.6), F is the desired $(1, f)$-odd factor of T. $\square$

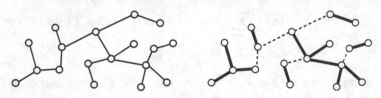

Fig. 6.2. A tree T of even order and its unique odd factor shown by bold edges; Every bold edge e satisfies $odd(T - e) = 2$.

Note that if a tree T of even order has a $(1, f)$-odd factor, then such a factor is unique in T (see Problem 6.1).

The next theorem is an easy consequence of the $(1, f)$-odd factor theorem since it is obtained only by defining $f(x)$ to be a large odd integer, where an **odd factor** and an **odd subgraph** mean a spanning subgraph and a subgraph with all degrees odd, respectively. Notice that every connected graph G has a vertex v such that $G - v$ is connected since this holds for a tree as well.

Theorem 6.7 (Problem 42 in Sec. 7 of [181]). *Let G be a connected general graph. Then the following statements hold.*
(i) If G has even order, then G has an odd factor.
(ii) If G has odd order, then G has an odd subgraph of order $|G| - 1$.

We are ready to give another proof for the $(1, f)$-odd factor theorem, which does not use the parity (g, f)-factor theorem.

Proof of the necessity of the $(1, f)$-odd factor theorem (2) ([14], [60]). We shall prove the theorem by induction on the size $\|G\|$ of G. We may assume that G is connected. Then G is of even order since $odd(G - \emptyset) = 0$ by (6.5). Moreover, by Theorem 6.6, we may assume that G is not a tree. It is clear that for every non-empty subset S of $V(G)$, we have

$$odd(G - S) \equiv |S| \equiv \sum_{x \in S} f(x) \pmod 2. \tag{6.7}$$

We consider the following two cases.

Case 1. $odd(G - X) < \sum_{x \in X} f(x)$ *for all* $\emptyset \neq X \subset V(G)$.

In this case, we have $odd(G - X) \leq \sum_{x \in X} f(x) - 2$ by (6.7). Since G is not a tree, we can take an edge e_1 of G such that $G - e_1$ is connected. For any non-empty subset $S \subset V(G - e_1) = V(G)$, we have

$$odd(G - e_1 - S) \leq odd(G - S) + 2 \leq \sum_{x \in S} f(x).$$

By the inductive hypothesis, $G - e_1$ has a $(1, f)$-odd factor, which is of course the desired $(1, f)$-odd factor of G.

Case 2. $odd(G - X) = \sum_{x \in X} f(x)$ *for some* $\emptyset \neq X \subset V(G)$.

Choose a maximal subset $S \subset V(G)$ that satisfies $odd(G - S) = \sum_{x \in S} f(x)$. Then

$$odd(G - Y) < \sum_{x \in Y} f(x) \qquad \text{for all} \quad S \subset Y \subseteq V(G). \qquad (6.8)$$

Claim 1. *Every even component of* $G - S$ *has a* $(1, f)$-*odd factor.*

Let D be an even component of $G - S$, and $\emptyset \neq X \subset V(D)$. Then by (6.8), we have

$$odd(G - S) + odd(D - X) = odd(G - (S \cup X)) < \sum_{x \in S \cup X} f(x),$$

which implies $odd(D - X) < \sum_{x \in X} f(x)$. Hence the claim holds by induction.

Claim 2. *For any odd component* C *of* $G - S$ *and any edge* $e = vw$ *joining* $v \in V(C)$ *to* $w \in S$, $C + e$ *has a* $(1, f)$-*odd factor, where* $f(w) = 1$ *(Fig. 6.3).*

Let $\emptyset \neq X \subset V(C + e) = V(C) \cup \{w\}$. If $X \neq \{w\}$, then by (6.8), we have

$$\sum_{x \in S \cup X} f(x) > odd(G - (S \cup T))$$

$$= odd(G - S) - 1 + odd\left((C + e) - (X \cup \{w\})\right)$$

$$\geq \sum_{x \in S} f(x) - 1 + odd((C + e) - X) - 1.$$

Thus $\sum_{x \in X} f(x) > odd((C+e)-X)-2$, which implies $\sum_{x \in X} f(x) \geq odd((C+e) - X)$ by (6.7). If $X = \{w\}$, then $C + e - \{w\} = C$ has odd order, and thus $odd((C+e)-\{w\}) = 1 = f(w)$. Hence $C+e$ satisfies (6.5), and has the desired $(1, f)$-odd factor by induction. Therefore Claim 2 is proved.

Let $C_1, C_2, \ldots, C_m$ be the odd components of $G - S$, where $m = odd(G - S) = \sum_{x \in S} f(x)$. We construct a bipartite graph B with bipartition $S \cup \{C_1, C_2, \ldots, C_m\}$ as follows: a vertex x of S and C_i are joined by an edge of B if and only if x and C_i are joined by at least one edge of G (see Fig. 6.3).

Claim 3. *The bipartite graph* B *has a factor* K *such that* $\deg_K(x) = f(x)$ *for* $x \in S$ *and* $\deg_K(C_i) = 1$ *for all* $1 \leq i \leq m$.

It follows from the connectedness of G that $N_B(S) = \{C_1, C_2, \ldots, C_m\}$. Assume that $|N_B(X)| < \sum_{x \in X} f(x)$ for some $\emptyset \neq X \subset S$. Then every vertex $C_i \in \{C_1, C_2, \ldots, C_m\} - N_B(X)$ is an isolated vertex of $B - (S - X)$, which implies that C_i is an odd component of $G - (S - X)$, and thus

$$odd(G - (S - X)) \geq |\{C_1, C_2, \ldots, C_m\} - N_B(X)|$$

$$> \sum_{x \in S} f(x) - \sum_{x \in X} f(x) = \sum_{x \in S - X} f(x).$$

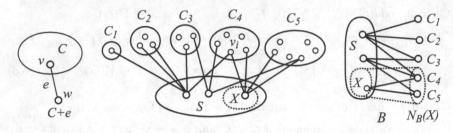

Fig. 6.3. A subgraph $C + e$, the odd components of $G - S$ with edges $v_i w_i$, and the bipartite graph B.

This contradicts (6.5). Therefore $|N_B(X)| \geq \sum_{x \in X} f(x)$ for all $X \subseteq S$, and so by the generalized marriage theorem (Theorem 2.10), B has the desired factor K.

Let K be the factor of B given in Claim 3. For every edge $w_i C_i$ of K with $w_i \in S$, choose a vertex $v_i \in V(C_i)$ that is adjacent to w_i in G, and take a $(1, f)$-odd factor $F(C_i + v_i w_i)$ of $C_i + v_i w_i$, whose existence is guaranteed by Claim 2. For every even component D_j of $G - S$, D_j has a $(1, f)$-odd factor $F(D_j)$ by Claim 1. By combining the above factors, we obtain the following desired $(1, f)$-odd factor F of G:

$$F = \bigcup_j F(D_j) \ \cup \ \bigcup_i F(C_i + v_i w_i).$$

Consequently the theorem is proved. $\square$

The next theorem is obtained by setting $f(x) = n$ for all vertices x, though the second proof of the $(1, f)$-odd factor theorem was first essentially given in [14].

Theorem 6.8 (Amahashi [14] (1985)). *Let G be a general graph and $n \geq 1$ be an odd integer. Then G has a $[1, n]$-odd factor if and only if*

$$odd(G - S) \leq n|S| \qquad for \ all \quad S \subseteq V(G). \tag{6.9}$$

6.2 $(1, f)$-odd subgraphs and structure theorem

Let G be a graph and $f : V(G) \to \{1, 3, 5, 7, \ldots\}$. Then a subgraph H of G is called a $(1, f)$**-odd subgraph** if

$$\deg_H(x) \in \{1, 3, \ldots, f(x)\} \qquad for \ all \quad x \in V(H). \tag{6.10}$$

Of course, a spanning $(1, f)$-odd subgraph is a $(1, f)$-odd factor. For a constant odd integer $n \geq 1$, a $[1, n]$-**odd subgraph** can be defined analogously.

By the $(1, f)$-odd factor theorem, we can expect that many results and properties on 1-factors and matchings can be extended to those on $(1, f)$-odd factors and $(1, f)$-odd subgraphs. In this section we show that some results on matchings, including the Gallai-Edmonds structure theorem, can be extended.

A $(1, f)$-odd subgraph H of G is said to be **maximum** if G has no $(1, f)$-odd subgraph H' such that $|H| < |H'|$. On the other hand, a $(1, f)$-odd subgraph K is called a **maximal** (with respect to vertex set) if G has no $(1, f)$-odd subgraph K' such that $V(K) \subset V(K')$. Note that a matching M is called maximal if there exists no matching M' such that $M \subset M'$, that is, "maximal" is defined by using edge sets instead of vertex sets. It is trivial that every maximum $(1, f)$-odd subgraph is a maximal $(1, f)$-odd subgraph, and the converse is also true. Namely, as we state in Problem 6.2, every maximal $(1, f)$-odd subgraph is a maximum $(1, f)$-odd subgraph.

Let G be a general graph, and H be a $(1, f)$-odd subgraph of G. Then for any cycle C of H, which may be a loop or a cycle consisting of two edges, $H - E(C)$ is also a $(1, f)$-odd subgraph with vertex set $V(H)$. By repeating this procedure until it has no cycle, we can obtain a $(1, f)$-odd subgraph that has no cycles and vertex set $V(H)$. In particular, a general graph G has a $(1, f)$-odd subgraph with vertex set W if and only if its underlying simple graph has one with vertex set W. Therefore, when we consider a $(1, f)$-odd subgraph or $(1, f)$-odd factor, we may restrict ourselves to simple graphs.

It is clear that the next theorem is a generalization of Theorem 4.11, which gives a formula for the order of a maximum matching.

Theorem 6.9 (Kano and Katona [120]). *Let G be a simple graph and let $f : V(G) \to \{1, 3, 5, \ldots\}$ be a function. Then the order of a maximum $(1, f)$-odd subgraph H of G is given by*

$$|H| = |G| - \max_{X \subseteq V(G)} \{odd(G - X) - \sum_{x \in X} f(x)\}. \qquad (6.11)$$

Proof. Let H be a maximum $(1, f)$-odd subgraph of G, and let

$$d = \max_{X \subseteq V(G)} \{odd(G - X) - \sum_{x \in X} f(x)\}.$$

Then $d \geq 0$ since $odd(G) \geq 0$, and $|G| + d$ is even since for every $X \subset V(G)$,

$$|G| \equiv odd(G - X) + |X| \equiv odd(G - X) - |X|$$
$$\equiv odd(G - X) - \sum_{x \in X} f(x) = d \pmod 2.$$

We first show that $|H| \leq |G| - d$. Let $S \subset V(G)$ such that

$$odd(G - S) - \sum_{x \in S} f(x) = d.$$

Then for any odd component C of $G - S$, if $V(C)$ is covered by H, then there exists at least one edge in H that joins C to S. Thus at most $\sum_{x \in S} \deg_H(x)$ odd components are covered by H. Since $\deg_H(x) \le f(x)$, at least $odd(G - S) - \sum_{x \in S} f(x)$ odd components of $G - S$ are not covered by H. This implies $|H| \le |G| - d$.

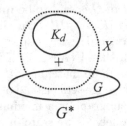

G^*

Fig. 6.4. $G^* = G + K_d$ and a subset X, $V(K_d) \subset X \subset V(G^*)$.

We next prove the reverse inequality. Let $G^* = G + K_d$ be the join of G and the complete graph K_d, and define $f' : V(G^*) \to \{1, 3, 5, \ldots\}$ by

$$f'(x) = \begin{cases} f(x) & \text{if } x \in V(G) \\ 1 & \text{if } x \in V(K_d). \end{cases}$$

Let $\emptyset \ne X \subset V(G^*)$. If $V(K_d) \not\subseteq X$, then $odd(G^* - X) \le 1 \le \sum_{x \in X} f'(x)$. If $V(K_d) \subseteq X$ (Fig. 6.4), then

$$odd(G^* - X) = odd(G - X \cap V(G)) \le \sum_{x \in X \cap V(G)} f(x) + d \le \sum_{x \in X} f'(x).$$

Hence by the $(1, f)$-odd factor theorem, G^* has a $(1, f')$-odd factor F'. Let $H' = F' - V(K_d)$. Since $\deg_{F'}(x) = f'(x) = 1$ for all $x \in V(K_d)$, H' has at most d even vertices. Therefore H' has at most d odd components.

By applying (i) or (ii) of Theorem 6.7 to each component of H' according to its parity, we obtain an odd subgraph M of H' such that $|M| \ge |H'| - d = |G| - d$. Since M is a $(1, f)$-odd subgraph of G, the proof is complete. $\qquad \square$

We now discuss barriers for $(1, f)$-odd factors. For a graph G and a function $f : V(G) \to \{1, 3, 5, \ldots\}$, a non-empty subset $S \subset V(G)$ is called a **barrier** for (G, f) if $odd(G - S) > \sum_{x \in S} f(x)$. A barrier S is said to be **minimal** if no proper subset of S is a barrier. One important feature of barriers is shown in the next theorem.

Theorem 6.10 (Topp and Vestergaard [224]). *Let G be a connected simple graph of even order, and let $f : V(G) \to \{1, 3, 5, \ldots\}$ be a function. Assume that G has no $(1, f)$-odd factor and let S be a minimal barrier for (G, f). Then*

(i) $odd(G - S) \geq \sum_{x \in S} f(x) + 2$;
(ii)   each vertex v of S is adjacent to at least $f(v) + 2$ distinct odd components of $G - S$;
(iii)   each vertex v of S is the center of an induced star $K(1, f(v)+2)$ in G; and
(iv) $\kappa(G) \leq |S| \leq \min\{(|G| - 2)/(f_0 + 1), (\alpha(G) - 2)/f_0, ||G|| - |G| + 2\}$, where $f_0 = \min\{f(x) : x \in V(G)\}$.

Proof. (i)   Let S be a minimal barrier for (G, f). Then $odd(G - S) > \sum_{x \in S} f(x)$, which implies (i) by (6.7).

(ii)   If $v \in S$ is adjacent to m distinct odd components of $G - S$, then since S is a minimal barrier and by (i), it follows that

$$\sum_{x \in S-v} f(x) \geq odd(G - (S - v))$$

$$\geq odd(G - S) - m \geq \sum_{x \in S} f(x) + 2 - m.$$

Hence we have $m \geq f(v) + 2$.

(iii)   Since each $v \in S$ is adjacent to at least $f(v) + 2$ odd components of $G - S$, we can find an induced star $K(1, f(v) + 2)$ with center v.

(iv)   Since S is a cut of G, $|S| \geq \kappa(G)$. By (i), we have

$$|G| \geq |S| + odd(G - S) \geq |S| + \sum_{x \in S} f(x) + 2 \geq |S| + f_0|S| + 2.$$

Hence $|S| \leq (|G| - 2)/(f_0 + 1)$.
By taking one vertex from each odd component of $G - S$, we get an independent set I of cardinality $odd(G - S)$. Therefore

$$\alpha(G) \geq |I| = odd(G - S) \geq \sum_{x \in S} f(x) + 2 \geq f_0|S| + 2,$$

and thus $|S| \leq (\alpha(G) - 2)/f_0$. We omit the proof of the last inequality.   □

Let $n \geq 1$ be an odd integer. By the above Theorem 6.10, it follows that if a connected graph G of even order has no induced star $K(1, n + 2)$, then G has a $[1, n]$-odd factor. This is a generalization of Theorem 2.43, which says that every connected $K(1, 3)$-free graph of even order has a 1-factor. Sumner extended this result to graphs with high connectivity, that is, he showed that if an n-connected graph of even order has no induced subgraph isomorphic to $K(1, n+1)$, then G has a 1-factor [218]. The following theorem is a generalization of this result.

Theorem 6.11 (Topp and Vestergaard [224]). *Let G be an n-connected simple graph of even order and let $f : V(G) \to \{1, 3, 5, \ldots\}$ be a function. If no vertex v of G is the center of an induced star $K(1, nf(v)+1)$, then G has a $(1, f)$-odd factor.*

Proof. Suppose that G does not have a $(1, f)$-odd factor. Let S be a minimal barrier for (G, f). Let $D_1, D_2, \ldots, D_t$ be the odd components of $G - S$. Since G is n-connected, for every $i \in \{1, \ldots, t\}$, the set $N_G(D_i) \cap S$ contains at least n distinct vertices, and so we can take a set E_i of n edges joining D_i to n distinct vertices of S. Then the set $E = \cup_{i=1}^t E_i$ has exactly nt edges. Since $t = odd(G - S) \geq \sum_{x \in S} f(x) + 2$ by Theorem 6.6 and (6.7), we have

$$|E| = nt \geq n(\sum_{x \in S} f(x) + 2) > \sum_{x \in S} nf(x).$$

This implies that some vertex $v \in S$ is incident with at least $nf(v) + 1$ edges of E. Certainly, if a vertex $v \in S$ is incident with at least $nf(v) + 1$ edges of E, then v the center of an induced star $K(1, nf(v) + 1)$ in G. Therefore the theorem is proved. □

It was shown in Theorem 2.41 that if a connected simple graph G of even order satisfies

$$N_G(S) = V(G) \quad \text{or} \quad |N_G(S)| > \frac{4}{3}|S| - 1 \quad \text{for all} \quad S \subset V(G),$$

then G has a 1-factor. This result is extended to $[1, n]$-odd factors in the next theorem. However, it is not known whether there exists a similar result for $(1, f)$-odd factors.

Theorem 6.12 ([60]). *Let G be a simple graph of even order and n an odd positive integer. If G satisfies*

$$N_G(S) = V(G) \quad or \quad |N_G(S)| > \left(1 + \frac{1}{3n}\right)|S| - \frac{1}{n} \tag{6.12}$$

for all $S \subseteq V(G)$, then G has a $[1, n]$-odd factor.

Proof. Suppose that G satisfies (6.12) but has no $[1, n]$-odd factor. Then by Theorem 6.8 and (6.7), there exists $\emptyset \neq S \subset V(G)$ such that $odd(G - S) \geq n|S| + 2$. Let m denote the number of isolated vertices of $G - S$, and let

$$t = 1 + \frac{1}{3n} \quad \text{and} \quad r = \frac{1}{n}.$$

We consider two cases.

Case 1. $m \geq 1$.

Since $N_G(V(G) - S) \neq V(G)$ (Fig. 6.5), by (6.12) we have

$$|N_G(V(G) - S)| > t|V(G) - S| - r = t|G| - t|S| - r.$$

It is clear that $|G| - m \geq |N_G(V(G) - S)|$. From these two inequalities, we obtain

$$\frac{t|S| + r - m}{t - 1} > |G|. \tag{6.13}$$

On the other hand, $G - S$ has at least $n|S| + 2 - m$ odd components with order at least three, and thus $m + 3(n|S| + 2 - m) \leq |G| - |S|$. Hence

$$(3n + 1)|S| + 6 - 2m \leq |G|. \tag{6.14}$$

Combining (6.13) and (6.14), we obtain

$$(3n + 1)|S| + 6 - 2m < \frac{t|S| + r - m}{t - 1}. \tag{6.15}$$

Substituting the values of t and r into (6.15), we obtain

$$(3n + 1)|S| + 6 - 2m < \frac{(1 + \frac{1}{3n})|S| + \frac{1}{n} - m}{\frac{1}{3n}}$$

$$(3n + 1)|S| + 6 - 2m < (3n + 1)|S| + 3 - 3nm$$

$$3 + (3n - 2)m < 0.$$

This is a contradiction.

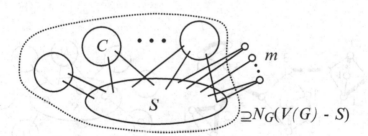

Fig. 6.5. A graph G with S and $N_G(V(G) - S)$.

Case 2. $m = 0$.

In this case, every odd component has at least three vertices. Take one odd component C of $G - S$, and let $X = V(G) - S - V(C)$ (Fig. 6.5). Then since $N_G(X) \neq V(G)$, it follows from (6.12) that $|N_G(X)| > t|X| - r$. It is obvious that $|N_G(X)| \leq |X| + |S|$. Thus $t|X| - r < |X| + |S|$ and hence

$$|X| < \frac{|S| + r}{t - 1}. \tag{6.16}$$

On the other hand, $|X| \geq 3(n|S|+1)$ as well. Thus combining this and (6.16), we obtain

$$3(n|S|+1) < \frac{|S|+r}{t-1}.$$

Substituting the values of t and r in the above inequality, we have

$$3(n|S|+1) < \frac{|S|+\frac{1}{n}}{\frac{1}{3n}}$$

$$3(n|S|+1) < 3n|S|+3$$

$$0 < 0.$$

This is a contradiction. Consequently the theorem is proved. □

Petersen proved that every 2-edge connected cubic graph has a 1-factor (see Theorem 2.36) and there are cubic graphs having no 1-factors. If we consider a special $(1,f)$-odd factor instead of a 1-factor, we can guarantee the existence of such a factor in every cubic graph as shown in the following theorem.

Theorem 6.13 ([129]). *Let G be a connected cubic multigraph, and define a function f as*

$$f(x) = \begin{cases} 3 & \text{if } x \text{ is a cut vertex,} \\ 1 & \text{otherwise.} \end{cases} \qquad (6.17)$$

Then G has a $(1,f)$-odd factor (see Fig. 6.6).

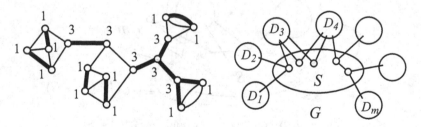

Fig. 6.6. (a) A cubic multigraph and its $(1,f)$-odd factor; numbers denote $f(x)$. (b) $G - S$ and the odd components of $G - S$.

Proof. Let $S \subset V(G)$ with $|S| \geq 2$, and let $D_1, D_2, \ldots, D_m$ be the odd components of $G - S$, where $m = odd(G - S)$ (Fig. 6.6). Note that if S consists of one vertex, then the required inequality (6.5) in the $(1,f)$-odd factor theorem easily holds. Hence we may assume that S contains at least two vertices. Moreover, if $m \leq 2$, then (6.5) also holds, and so we may assume $m \geq 3$. For every $D = D_i$, it follows that

$$1 \equiv 3|D| = \sum_{x \in V(D)} \deg_G(x) = 2||D|| + e_G(D, S) \equiv e_G(D, S) \quad (\text{mod } 2).$$

Hence $e_G(D, S) = 1$ or $e_G(D, S) \geq 3$, moreover, if $e_G(D, S) = 1$ then D must be adjacent to a cut vertex of S. Let

$$\Omega = \{D_i : e_G(D_i, S) = 1, \ 1 \leq i \leq m\} \qquad \text{and} \qquad t = |\Omega|.$$

Let $Cutv(S)$ denote the set of cut vertices of S. Since G is connected and $m \geq 3$, for every cut vertex v of S, at most two components in Ω are adjacent to v (Fig. 6.6), and thus $t \leq 2|Cutv(S)|$. It follows from $e_G(D, S) = 1$ or $e_G(D, S) \geq 3$ that

$$t + 3(m - t) \leq e_G(D_1 \cup D_2 \cup \cdots \cup D_m, S)$$
$$\leq \sum_{x \in S} \deg_G(x) = 3|S| = \sum_{x \in S - Cutv(S)} 3f(x) + \sum_{x \in Cutv(S)} f(x).$$

Since $t \leq 2|Cutv(S)| = (2/3)\sum_{x \in Cutv(S)} f(x)$, it follows from the above inequality that

$$m \leq \sum_{x \in S - Cutv(S)} f(x) + \frac{1}{3}\sum_{x \in Cutv(S)} f(x) + \frac{2}{3}t$$
$$\leq \sum_{x \in S - Cutv(S)} f(x) + \left(\frac{1}{3} + \frac{4}{9}\right)\sum_{x \in Cutv(S)} f(x)$$
$$\leq \sum_{x \in S} f(x).$$

Consequently, by the $(1, f)$-odd factor theorem, G has the desired $(1, f)$-odd factor. $\square$

Here we give some remarks on the following open problem, which one could easily surmise from Theorems 6.5 and 6.6. Note that if G satisfies

$$odd(G - S) \leq \sum_{x \in S} f(x) \qquad \text{for all} \quad S \subset V(G),$$

then by taking a minimal $(1, f)$-odd factor, which is a forest, and by adding some edges to it, we can obtain a spanning tree T. This spanning tree T satisfies $odd(T - v) \leq f(v)$ for all $v \in V(T)$.

Problem 6.14. Let G be a connected simple graph and $h : V(G) \to \{2, 4, 6, \ldots\}$.
   (1) If G satisfies

$$odd(G - S) \leq 2|S| \qquad \text{for all} \quad \emptyset \neq S \subset V(G),$$

what factor or property does G have?

(2) If G satisfies

$$odd(G - S) \leq \sum_{x \in S} h(x) \qquad \text{for all} \quad \emptyset \neq S \subset V(G),$$

what factor or property does G have?

(3) If G satisfies

$$odd(G - S) \leq \sum_{x \in S} h(x) \qquad \text{for all} \quad \emptyset \neq S \subset V(G),$$

then does G has a spanning tree T such that

$$odd(T - v) \leq h(v) \qquad \text{for all} \quad v \in V(T)?$$

For problem (3) above, the following are some results hinting that no such spanning tree is likely to exist.

Remark 6.15. Let $m \geq 2$ be an even integer and $G = K(2, 2m)$. Then G has even order and satisfies $odd(G - S) \leq m|S|$ for all $S \subset V(G)$. However, G has no spanning tree T such that $odd(T - v) \leq m$ for all $v \in (G)$.

Remark 6.16. Let n be a positive odd integer and G be the cycle C_n of order n. Then G has odd order and satisfies $odd(G - S) \leq |S|$ for all non-empty subsets S of $V(G)$, but G has no spanning tree T such that $odd(T - v) \leq 1$ for all $v \in V(G)$.

Remark 6.17 (Saito [215]). Let $m \geq 2$ be an even integer and $G = K(3, 3m)$. Then G has odd order and satisfies $odd(G - S) \leq m|S|$ for all non-empty subsets S of $V(G)$. However, G has no spanning tree T such that $odd(T - S) \leq m|S|$ for all non-empty subsets S of $V(G)$, although G contains a spanning tree R which satisfies $odd(R - v) \leq m$ for all $v \in V(G)$.

We study some other properties of maximum $(1, f)$-odd subgraphs, which are extensions of properties of maximum matchings. In particular, we obtain a structure theorem on $(1, f)$-odd subgraphs. In order to give short proofs for these results, we introduce a new method based on Kano, Katona and Szabó [123]. Moreover, by this method, we can generalize some properties of elementary graphs to those of elementary graphs with respect to $(1, f)$-odd factors [123], which are not explained here. For elementary graphs, we refer the reader to Lovász and Plummer [182] and Yu and Liu [254].

For a simple graph G and a function $f : V(G) \to \{1, 3, 5, \ldots\}$, define the new graph G^f as follows: replace every vertex $v \in V(G)$ by the complete graph $K_{f(v)}$ on $f(v)$ vertices, and for every edge xy of G, join every vertex of $K_{f(x)}$ to every vertex of $K_{f(y)}$. In particular, there are $f(x)f(y)$ edges between $K_{f(x)}$ and $K_{f(y)}$ in G^f (see Fig. 6.7). Then

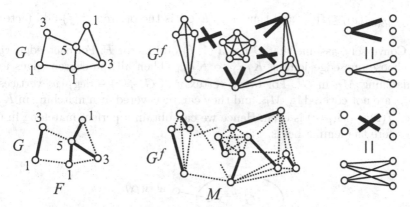

Fig. 6.7. A graph G and G^f; where the numbers denote $f(v)$; a $(1, f)$-odd factor F of G and a perfect matching M in G^f, where $F = M \cap E(G)$.

$$|G^f| = \sum_{x \in V(G)} f(x) \quad \text{and} \quad \|G^f\| = \sum_{xy \in E(G)} f(x)f(y).$$

There is a strong relationship between maximum matchings in G^f and maximum $(1, f)$-odd subgraphs of G. Before discussing this, we make a remark on matchings in G^f. If a matching M_1 in G^f contains two edges joining $K_{f(u)}$ to $K_{f(w)}$, then remove these two edges and add two edges, one of $K_{f(u)}$ and the other of $K_{f(w)}$, to M_1 so that the two added edges cover the four endpoints of the two removed edges. By repeating this procedure, we can obtain a matching M_2 such that $V(M_2) = V(M_1)$ and M_2 contains at most one edge between $K_{f(x)}$ and $K_{f(y)}$ for all edges xy of G. Therefore, we can obtain a matching in G^f possessing the following property.

Property (a) A matching in G^f that contains at most one edge between $K_{f(x)}$ and $K_{f(y)}$ for every edge xy of G.

For a matching M in G^f having property (a), let $M \cap E(G)$ denote the subgraph of G induced by

$$\{xy \in E(G) : M \text{ contains one edge joining } K_{f(x)} \text{ to } K_{f(y)}\}. \qquad (6.18)$$

The next lemma is a starting point of the following results on maximum $(1, f)$-odd subgraphs.

Lemma 6.18. *Let G be a simple graph and $f : V(G) \rightarrow \{1, 3, 5, \ldots\}$. Then G has a $(1, f)$-odd factor if and only if G^f has a perfect matching (Fig. 6.7).*

Proof. Suppose that G^f has a perfect matching M with property (a). Then for every vertex v of G, $\deg_{M \cap E(G)}(v)$ is odd since an even number of vertices of $K_{f(v)}$ are saturated by the edges of $M \cap E(K_{f(v)})$. It is clear that

$\deg_{M \cap E(G)}(v) \leq f(v)$, and thus $M \cap E(G)$ is the desired $(1, f)$-odd factor of G.

Conversely, assume that G has a $(1, f)$-odd factor F. For each edge xy of F, choose one edge joining $K_{f(x)}$ to $K_{f(y)}$. Then all the chosen edges form a matching M_F in G^f. For every vertex v of G, $f(v) - \deg_F(v)$ vertices of $K_{f(v)}$ are not covered by M_F, and they can be covered by a matching in $K_{f(v)}$ since $f(v) - \deg_F(v)$ is even. Hence we can obtain a perfect matching in G^f. Therefore the lemma holds. □

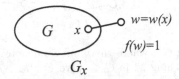

Fig. 6.8. A graph G and G_x for a vertex x.

For a vertex x of G, we denote by G_x the graph obtained from G by adding a new vertex $w = w(x)$ together with a new edge wx, and define $f(w) = 1$ (Fig. 6.8). Let

$$\tau_f(G) = \text{the order of a maximum } (1, f)\text{-odd subgraph of } G.$$

Note that $\tau(G) = \tau_1(G)$ denotes the order of a maximum matching in G, where $1 : V(G) \rightarrow \{1\}$. Let

$$D_f(G) = \{x \in V(G) : \tau_f(G_x) = \tau_f(G) + 2\}.$$

Namely, $D_f(G)$ consists of vertices x such that G_x has a maximum $(1, f)$-odd subgraph of order $\tau_f(G) + 2$, which must contain the edge $xw(x)$. Recall that $D(G)$ consists of all the vertices v of G such that v is not saturated by some maximum matching in G. It is immediate that $D(G) = D_1(G)$, where $1 : V(G) \rightarrow \{1\}$.

Lemma 6.19. *Let G be a simple graph and $f : V(G) \rightarrow \{1, 3, 5, \ldots\}$. Then $V(K_{f(v)}) \cap D(G^f) \neq \emptyset$ implies $V(K_{f(v)}) \subseteq D(G^f)$. Moreover, it follows that*

$$|G| - \tau_f(G) = |G^f| - \tau(G^f), \quad \text{and} \tag{6.19}$$
$$V(K_{f(v)}) \subseteq D(G^f) \quad \text{if and only if} \quad v \in D_f(G). \tag{6.20}$$

Proof. If there exists a maximum matching in G^f that does not saturate a vertex in $K_{f(v)}$, then obviously for each vertex x of $K_{f(v)}$, G^f has a maximum matching that does not saturate x. Hence $V(K_{f(v)}) \cap D(G^f) \neq \emptyset$ implies $V(K_{f(v)}) \subseteq D(G^f)$.

We next prove (6.19). Let $m = |G^f| - \tau(G^f)$. Then every maximum matching in G^f does not saturate exactly m vertices of G^f. By Lemma 6.18, if $m = 0$ then (6.19) holds. So we may assume $m \geq 1$.

For convenience, we say that a matching M in G^f **saturates** $K_{f(v)}$ if M saturates all the vertices of $K_{f(v)}$, otherwise M does not saturate $K_{f(v)}$.

Let M be a maximum matching of G^f having property (a). Then for any vertex v of G, $\deg_{M \cap E(G)}(v)$ is even if and only if M does not saturate $K_{f(v)}$. Thus there are exactly m vertices in G whose degrees in $M \cap E(G)$ are even. Since $M \cap E(G)$ contains at most m odd components, by Theorem 6.7 $M \cap E(G)$ contains a $(1, f)$-odd subgraph H with order at least $|G| - m$, which implies $|G| - \tau_f(G) \leq m$.

Conversely, let H^* be a maximum $(1, f)$-odd subgraph of G. Then G^f has a maximal matching M^* such that $M^* \cap E(G) = E(H^*)$, that is, M^* saturates $K_{f(v)}$ if and only if H^* contains v. Hence M^* satisfies $|G^f| - |M^*| = |G| - |H^*|$. Therefore

$$|G| - \tau_f(G) = |G| - |H^*| = |G^f| - |M^*| \geq |G^f| - \tau(G^f) = m.$$

Hence equality (6.19) holds.

Finally, we prove (6.20). Assume that a vertex v of G is contained in $D_f(G)$. Let H_v be a maximum $(1, f)$-odd subgraph of $G_v = G + vw$, where $w = w(v)$. Since $\tau_f(G_v) = \tau_f(G) + 2$, H_v must contain w, and so $H_v - w$ has even degree only at v and does not contain exactly $|G_v| - \tau_f(G_v)$ vertices of G except at v. Thus there exists a matching M in G^f such that $M \cap E(G) = E(H_v - w)$ and M does not saturate the following number of vertices of G^f since $|H_v| = \tau_f(G_v) = \tau_f(G) + 2$.

$$|G^f| - |M| = |G_v| - |H_v| + 1 = |G| - \tau_f(G).$$

Hence by (6.19), M is a maximum matching in G^f and does not saturate $K_{f(v)}$. Hence $V(K_{f(v)}) \subseteq D(G^f)$.

Conversely, assume that $V(K_{f(v)}) \subseteq D(G^f)$ for some vertex v of G. Then G^f has a maximum matching M with property (a) that does not saturate $K_{f(v)}$. Then $M \cap E(G)$ has even degree at v, and a vertex u of G has odd degree in $M \cap E(G)$ if and only if $V(K_{f(u)}) \subseteq V(M)$. Thus $M \cap E(G)$ has even degree at exactly $|G^f| - \tau(G^f)$ vertices of G including v. Hence a subgraph $(M \cap E(G)) + vw$ of $G_v = G + vw$ has precisely $|G^f| - \tau(G^f) - 1$ vertices of even degree, which implies by Theorem 6.7 that $(M \cap E(G)) + vw$ contains a $(1, f)$-odd subgraph H with order at least $|G_v| - (|G^f| - \tau(G^f) - 1)$. Hence by (6.19), we have

$$|G_v| - |H| \leq |G^f| - |\tau(G^f)| - 1 = |G| - \tau_f(G) - 1.$$

Since $|G_v| = |G| + 1$, we have $\tau_f(G) + 2 \leq |H|$, and thus $\tau_f(G_v) = \tau_f(G) + 2$ since $|H| \leq \tau_f(G_v) \leq \tau_f(G) + 2$. Therefore $v \in D_f(G)$. Consequently the lemma is proved. $\square$

The proof of Lemma 6.19 includes the following lemma.

Lemma 6.20. *Let G be a simple graph and $f : V(G) \to \{1, 3, 5, \ldots\}$. If M is a maximum matching in G^f with property (a), then $M \cap E(G)$ is a maximum $(1, f)$-odd subgraph of G. Conversely, if H is a maximum $(1, f)$-odd subgraph of G, then there exists a maximum matching M^* in G^f with property (a) such that $M^* \cap E(G) = E(H)$.*

Recall that a graph G is said to be **factor-critical** if $G - v$ has a 1-factor for every vertex v of G. From this definition, when $f : V(G) \to \{1, 3, 5, \ldots\}$ is given, a graph G is said to be **critical with respect to $(1, f)$-odd factors** if $G_v = G + vw$ has a $(1, f)$-odd factor for all vertices v of G, where $w = w(v)$.

Lemma 6.21. *Let G be a simple graph and $f : V(G) \to \{1, 3, 5, \ldots\}$. Then G is critical with respect to $(1, f)$-odd factors if and only if G^f is factor-critical.*

Proof. Assume that G is critical with respect to $(1, f)$-odd factors. Then for an arbitrary vertex v of G, $G_v = G + vw$ has a $(1, f)$-odd factor F_v. Then $F_v - w$ is a spanning subgraph of G with only one vertex v of even degree. Thus there exists a matching M in G^f such that $M \cap E(G) = E(F_v - w)$ and M does not saturate exactly one vertex of G^f, which is contained in $K_{f(v)}$. Hence M is a maximum matching in G^f, and it is immediate that for any vertex x of $K_{f(v)}$, $G^f - x$ has a 1-factor. Hence G^f is factor-critical since v is arbitrarily chosen.

Next suppose G^f is factor-critical. Let v be an arbitrary vertex of G, and let x be a vertex of $K_{f(v)}$. Then $G^f - x$ has a 1-factor M, and so $M \cap E(G)$ is a spanning subgraph of G having exactly one vertex v of even degree. Hence $(M \cap E(G)) + vw$ is a $(1, f)$-odd factor of G_v. Hence G is critical with respect to $(1, f)$-odd factors. □

We are ready to give a structure theorem on $(1, f)$-odd subgraphs. Let G be a simple graph and $f : V(G) \to \{1, 3, 5, \ldots\}$. Define

$$D_f(G) = \{x \in V(G) : \tau(G_x) = \tau(G) + 2\},$$
$$A_f(G) = N_G(D_f(G)) \setminus D_f(G), \quad \text{and}$$
$$C_f(G) = V(G) - D_f(G) - A_f(G).$$

Namely, $A_f(G)$ is the set of vertices of $V(G) - D_f(G)$ that are adjacent to at least one vertex in $D_f(G)$, and $V(G)$ is decomposed into three disjoint subsets

$$V(G) = D_f(G) \cup A_f(G) \cup C_f(G).$$

Note that if $f(x) = 1$ for all vertices x of G, then the above decomposition is equivalent to the Gallai-Edmonds decomposition $D(G) \cup A(G) \cup C(G)$. The Gallai-Edmonds Structure Theorem 2.47 on matchings can be extended to a structure theorem on $(1, f)$-odd subgraphs as follows.

Theorem 6.22 (Structure Theorem on $(1, f)$-odd Subgraphs, [121]). *Let G be a simple graph and $f : V(G) \to \{1, 3, 5, \ldots\}$ be a function. Let $V(G) = D_f(G) \cup A_f(G) \cup C_f(G)$ be the decomposition defined above. Then the following statements hold (see Fig. 6.9) :*

(i) Every component of $\langle D_f(G) \rangle_G$ is critical with respect to $(1, f)$-odd factors.
(ii) $\langle C_f(G) \rangle_G$ has a $(1, f)$-odd factor.
(iii) Every maximum $(1, f)$-odd subgraph H of G covers $C_f(G) \cup A_f(G)$, and for every vertex $u \in A_f(G)$, $\deg_H(u) = f(u)$ and every edge of H incident with u joins u to a vertex in $D_f(G)$.
(iv) The order of a maximum $(1, f)$-odd subgraph H is given by

$$|H| = |G| - \omega(\langle D_f(G) \rangle_G) + \sum_{x \in A_f(G)} f(x), \qquad (6.21)$$

where $\omega(\langle D_f(G) \rangle_G)$ denotes the number of components of $\langle D_f(G) \rangle_G$.

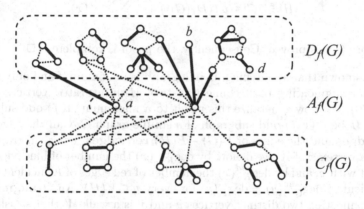

$f(x) = 1$ for $x \in \{a, b, c, d\}$, and $f(x) = 3$ otherwise.

Fig. 6.9. A graph G and its decomposition $D_f(G) \cup A_f(G) \cup C_f(G)$.

Proof. By using the previous lemmas, we shall derive this theorem from the Gallai-Edmonds Structure Theorem 2.47, referred to as the structure theorem in this proof.

By Lemma 6.19 and the construction of G^f, it follows that

$$V(K_{f(v)}) \subseteq A(G^f) \qquad \text{if and only if} \qquad v \in A_f(G); \quad \text{and}$$
$$V(K_{f(v)}) \subseteq C(G^f) \qquad \text{if and only if} \qquad v \in C_f(G).$$

Since each component of $C(G^f)$ has a 1-factor by the structure theorem, each component of $C_f(G)$ has a $(1, f)$-odd factor by Lemma 6.18. Thus (ii) holds. Similarly, by the structure theorem and by Lemma 6.21, statement (i) holds.

Let H be a maximum $(1, f)$-odd subgraph of G. Then G^f has a maximum matching M with property (a) such that $M \cap E(G) = E(H)$ by Lemma 6.20. By the structure theorem, every edge of M incident with a vertex u of $A(G^f)$ joins u to $D(G^f)$. Hence every edge of H incident with a vertex of $A_f(G)$ joins $A_f(G)$ to $D_f(G)$. Hence statement (iii) follows.

By the structure theorem, it follows that

$$|M| = |G^f| - \omega(\langle D(G^f) \rangle) + |A(G^f)|.$$

Since $|G| - |H| = |G^f| - |M|$ by (6.19) and since

$$\omega(\langle D(G^f) \rangle) = \omega(\langle D_f(G) \rangle), \qquad |A(G^f)| = \sum_{x \in A_f(G)} f(x),$$

we have

$$|H| = |G| - \omega(\langle D_f(G) \rangle) + \sum_{x \in A_f(G)} f(x).$$

Therefore (iv) is proved. Consequently, the proof is complete.   □

It is known that a matching M in a graph is maximum if and only if there exists no augmenting path connecting two M-unsaturated vertices (Theorem 2.20). We now generalize this result to a maximum $(1, f)$-odd subgraph.

Let H be a $(1, f)$-odd subgraph of a simple graph G. Call the edges of H **blue edges** and the edges of $E(G) - E(H)$ **red edges**. For a subgraph K of G and a vertex x of K, we denote by $\deg_K^{blue}(x)$ the number of blue edges of K incident with x, and by $\deg_K^{red}(x)$ the number of red edges of K incident with x. In particular, $\deg_G^{blue}(y) = \deg_H(y)$ for every $y \in V(H)$. An H-**augmenting walk** connecting two distinct vertices u and v is a walk W that satisfies

(i) $\deg_W^{blue}(u) = \deg_W^{blue}(v) = 0$,
(ii) $\deg_W^{red}(u) = \deg_W^{red}(v) = 1$,   and
(iii) $\deg_W^{red}(x) - \deg_W^{blue}(x) \le f(x) - \deg_H(x)$ for all $x \in V(W) - \{u, v\}$.

It is easy to see that if $f : V(G) \to \{1\}$ and H is a matching, then an H-augmenting walk is nothing but an augmenting path.

Theorem 6.23 ([120]). *Let G be a simple graph, $f : V(G) \to \{1, 3, 5, \dots\}$ a function, and H a $(1, f)$-odd subgraph of G. Then H is a maximum $(1, f)$-odd subgraph of G if and only if G has no H-augmenting walk.*

Proof. Suppose that H is a $(1, f)$-odd subgraph of G and there is an H-augmenting walk W connecting two distinct vertices u and v, where $u, v \notin V(H)$. Then $W \triangle H$, which is the subgraph of G induced by

$(E(H) \cup E(W)) - (E(H) \cap E(W))$, is a $(1, f)$-odd subgraph since for every vertex $x \in V(W) - \{u, v\}$,

$$\deg_{W \triangle H}(x) = \deg_H(x) - \deg_W^{blue}(x) + \deg_W^{red}(x)$$
$$= \deg_H(x) + \deg_W(x) - 2\deg_W^{blue}(x)$$
$$\equiv \deg_H(x) \pmod 2 \qquad (\text{as } \deg_W(x) \text{ is even})$$

and $\deg_{W \triangle H}(x) \leq f(x)$ by (iii). Furthermore, $W \triangle H$ covers all the vertices of H and $\{u, v\}$, therefore H is not a maximum $(1, f)$-odd subgraph.

Assume that a $(1, f)$-odd subgraph H of G is not maximum. Call the edges of H blue edges and the edges not in H red edges. Then there exists a matching M in G^f with property (a) such that $M \cap E(G) = E(H)$ and M does not saturate at most one vertex in each $K_f(x)$. By Lemma 6.20, M is not a maximum matching in G^f. Hence there exists an augmenting path P connecting two distinct M-unsaturated vertices x_1 of $K_{f(u)}$ and y_1 of $K_{f(v)}$. By taking a shortest augmenting path, we may assume that P passes through $K_{f(u)}$ and $K_{f(v)}$ exactly once.

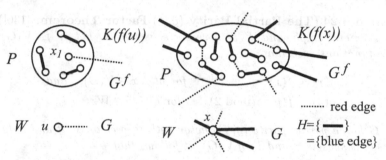

Fig. 6.10. An augmenting path P of G^f and a walk $W = E(P) \cap E(G)$ of G, which is an H-augmenting walk.

We now show that $W = E(P) \cap E(G)$ is an H-augmenting walk of G (Fig. 6.10). Since all the vertices of $V(K(u)) - x_1$ are covered by $M \cap E(K_{f(u)})$, we have

$$\deg_W^{blue}(u) = \deg_W^{blue}(v) = 0 \quad \text{and} \quad \deg_W^{red}(u) = \deg_W^{red}(v) = 1.$$

For any vertex $x \in V(G) - \{u, v\}$, it follows that

$$\deg_W^{red}(x) - \deg_W^{blue}(x)$$
$$\leq \text{the number of blue edges in } K_{f(x)}$$
$$\leq f(x) - \deg_H(x)$$

Hence W is the desired H-augmenting walk. $\square$

Other results on $(1, f)$-odd subgraphs and $(1, f)$-odd factors can be found in [120], [121] and [123].

6.3 Partial parity (g, f)-factors and coverings

For a given vertex set W of a graph G, consider two functions $g, f : V(G) \to \mathbb{Z}$ such that

$$g(x) \le f(x) \qquad \text{for all } x \in V(G), \text{ and}$$
$$g(y) \equiv f(y) \pmod 2 \qquad \text{for all } y \in W.$$

Then a spanning subgraph F of G is called a **partial parity** (g, f)-**factor with respect to** W if

$$g(x) \le \deg_F(x) \le f(x) \qquad \text{for all } x \in V(G), \text{ and}$$
$$\deg_F(y) \equiv f(y) \pmod 2 \qquad \text{for all } y \in W.$$

Note that if $W = \emptyset$, then a partial parity (g, f)-factor is a (g, f)-factor and, if $W = V(G)$, then a partial parity (g, f)-factor is a parity (g, f)-factor. We begin with a necessary and sufficient condition for a graph to have a partial parity (g, f)-factor.

Theorem 6.24 (The Partial Parity (g, f)-Factor Theorem, [130]). *Let G be a general graph, and W be a vertex subset of G. Let $g, f : V(G) \to \mathbb{Z}$ be two functions such that*

$$g(x) \le f(x) \qquad \text{for all } x \in V(G), \text{ and}$$
$$g(y) \equiv f(y) \pmod 2 \qquad \text{for all } y \in W.$$

Then G has a partial parity (g, f)-factor with respect to W if and only if for all disjoint subsets S and T of $V(G)$, it follows that

$$\eta_2(S, T) = \sum_{x \in S} f(x) + \sum_{x \in T} (\deg_G(x) - g(x))$$
$$-e_G(S, T) - q_2(W; S, T) \ge 0, \qquad (6.22)$$

where $q_2(W; S, T)$ denotes the number of components C of $G - (S \cup T)$ such that

$$g(x) = f(x) \qquad \text{for all } x \in V(C) \setminus W, \text{ and}$$
$$\sum_{x \in V(C)} f(x) + e_G(C, T) \equiv 1 \pmod 2. \qquad (6.23)$$

Proof. We first construct a new graph G^* from G by adding $(f(y) - g(y))/2$ new loops to every vertex $y \in W$ (see Fig. 6.11). Then define

$$g^*(x) = \begin{cases} f(x) & \text{if } x \in W, \\ g(x) & \text{otherwise.} \end{cases}$$

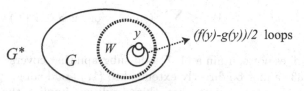

Fig. 6.11. The new graph G^* constructed from G.

It is easy to see that G has the desired partial parity (g, f)-factor with respect to W if and only if G^* has a (g^*, f)-factor. By the (g, f)-factor theorem, G^* has a (g^*, f)-factor if and only if for all disjoint subsets S and T of $V(G^*) = V(G)$, it follows that

$$\sum_{x \in S} f(x) + \sum_{x \in T} (\deg_{G^*}(x) - g^*(x)) - e_{G^*}(S, T) - q_{G^*}(S, T) \geq 0. \qquad (6.24)$$

It is clear that

$$\deg_{G^*}(x) - g^*(x) = \deg_G(x) - g(x) \qquad \text{for all} \quad x \in V(G).$$

Moreover, a component C of $G^* - (S \cup T)$ is a (g^*, f)-odd component if and only if

$$g^*(x) = f(x) \qquad \text{for all} \quad x \in V(C), \text{ and}$$

$$\sum_{x \in V(C)} f(x) + e_{G^*}(C, T) \equiv 1 \quad (\text{mod } 2).$$

The above conditions are equivalent to

$$g(x) = f(x) \qquad \text{for all} \quad x \in V(C) \setminus W, \text{ and}$$

$$\sum_{x \subset V(C)} f(x) + e_G(C, T) \equiv 1 \quad (\text{mod } 2).$$

Hence $q_{G^*}(S, T) = q_2(W; S, T)$. Since $e_{G^*}(S, T) = e_G(S, T)$, (6.24) is equivalent to (6.22). Consequently, the theorem is proved. $\square$

Before giving an application of the partial parity (g, f)-factor theorem, we make a remark about matchings. It is shown in Theorem 2.33 that for a graph G and its vertex subset W, G has a matching that covers W if and only if

$$odd(G - S \,|\, W) \leq |S| \qquad \text{for all} \quad S \subset V(G),$$

where $odd(G - S \,|\, W)$ denotes the number of odd components of $G - S$, all of whose vertices are contained in W. However, this result cannot be directly extended to $(1, f)$-odd factors. For example, consider the tree T given in Fig. 6.12. Let $W = \{\bullet\}$, and define the functions f as in Fig. 6.12, where numbers denote $f(v)$. Then T satisfies

$$odd(T - S \mid W) \le \sum_{x \in S} f(x) \qquad \text{for all } S \subset V(T).$$

However, T does not contain a $(1, f)$-odd subgraph that covers W. Therefore, Theorem 2.33 cannot be directly extended to $(1, f)$-odd subgraphs.

However, we may expect that this condition implies the existence of another subgraph with some other property. The next theorem provides an answer to this question, and it can be proved by using the partial parity (g, f)-factor theorem. Moreover, if f is defined as $f(x) = 1$ for all vertices x, then the following theorem implies the previous result on matchings.

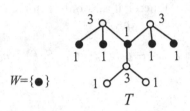

$W=\{\bullet\}$

Fig. 6.12. A tree that has no $(1, f)$-odd subgraph containing W.

Theorem 6.25 ([130]). *Let G be a general graph, W a non-empty subset $V(G)$, and $f : V(G) \to \mathbb{Z}^+$ a function such that $f(y)$ is an odd integer for all $y \in W$. Then G has a subgraph H covering W such that*

$$1 \le \deg_H(x) \le f(x) \qquad \text{for all } x \in V(H), \text{ and} \qquad (6.25)$$
$$\deg_H(y) \equiv 1 \pmod 2 \qquad \text{for all } y \in W \qquad (6.26)$$

if and only if

$$odd(G - S \mid W) \le \sum_{x \in S} f(x) \qquad \text{for all } S \subset V(G), \qquad (6.27)$$

where $odd(G - S \mid W)$ denotes the number of odd components C of $G - S$ with $V(C) \subseteq W$.

Proof. By adding the vertices of $V(G) - V(H)$ to H, we obtain a factor H' that satisfies the following inequality instead of (6.25).

$$0 \le \deg_{H'}(x) \le f(x) \qquad \text{for all } x \in V(G). \qquad (6.28)$$

On the other hand, we can remove the isolated vertices from a factor H' satisfying (6.28) and (6.26) in order to obtain a subgraph H satisfying (6.25) and (6.26). Thus we can regard H as a spanning subgraph satisfying (6.28) and (6.26).

Assume that G has a factor H that satisfies (6.28) and (6.26). Then for any odd component C of $G - S$ with $V(C) \subseteq W$, at least one edge of H joins C to S by (6.26). Hence

$$odd(G - S \,|\, W) \leq \sum_{x \in S} \deg_H(x) \leq \sum_{x \in S} f(x).$$

We next prove the sufficiency. Let N be a sufficiently large odd integer. Define a function $g : V(G) \to \mathbb{Z}$ by

$$g(x) = -N \qquad \text{for all} \quad x \in V(G).$$

Then a partial parity (g, f)-factor with respect to W is the desired factor satisfying (6.28) and (6.26). Therefore, it suffices to show that G, g and f satisfy condition (6.22) in the partial parity (g, f)-factor theorem.

Let S and T be two disjoint subsets of $V(G)$. If $T \neq \emptyset$, then (6.22) holds since

$$\deg_G(x) - g(x) = \deg_G(x) + N \qquad \text{for every} \quad x \in T.$$

Thus we may assume that $T = \emptyset$. It is immediate that a component C of $G - S$ satisfying (6.23) is covered by W and has odd order. Therefore, it follows from (6.27) that

$$\eta_2(S, \emptyset) = \sum_{x \in S} f(x) - q_2(W; S, \emptyset) = \sum_{x \in S} f(x) - odd(G - S \,|\, W) \geq 0.$$

Consequently, G has a partial parity (g, f)-factor with respect to W, and the theorem is proved. $\square$

We now give some results on covering a vertex subset with paths and cycles, which are applications of the parity (g, f)-factor theorem.

Theorem 6.26 (Kano and Matsuda [130]). *Let G be a simple graph and W be a set of an even number of vertices of G. Then G has a set of vertex disjoint paths such that the set of their end-vertices is precisely equal to W (Fig. 6.13) if and only if*

$$odd(W; G - S) \leq |S \cap W| + 2|S \setminus W| \qquad \text{for all} \quad S \subseteq V(G), \qquad (6.29)$$

where $odd(W; G - S)$ denotes the number of components D of $G - S$ such that $|V(D) \cap W|$ is odd (Fig. 6.13).

Proof. Let H be a subgraph of G that satisfies

$$\deg_H(x) = 1 \qquad \text{for all} \quad x \in W, \text{ and}$$
$$\deg_H(y) = 2 \qquad \text{for all} \quad y \in V(H) \setminus W. \qquad (6.30)$$

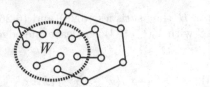

Fig. 6.13. A set of vertex disjoint paths whose end-vertices are exactly W; D is a component of $G - S$.

Then each component of H is a path or a cycle, and the set of paths in H is the desired set of disjoint paths of G. Hence G has the desired set of disjoint paths if and only if G has a subgraph H satisfying (6.30). We show that the existence of such a subgraph H is equivalent to (6.29).

Suppose that G has a subgraph H satisfying (6.30). For each component D of $G - S$ such that $|V(D) \cap W|$ is odd, at least one edge of H joins D to S. Thus we obtain

$$odd(W; G - S) \leq \sum_{x \in S} \deg_H(x) = |S \cap W| + 2|S \setminus W|.$$

Hence (6.29) follows.

Conversely assume that (6.29) holds. Let N be a sufficiently large integer, and define two functions $g, f : V(G) \to \mathbb{Z}$ by

$$g(x) = \begin{cases} -2N - 1 & \text{if } x \in W \\ -2N & \text{otherwise,} \end{cases} \quad \text{and} \quad f(x) = \begin{cases} 1 & \text{if } x \in W \\ 2 & \text{otherwise.} \end{cases}$$

Then the desired subgraph H is a parity (g, f)-factor F of G, which may have degree 0 for some vertices of $V(G) - W$ but must have degree 1 for every vertex of W. More precisely, the subgraph of G induced by $E(F)$ is the desired subgraph H. Hence it suffices to show that G, g and f satisfy (6.2) in the parity (g, f)-factor theorem.

Since $odd(W; G) = 0$, every component of G contains an even number of vertices in W. Thus it follows that $\eta(\emptyset, \emptyset) = -q(\emptyset, \emptyset) = 0$. Let S and T be disjoint subsets of $V(G)$ such that $S \cup T \neq \emptyset$. If $T \neq \emptyset$, then since $-g(x) \geq 2N$ is sufficiently large, we have

$$\eta(S, T) = \sum_{x \in S} f(x) + \sum_{x \in T} (\deg_G(x) - g(x)) - e_G(S, T) - q(S, T) \geq 0.$$

Hence we may assume that $T = \emptyset$. Then $q(S, \emptyset) = odd(W; G - S)$ and thus

$$\eta(S, \emptyset) = \sum_{x \in S} f(x) - q(S, \emptyset) = 2|S \setminus W| + |S \cap W| - odd(W; G - S) \geq 0,$$

as desired. $\square$

Theorem 6.27 ([130]). *Let G be a simple graph and W be a vertex subset of G. Then G has a set of vertex disjoint cycles that cover W (Fig. 6.14) if and only if for all disjoint subsets $S \subseteq V(G)$ and $T \subseteq W$, it follows that*

$$2|S| + \sum_{x \in T} (\deg_G(x) - 2) - e_G(S, T) - q_3(S, T) \geq 0, \qquad (6.31)$$

where $q_3(S, T)$ denotes the number of component C of $G - (S \cup T)$ such that $e_G(C, T) \equiv 1 \pmod 2$.

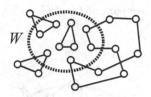 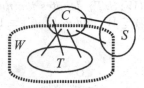

Fig. 6.14. A set of vertex disjoint cycles that cover W; a component C of $G - (S \cup T)$ counted in $q_3(S, T)$.

Proof. Let N be a sufficiently large number, and define two functions $g, f : V(G) \to \mathbb{Z}$ by

$$g(x) = \begin{cases} 2 & \text{if } x \in W \\ -2N & \text{otherwise,} \end{cases} \qquad \text{and} \qquad f(x) = 2 \quad \text{for all} \quad x \in V(G).$$

Then G has a parity (g, f)-factor, which may have degree 0 for some vertices of $V(G) - W$, if and only if G has the desired set of disjoint cycles. Thus it suffices to show that (6.2) and (6.31) are equivalent for the two functions g and f.

Suppose that (6.2) holds. Let $S \subseteq V(G)$ and $T \subseteq W$ such that $S \cap T = \emptyset$. Then by (6.2), we have

$$\eta(S, T) = 2|S| + \sum_{x \in T} (\deg_G(x) - 2) - e_G(S, T) - q(S, T) \geq 0.$$

Moreover, every component D of $G - (S \cup T)$ satisfying (6.3) satisfies $e_G(T, D) \equiv 1 \pmod 2$. Thus $q(S, T) = q_3(S, T)$, and hence (6.31) holds.

Conversely, assume that (6.31) holds. It follows that $\eta(\emptyset, \emptyset) = -q(\emptyset, \emptyset) = 0$ since $f(x) = 2$ for all $x \in V(G)$. Let S and T be disjoint subsets of $V(G)$ such that $S \cup T \neq \emptyset$. If $T \setminus W \neq \emptyset$, then since $-g(x) = 2N$ is sufficiently large for $x \in T \setminus W$, we have $\eta(S, T) \geq 0$. Thus we may assume that $T \subseteq W$. It follows from $f(x) = 2$ that $q(S, T) = q_3(S, T)$. Hence by (6.31), we obtain

$$\eta(S, T) = 2|S| + \sum_{x \in T} (\deg_G(x) - 2) - e_G(S, T) - q(S, T) \geq 0.$$

Therefore (6.2) holds. Consequently, the theorem is proved. □

Theorem 6.28 ([130]). *Let G be a graph and W a set of vertices of G. Then G has a set of vertex disjoint cycles and paths covering W such that all the vertices of the cycles and all the inner vertices of the paths are contained in W, and all the end-vertices of the paths are contained in $V(G)-W$ (Fig. 6.15) if and only if for all disjoint subsets $S \subseteq V(G)$ and $T \subseteq W$, it follows that*

$$|S \setminus W| + 2|S \cap W| + \sum_{x \in T}(\deg_G(x) - 2)$$
$$-e_G(S,T) - q_4(W;S,T) \geq 0, \qquad (6.32)$$

where $q_4(W;S,T)$ denotes the number of component C of $G - (S \cup T)$ such that $V(C) \subseteq W$ and $e_G(T,C) \equiv 1 \pmod 2$.

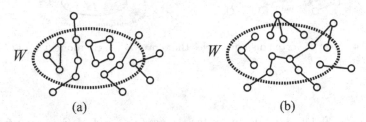

Fig. 6.15. (a) A set of vertex disjoint cycles and paths that have the properties given in Theorem 6.28; (b) A $[1,3]$-subgraph covering W.

Proof. Let N be a sufficiently large integer, and define two functions $g, f : V(G) \to \mathbb{Z}$ by

$$g(x) = \begin{cases} 2 & \text{if } x \in W \\ -N & \text{otherwise,} \end{cases} \quad \text{and} \quad f(x) = \begin{cases} 2 & \text{if } x \in W \\ 1 & \text{otherwise} \end{cases}$$

Then G has a (g,f)-factor, which may have degree 0 for some vertices of $V(G)-W$, if and only if G has the desired set of cycles and paths. Hence it suffices to show that (6.32) and (4.1) in the (g,f)-factor theorem are equivalent.

Suppose that (4.1) holds. Let $S \subseteq V(G)$ and $T \subseteq W$ such that $S \cap T = \emptyset$. Then

$$\gamma(S,T) = |S \setminus W| + 2|S \cap W| + \sum_{x \in T}(\deg_G(x) - 2)$$
$$-e_G(S,T) - q^*(S,T) \geq 0.$$

Since every component D of $G - (S \cup T)$ counted in $q^*(S,T)$ satisfies $f(x) = g(x)$ for all $x \in V(D)$ and $\sum_{x \in V(D)} f(x) + e_G(T,D) \equiv 1 \pmod 2$, we have $V(D) \subseteq W$ and $q^*(S,T) = q_4(W;S,T)$. Hence (6.32) follows.

Conversely assume that (6.32) holds. It follows that $\gamma_G(\emptyset, \emptyset) = -q^*(\emptyset, \emptyset) = 0$ since $g(x) < f(x)$ for all $x \in V(G) - W$ and $f(y) = 2$ for all $y \in W$. Let S and T be disjoint subsets of $V(G)$ such that $S \cup T \neq \emptyset$. If $T \setminus W \neq \emptyset$, then $\gamma(S, T) \geq 0$ since $-g(x) = N$ is sufficiently large for $x \in T \setminus W$. Thus we may assume that $T \subseteq W$. It follows immediately that $q^*(S, T) = q_4(W; S, T)$, and we obtain the following inequality from (6.32):

$$\gamma(S, T) = |S \setminus W| + 2|S \cap W| + \sum_{x \in T}(\deg_G(x) - 2)$$
$$-e_G(S, T) - q^*(S, T) \geq 0.$$

Therefore (6.32) and (4.1) are equivalent, and the theorem is proved. $\square$

We finally consider a problem of covering a given vertex subset with a $[1, n]$-subgraph, each of whose vertices has degree between 1 and n. The condition for the existence of such a subgraph, which is given in the following theorem, is a natural extension of the criterion for the existence of a $[1, n]$-factor since the 1-factor theorem is generalized to the theorem of a matching which covers a given subset in a graph G.

Theorem 6.29. Let G be a simple graph, W a subset of $V(G)$ and $n \geq 2$ be an integer. Then G has a $[1, n]$-subgraph that covers W (Fig. 6.15) if and only if

$$iso(G - S \,|\, W) \leq n|S| \qquad \text{for all} \quad S \subseteq V(G), \tag{6.33}$$

where $iso(G - S \,|\, W)$ denotes the number of isolated vertices of $G - S$ contained in W.

Proof. Suppose first that G has a $[1, n]$-subgraph H which covers W. Then for every isolated vertex v of $G - S$ contained in W, H has at least one edge joining v to S. Hence

$$iso(G - S \,|\, W) \leq \sum_{x \in S} \deg_H(x) \leq n|S|.$$

Next assume that (6.33) holds. Let N be a sufficiently large integer, and define two functions g and f as follows:

$$g(x) = \begin{cases} 1 & \text{if } x \in W \\ -N & \text{otherwise,} \end{cases} \qquad \text{and} \qquad f(x) = n \quad \text{for all} \quad x \in V(G).$$

Then a (g, f)-factor of G is the desired subgraph. So it suffices to show that G satisfies condition (4.1) in the (g, f)-factor theorem.

Let S and T be two disjoint subsets of $V(G)$. If $T \setminus W \neq \emptyset$, then $\gamma(S, T) \geq 0$ since $-g(x) = N$ is sufficiently large for every $x \in T \setminus W$. Thus we may assume that $T \subseteq W$. Note that $q^*(S, T) = 0$ since $g(x) < f(x)$ for all $x \in V(G)$. Since $\deg_{G-S}(x) = 0$ if and only if x is an isolated vertex of $G - S$, we obtain the following from $T \subseteq W$ and (6.33).

$$\gamma(S,T) = n|S| + \sum_{x\in T}(\deg_G(x) - 1) - e_G(S,T)$$

$$= n|S| + \sum_{x\in T}(\deg_{G-S}(x) - 1)$$

$$\geq n|S| - iso(G - S\,|\,W) \geq 0.$$

Therefore the theorem is proved. □

6.4 $\mathcal{H}$-factors

We introduce the most general degree factor, which is called an $\mathcal{H}$-factor, and give some remarks on it without proofs. For a vertex v of a graph G, let $\mathcal{H}_v$ denote a non-empty subset of $\{0,1,2,3,\ldots\}$. Then a spanning subgraph K of G is called an $\mathcal{H}$-**factor of** G if

$$\deg_K(x) \in \mathcal{H}_x \qquad \text{for all} \quad x \in V(G).$$

For a spanning subgraph F of G and for a vertex v of G, define

$$\delta(\mathcal{H};F,v) = \min\{|\deg_F(v) - i| \,:\, i \in H_v\},$$

and let

$$\delta(\mathcal{H};F) = \sum_{x\in V(G)} \delta(\mathcal{H};F,x).$$

Thus a spanning subgraph F is an $\mathcal{H}$-factor if and only if $\delta(\mathcal{H};F) = 0$. The minimum $\delta(\mathcal{H};F)$ among all the spanning subgraphs F of G is denoted by $\delta_{\mathcal{H}}(G)$, namely,

$$\delta_{\mathcal{H}}(G) = \min\{\delta(\mathcal{H};F) \,:\, F \text{ are the spanning subgraphs of } G.\}$$

A spanning subgraph F is called $\mathcal{H}$-**optimal** if

$$\delta(\mathcal{H};F) = \delta_{\mathcal{H}}(G).$$

The **degree prescribed subgraph problem** is to determine the value of $\delta_{\mathcal{H}}(G)$.

An integer h is called a **gap** of $\mathcal{H}_v$ if $h \notin \mathcal{H}_v$ but $\mathcal{H}_v$ contains an element less than h and an element greater than h.

Lovász [176] gave a structural description on the degree prescribed subgraph problem in the case where $\mathcal{H}_v$ has no two consecutive gaps for all $v \in V(G)$. He showed that the problem is NP-complete without this restriction. The first polynomial time algorithm was given by Cornuéjols [57]. It is implicit in Cornuéjols [57] that this algorithm implies a Gallai–Edmonds type structure theorem for the degree prescribed subgraph problem, which is similar to, but in some respects more compact, than that of Lovász.

Of course, for a function $f : V(G) \rightarrow \{1, 3, 5, \ldots\}$, if we define $\mathcal{H}_v = \{1, 3, \ldots, f(v)\}$ for every vertex v of G, then a $(1, f)$-odd factor is an $\mathcal{H}$-factor. Similarly, a (g, f)-factor and a parity (g, f)-factor are special $\mathcal{H}$-factors. Thus these results are generalizations of previously obtained results, though they are complicated. For details and works on $\mathcal{H}$-factors, we refer the reader to the Doctoral Thesis [220] by Szabó and the book [254] by Yu and Liu.

Problems

6.1. Let T be a tree of even order and $f : V(T) \rightarrow \{1, 3, 5, \ldots\}$. Show that if T has a $(1, f)$-odd factor, then such a factor is unique in T.

6.2. Let G be a simple graph and $f : V(G) \rightarrow \{1, 3, 5, \ldots\}$. Then a $(1, f)$-odd subgraph is called a **maximal $(1, f)$-odd subgraph** if G has no $(1, f)$-odd subgraph H' such that $V(H) \subset V(H')$. Show that every maximal $(1, f)$-odd subgraph is a maximum $(1, f)$-odd subgraph.

6.3. Let G be a simple graph, $f : V(G) \rightarrow \{1, 3, 5, \ldots\}$, and B and R vertex subsets of G with $|B| < |R|$. Prove the following two statements by using the structure theorem on $(1, f)$-odd subgraphs.

(1) If there exists a $(1, f)$-odd subgraph which covers B and another one which covers R, then there exists a $(1, f)$-odd subgraph that covers B and at least one vertex of $R \backslash B$ ([120]).

(2) If there exists a maximum $(1, f)$-odd subgraph which avoids B and another one which avoids R, then there exists a maximum $(1, f)$-odd subgraph which avoids B and at least one vertex of $R \setminus B$. Note that a $(1, f)$-odd subgraph H **avoids** a vertex subset X if H covers no vertex of X.

6.4. Prove the necessity of the partial parity (g, f)-factor theorem without using the parity (g, f)-factor theorem.

7

Component Factors

In this chapter we investigate a **component factor**, which is a spanning subgraph having specified components. For example, a K_2-factor, each of whose components is the complete graph K_2, is the same as a 1-factor. For a family S of connected graphs, a subgraph H of a graph G is called an S-**subgraph** if each component of H is isomorphic to an element of S, and a spanning S-subgraph is called an S-**factor**. If S consists of exactly one graph K, then a $\{K\}$-factor is often called a K-factor.

7.1 Path factors and star factors

We first consider conditions for a graph to have a path factor or a star factor, whose components are paths or stars, respectively.

We need some more definitions on S-subgraphs, where S denotes a family of connected graphs. An S-subgraph H of a graph G is said to be **maximum** if G has no S-subgraph K with $|K| > |H|$, and H is called **maximal** if G has no S-subgraph K' with $V(K') \supset V(H)$.

We start with a component factor contained in a graph G that satisfies $iso(G - S) \leq |S|$ for all $S \subset V(G)$. The following lemma shows that the above condition on the isolated vertices of $G - S$ can be expressed by using independent sets and their neighborhoods.

Lemma 7.1. *Let G be a simple graph and λ a positive real number. Then the following two statements are equivalent:*
  (i) $iso(G - S) \leq \lambda|S|$ for all $S \subset V(G)$.
  (ii) $\lambda|N_G(X)| \geq |X|$ for all independent set X of G.
Moreover, when $\lambda = 1$, (i) and (ii) are equivalent to
  (iii) $|N_G(X)| \geq |X|$ for all $X \subseteq V(G)$.

Proof. Assume that statement (i) holds. Then G has no trivial components, and for any independent set X of G, we have

J. Akiyama and M. Kano, *Factors and Factorizations of Graphs*,
Lecture Notes in Mathematics 2031, DOI 10.1007/978-3-642-21919-1_7,
© Springer-Verlag Berlin Heidelberg 2011

$$|X| \leq iso(G - N_G(X)) \leq \lambda |N_G(X)|.$$

Hence (ii) follows.

Conversely, assume that (ii) holds. Then G has no trivial components, and for any vertex set S of G, $Iso(G-S)$ is an independent set of G, and so we have

$$iso(G - S) = |Iso(G - S)| \leq \lambda |N_G(Iso(G - S))| = \lambda |S|.$$

Therefore (i) holds.

Assume that (ii) with $\lambda = 1$ holds. Then G has no trivial components, and so $N_G(V(G)) = V(G)$. Let $\emptyset \neq X \subset V(G)$, and $Z = X \cap N_G(X)$. Then $X - Z$ is an independent set of G, $N_G(Z) \supseteq Z$ and $N_G(X-Z) \cap Z = \emptyset$. By (ii) we have

$$|N_G(X)| \geq |N_G(X - Z)| + |Z| \geq |X - Z| + |Z| = |X|.$$

Hence (iii) holds. Trivially (iii) implies (ii). Therefore the lemma is proved. $\square$

G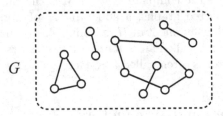

Fig. 7.1. A $\{K_2, C_n : n \geq 3\}$-factor of a graph G.

Theorem 7.2 (Tutte [227]). *Let G be a simple graph. Then G has a $\{K_2, C_n : n \geq 3\}$-factor (Fig. 7.1) if and only if*

$$iso(G - S) \leq |S| \qquad \text{for all} \quad S \subset V(G). \tag{7.1}$$

Proof (1). Assume first that G has a $\{K_2, C_n : n \geq 3\}$-factor F. Then for any subset $\emptyset \neq S \subset V(G)$, we have

$$iso(G - S) \leq iso(F - S) \leq |S|$$

since each component of F is K_2 or a cycle.

We next prove the sufficiency by reducing the problem to a problem of perfect matching in a special bipartite graph.

Construct a bipartite graph H with bipartition (A, B), $A = V(G)$, $B = V(G)$, in which two vertices $x \in A$ and $y \in B$ are adjacent if and only if x and y are adjacent in G (see Fig. 7.2). For a subset $S \subseteq V(G)$,

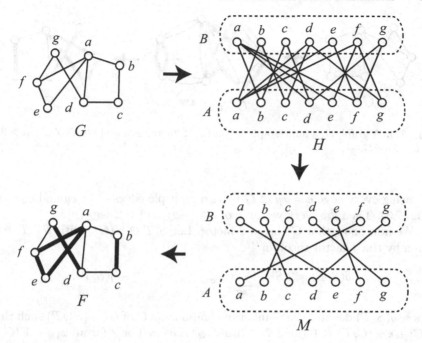

Fig. 7.2. A graph G, a bipartite graph H, a perfect matching M of H, and a $\{K_2, C_n : n \geq 3\}$-factor F of G.

S_A and S_B denote the subsets of A and B, respectively, corresponding to S. Conversely, for a subset $S_A \subseteq A$, S denotes the corresponding vertex set of G.

Let $\emptyset \neq X_A \subseteq A$. Then $X \subseteq V(G)$, and by Lemma 7.1 we have

$$|N_H(X_A)| = |N_G(X)| \geq |X| = |X_A|.$$

Therefore, H has a perfect matching M by the marriage theorem (Theorem 2.1) (see Fig. 7.2). From the perfect matching M of H, we obtain a subgraph F of G induced by the edge set

$$E(F) = \{xy \in E(G) : xy \in M\}.$$

Then for every vertex v of G, there exist two edges vy ($v \in A, y \in B$) and xv ($x \in A$ and $v \in B$) in M. If $x = y$, then F contains one edge $vy = xv$, which forms K_2 in F. If $x \neq y$, then $\deg_F(v) = 2$. Hence these edges form disjoint cycles in F. Consequently, F is the desired $\{K_2, C_n : n \geq 3\}$-factor of G. $\square$

We next give another proof of Theorem 7.2 by using the f-factor theorem (Theorem 3.2).

Proof (2). We give another proof of the sufficiency. We may assume that G is connected. Construct a multigraph G^* with vertex set $V(G)$ from G by

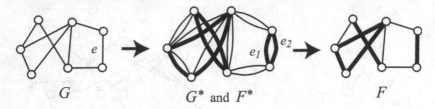

Fig. 7.3. A graph G, a multigraph G^* and its 2-factor F^*, and a $\{K_2, C_n : n \geq 3\}$-factor F of G.

replacing every edge $e = xy$ of G by two multiple edges $e_1 = xy$ and $e_2 = xy$ (Fig. 7.3). Then for every vertex v of G^*, $\deg_{G^*}(v) = 2\deg_G(v)$.

We now show that G^* has a 2-factor. Let $S, T \subset V(G^*)$ with $S \cap T = \emptyset$. Then by the f-factor theorem

$$\delta(S, T) = 2|S| + \sum_{x \in T} (\deg_{G^* - S}(x) - 2) - q(S, T),$$

where $q(S, T)$ denotes the number of components C of $G^* - (S \cup T)$ such that $2|C| + e_{G^*}(C, T) \equiv 1 \pmod 2$. Since $e_{G^*}(x, y) = 0$ or 2 for all $x, y \in V(G^*)$, it follows that $q(S, T) = 0$. Let

$$X = \{x \in T : N_G(x) \not\subseteq S\}, \quad \text{and} \quad Y = \{x \in T : N_G(x) \subseteq S\}.$$

Then $T = X + Y$, $\deg_{G^* - S}(x) \geq 2$ for all $x \in X$, $\deg_{G^* - S}(y) = 0$ for all $y \in Y$ and $Y \subseteq Iso(G - S)$. Thus we have

$$\delta(S, T) = 2|S| + \sum_{x \in X \cup Y} (\deg_{G^* - S}(x) - 2)$$
$$\geq 2|S| - 2|Y|$$
$$\geq 2(|S| - iso(G - S)) \geq 0.$$

Hence G^* has a 2-factor F^*.

Let F be a spanning subgraph of G induced by an edge set

$$E(F) = \{e \in E(G) : e_1 \text{ or } e_2 \text{ are contained in } F^*.\}$$

Then F becomes the desired $\{K_2, C_n : n \geq 3\}$-factor of G (Fig. 7.3). □

A **path factor** of a graph G is a spanning subgraph, each of whose components is a path of order at least two (Fig. 7.5). The next theorem gives a criterion for a graph to have a path factor. It is obvious that if a graph has a $\{K_2, C_n : n \geq 3\}$-factor, then it also has a path-factor. Hence the condition of the next theorem is weaker than that of Theorem 7.2.

Theorem 7.3 (The Path Factor Theorem, Akiyama, Avis and Era [3]**).** *A simple graph G has a path factor if and only if*

$$iso(G - S) \le 2|S| \qquad for\ all \quad S \subset V(G). \tag{7.2}$$

The proof of Theorem 7.3 is given later. We give an application of the above theorem, but first we require the following lemma.

Lemma 7.4. *Let G be a connected plane simple graph, and let r be the number of regions of G. Then*

$$r \le 2|G| - 4.$$

Proof. Let $p = |G|$ and $q = ||G||$. By substituting $q \le 3p - 6$ of Theorem 1.10 into $p - q + r = 2$ of Euler's formula (Theorem 1.9), we obtain $r \le 2p - 4$. □

Theorem 7.5 ([9]). *Every connected maximal plane simple graph has a path factor.*

Proof. Let G be a connected maximal plane simple graph, and let $\emptyset \ne S \subset V(G)$. It is obvious that $N_G(Iso(G - S)) \subseteq S$, and every isolated vertex of $G - S$ corresponds to an inner region of an induced subgraph $H = \langle S \rangle_G$ since G is a maximal plane graph (see Fig. 7.4). Let $Comp(H)$ denote the set of components of H. For a component D of H containing at least two regions, we denote by $reg(D)$ the number of regions in D. By Lemma 7.4,

$$reg(D) \le 2|D| - 4 \le 2|D|.$$

Thus we have

$$iso(G - S) \le \sum_{D \in Comp(H)} reg(D) \le \sum_{D \in Comp(H)} 2|D| \le 2|H| = 2|S|.$$

Therefore, by the path factor theorem, G has a path-factor. □

Since every path of order at least four has a $\{P_2, P_3\}$-factor, a graph G has a path factor if and only if G has a $\{P_2, P_3\}$-factor. Moreover, since $P_2 = K_{1,1}$ and $P_3 = K_{1,2}$, the path factor theorem is included in the following theorem. However, the proof of the following theorem given in Amahashi and Kano [15], which uses alternating paths, is basically the same as the proof of the path factor theorem given in [3]. On the other hand, the proof by Las Vergnas uses the (g, f)-factor theorem, and is explained later.

Theorem 7.6 (The Star Factor Theorem [165], **and** [15] **).** *Let $n \ge 2$ be an integer and G be a simple graph. Then G has a $\{K_{1,1}, K_{1,2}, \ldots, K_{1,n}\}$-factor (Fig. 7.5) if and only if*

$$iso(G - S) \le n|S| \qquad for\ all \quad S \subset V(G). \tag{7.3}$$

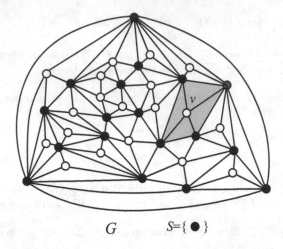

Fig. 7.4. (1) A connected maximal plane simple graph G, a vertex set S, an isolated vertex v of $G - S$, and the region of $H = \langle S \rangle_G$ corresponding to v.

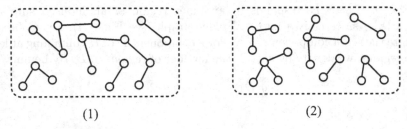

(1)   (2)

Fig. 7.5. (1) A path factor; and (2) $\{K_{1,1}, K_{1,2}, K_{1,3}\}$-factor of a graph.

Proof of the necessity. Suppose that G has a $\{K_{1,1}, K_{1,2}, \ldots, K_{1,n}\}$-factor F. Let $S \subset V(G)$. Then for each component D of F, we have $iso(D - S \cap V(D)) \le n|S \cap V(D)|$. Hence

$$iso(G - S) \le iso(F - S) \le n|S|.$$

Therefore the necessity is proved.   □

Proof of the sufficiency (1). We prove the sufficiency by induction on the size $\|G\|$. We may assume that G is connected since otherwise, by applying the induction hypothesis to each component of G, we can obtain the desired factor of G. Let

$$\beta = \min \big\{ n|X| - iso(G - X) : X \subseteq V(G) \text{ and } iso(G - X) \ge 1 \big\}.$$

Then $\beta \ge 0$ as $iso(G - X) \le n|X|$, and

$$iso(G - X) \le n|X| - \beta \tag{7.4}$$

for all $X \subseteq V(G)$ with $iso(G - X) \geq 1$. Take a maximal subset S of $V(G)$ such that $n|S| - iso(G - S) = \beta$ and $iso(G - S) \geq 1$.

Claim 1. *Every nontrivial component of $G - S$ has a $\{K_{1,1}, \ldots, K_{1,n}\}$-factor.*

Let D be a nontrivial component of $G - S$, and let $\emptyset \neq X \subset V(D)$. Then by (7.4), we have

$$iso(G - S) + iso(D - X) = iso(G - (S \cup X))$$
$$\leq n|S \cup X| - \beta = n|S| + n|X| - \beta.$$

Thus $iso(D - X) \leq n|X|$, which implies that D has a $\{K_{1,1}, \ldots, K_{1,n}\}$-factor by the induction hypothesis.

We construct a bipartite graph B with bipartition $S \cup Iso(G - S)$ in which two vertices $x \in S$ and $y \in Iso(G - S)$ are adjacent if and only if x and y are adjacent in G.

Claim 2. *For every $\emptyset \neq Y \subseteq S$, it follows that $|N_B(Y)| \geq n|Y| - \beta$, and $|N_B(S)| = n|S| - \beta$.*

It follows from the choice of S that $|N_B(S)| = |Iso(G - S)| = n|S| - \beta$. Let $\emptyset \neq Y \subset S$. Then $Iso(G - S) - N_B(Y)$ is a set of isolated vertices of $G - (S - Y)$, and so

$$n|S - Y| \geq iso(G - (S - Y)) \geq |Iso(G - S) - N_B(Y)|$$
$$= iso(G - S) - |N_B(Y)| = n|S| - \beta - |N_B(Y)|.$$

Hence $|N_B(Y)| \geq n|Y| - \beta$, and so Claim 2 holds.

Claim 3. $0 \leq \beta \leq 1$.

Assume that $\beta \geq 2$. Take one edge e, and let $\emptyset \neq X \subset V(G)$. If $iso(G - X) = 0$, then $iso(G - e - X) \leq 2 \leq n|X|$. If $iso(G - X) \geq 1$, then by (7.4) we have

$$iso(G - e - X) \leq iso(G - X) + 2 \leq n|X| - \beta + 2 \leq n|X|.$$

Hence by the induction hypothesis, $G - e$ has a $\{K_{1,1}, \ldots, K_{1,n}\}$-factor, which is the desired factor of G. Therefore, we may assume that Claim 3 holds.

If $\beta = 0$, then by Claim 2, B has a $K_{1,n}$-factor by the generalized marriage theorem (Theorem 2.10). By combining this factor and $\{K_{1,1}, \ldots, K_{1,n}\}$-factors of non-trivial components of $G - S$ guaranteed by Claim 1, we obtain the desired factor of G.

Assume $\beta = 1$. By adding one vertex u to B, and by joining u to every vertex of S, we obtain a bipartite graph B^*. Then by Claim 2,

$$N_{B^*}(S) = Iso(G - S) \cup \{u\}, \quad \text{and}$$
$$|N_{B^*}(X)| = |N_B(X)| + 1 \geq n|X| \quad \text{for all} \quad X \subseteq S.$$

Again by the generalized marriage theorem, B^* has a $K_{1,n}$-factor F^*, whose centers are S. It is obvious that $F^* - \{u\}$ is a $\{K_{1,n-1}, K_{1,n}\}$-factor of B. By combining $\{K_{1,1}, \ldots, K_{1,n}\}$-factors of all nontrivial components of $G - S$ and the above $\{K_{1,n-1}, K_{1,n}\}$-factor of B, we obtain the desired $\{K_{1,1}, \ldots, K_{1,n}\}$-factor of G. $\square$

We next give another proof using alternating paths, whose proof is given in [15] and similar to the proof in [3]. Furthermore, an algorithm for finding a star factor, if any exist, is obtained from this proof.

Proof of the sufficiency (2). Assume that G satisfies the condition (7.3) but has no $\{K_{1,1}, \ldots, K_{1,n}\}$-factor. We shall derive a contradiction. Let H be a maximal $\{K_{1,1}, \ldots, K_{1,n}\}$-subgraph of G. So G has no $\{K_{1,1}, \ldots, K_{1,n}\}$-subgraph H' with $V(H') \supset V(H)$. Since G has no $\{K_{1,1}, \ldots, K_{1,n}\}$-factor, there exists a vertex $v \in V(G) - V(H)$. Moreover, it is clear that no two vertices of $V(G) - V(H)$ are joined by an edge of G. An H-**alternating path** of G is a path whose edges are alternately in $E(H)$ and not in $E(H)$ (Fig. 7.6).

Claim 1. *For every H-alternating path $P = (v, x_1, y_1, x_2, y_2, \ldots, x_r$ or $y_r)$ starting at v, it follows that $\deg_H(x_i) = n$ and $\deg_H(y_i) = 1$ for every i.*

Since $vx_1 \notin E(H)$ and P is an H-alternating path, we have $x_i y_i \in E(H)$ and $y_i x_{i+1} \notin E(H)$. Let D_i denote a component of H containing x_i. If $\deg_H(x_1) = 1$ and $D_1 \cong K_{1,1}$, then $H + vx_1$ is a $\{K_{1,1}, \ldots, K_{1,n}\}$-subgraph of G with vertex set $V(H) \cup \{v\}$, which contradicts the maximality of H. If $\deg_H(x_1) = 1$ and $D_1 \not\cong K_{1,1}$, then x_1 is an end-vertex of D_1, and so by removing an edge of D_1 incident with x_1 and by adding vx_1, we obtain a $\{K_{1,1}, \ldots, K_{1,n}\}$-subgraph with vertex set $V(H) \cup \{v\}$, a contradiction (Fig. 7.6). If $2 \leq \deg_H(x_1) < n$, then x_1 is the center of D_1, and $H + vx_1$ is a $\{K_{1,1}, \ldots, K_{1,n}\}$-subgraph, a contradiction. Hence $\deg_H(x_1) = n$, which implies $\deg_H(y_1) = 1$.

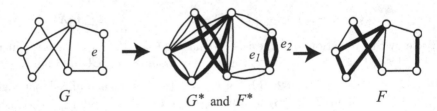

$$G \qquad\qquad G^* \text{ and } F^* \qquad\qquad F$$

Fig. 7.6. An H-alternating path $P = (v, x_1, y_1, x_2, y_2, \ldots)$ with $n = 3$, where $v, x_i, y_i \in V(G)$ and H is a $\{K_{1,1}, K_{1,2}, K_{1,3}\}$-subgraph.

If either $\deg_H(x_2) = 0$ or $\deg_H(x_2) = 1$ and $D_2 \cong K_{1,1}$, then $H - x_1 y_1 + vx_1 + y_1 x_2$ is a $\{K_{1,1}, \ldots, K_{1,n}\}$-subgraph with vertex set

$V(H) \cup \{v\}$, a contradiction. If $\deg_H(x_2) = 1$ and $D_2 \not\cong K_{1,1}$, then by removing two edges x_1y_1 and an edge of D_2 incident with x_2 and by adding vx_1 and y_1x_2, we obtain a larger $\{K_{1,1}, \ldots, K_{1,n}\}$-subgraph, a contradiction. If $2 \le \deg_H(x_2) < n$, then x_2 is the center of D_2, and $H - x_1y_1 + vx_1 + y_1x_2$ is a $\{K_{1,1}, \ldots, K_{1,n}\}$-subgraph, a contradiction (Fig. 7.6). Hence $\deg_H(x_2) = n$ and $\deg_H(y_2) = 1$. By repeating the above argument, we can prove Claim 1.

Define three vertex sets $Alt(v)$, $Odd(v)$ and $Even(v)$ of G as follows:

$$Alt(v) = \{z \in V(G) : \exists \text{ an } H\text{-alternating } v\text{-}z \text{ path}\},$$
$$Odd(v) = \{x \in V(G) : \exists \text{ an } H\text{-alternating } v\text{-}x \text{ path of odd length}\},$$
$$Even(v) = \{y \in V(G) : \exists \text{ an } H\text{-alternating } v\text{-}y \text{ path of even length}\}.$$

Claim 2. *(i) $Alt(v) - \{v\} \subseteq V(H)$, $Odd(v) \cap Even(v) = \emptyset$, and $Alt(v)$ is a disjoint union of $\{v\}$, $Odd(v)$ and $Even(v)$.*
(ii) $|Even(v)| = n|Odd(v)|$.
(iii) If a vertex z is adjacent to $Even(v)$ in G, then $z \in Odd(v)$.

By Claim 1, $x_i, y_i \in V(H)$ and so $Alt(v) - \{v\} \subseteq V(H)$. By Claim 1, if $x \in Odd(v)$, then $\deg_H(x) = n \ge 2$, and if $y \in Even(v)$, then $\deg_H(y) = 1$. Hence $Odd(v) \cap Even(v) = \emptyset$, and (i) holds.

Consider a subgraph of G induced by an edge set $Odd(v) \cup Even(v)$. Then by Claim 1, every component is $K_{1,n}$, its center is contained in $Odd(v)$, and its end-vertices are contained in $Even(v)$. Hence $|Even(v)| = n|Odd(v)|$.

Assume that a vertex z is adjacent to a vertex y of $Even(v)$ in G. Then $\deg_H(y) = 1$ and there exists an H-alternating v-y path $P = (v, x_1, y_1, \ldots, x_r, y_r = y)$. If $z \notin V(P)$, then $yz \notin E(H)$ and $P + yz$ is an H-alternating v-z path of odd length, and so $z \in Odd(v)$. If $z \in V(P)$ and $z \notin Odd(v)$, then $z = y_j$ and there exists an H-alternating v-y path $(v, x_1, y_1, \ldots, x_j, y_j = z, y_r = y)$ of odd length, which implies $y \in Odd(v)$ and contradicts (i). Therefore (iii) holds.

By Claim 2, we have

$$\{v\} \cup Even(v) \subseteq Iso(G - Odd(v)), \quad \text{and so}$$
$$1 + n|Odd(v)| \le iso(G - Odd(v)).$$

This contradicts (7.3). Consequently the proof is complete. $\square$

Corollary 7.7 (Payan). *Let $n \ge 2$ be an integer and G be a simple graph. If $\Delta(G)/\delta(G) \le n$, then G has a $\{K_{1,1}, K_{1,2}, \ldots, K_{1,n}\}$-factor.*

Proof. Let $\emptyset \ne S \subset V(G)$. Then by counting the number of edges joining $Iso(G - S)$ to S, we have

$$iso(G - S)\delta(G) \le e_G(Iso(G - S), S) \le \Delta(G)|S|.$$

Hence

$$iso(G - S) \leq \frac{\Delta(G)}{\delta(G)} |S| \leq n|S|.$$

Therefore, by the star factor theorem G has a $\{K_{1,1}, \ldots, K_{1,n}\}$-factor.   □

A $\{K_{1,1}, \ldots, K_{1,n}\}$-factor can be generalized as follows. Let G be a graph and let $f : V(G) \to \{2, 3, 4, \ldots\}$ be a function defined on $V(G)$. Then a subgraph H is called an f-**star subgraph** of G if each component of H is a star and its center x satisfies $\deg_H(x) \leq f(x)$. A spanning f-star subgraph is called an f-**star factor**. So if $f(x) = n$ for all $x \in V(G)$, then an f-star factor is nothing but a $\{K_{1,1}, \ldots, K_{1,n}\}$-factor. The following theorem gives a criterion for a graph to have an f-star factor, and it is a corollary of the (g, f)-factor theorem with $g \leq 1$. Note that it can also be proved in a similar way as the star factor theorem.

Theorem 7.8 (The f-Star Factor Theorem, Las Vergnas [165]**).** *Let* G *be a simple graph and* $f : V(G) \to \{2, 3, 4, \ldots\}$. *Then* G *has an* f-*star factor if and only if*

$$iso(G - S) \leq \sum_{x \in S} f(x) \qquad \text{for all} \quad S \subset V(G). \tag{7.5}$$

Proof. Suppose that G has an f-star factor F. Then for any subset $S \subset V(G)$, it follows that

$$iso(G - S) \leq iso(F - S) \leq \sum_{x \in S} f(x)$$

since each component D of F is a star and satisfies $iso(D - X) \leq \sum_{x \in X} f(x)$, where $X = V(D) \cap S$. Therefore the necessity is proved.

The sufficiency of the theorem is an immediate consequence of the $(1, f)$-factor theorem with $g \leq 1$ (Theorem 4.4). Define $g(x) = 1$ for all $x \in V(G)$. Let $S \subset V(G)$. Then since $f(x) \geq 2$ for all $x \in V(G)$, we have $odd(g; G - S) = iso(G - S)$ and thus

$$odd(g; G - S) = iso(G - S) \leq \sum_{x \in S} f(x).$$

Hence G has a $(1, f)$-factor by the $(1, f)$-factor theorem with $g \leq 1$. Let F be a minimal $(1, f)$-factor of G with respect to edge set. Namely, F is a $(1, f)$-factor of G but for any edge e of F, $F - e$ is not a $(1, f)$-factor anymore. It is obvious that F has no cycle, and so F is a forest. If F contains a path of order four as a subgraph, then we can remove its central edge, and so F does not contain P_4 as a subgraph, which implies that each component of F is a star. Since F is a $(1, f)$-factor, the center x of each component satisfies $\deg_F(x) \leq f(x)$. Therefore F is the desired f-star factor, and theorem is proved.   □

A formula for the order of a maximum f-star subgraph of a graph is easily obtained as a maximum matching, which is given in the following theorem.

Theorem 7.9. *Let G be a simple graph and $f : V(G) \to \{2, 3, 4, \ldots\}$. Then the order of a maximum f-star subgraph H of G is given by*

$$|H| = |G| - \max_{X \subset V(G)} \{iso(G - X) - \sum_{x \in X} f(x)\}. \qquad (7.6)$$

Proof. We may assume that G is connected. Let

$$d = \max_{X \subset V(G)} \{iso(G - X) - \sum_{x \in X} f(x)\}. \qquad (7.7)$$

Then $d \geq 0$ by considering the case $X = \emptyset$. Moreover, if $d = 0$, then the theorem follows from the f-star factor theorem. Hence we may assume $d \geq 1$. Let S be a subset $V(G)$ such that

$$iso(G - S) - \sum_{x \in S} f(x) = d.$$

Then by considering the subgraph $\langle S \cup Iso(G - S)\rangle_G$, we have that every f-star subgraph of G cannot cover at least $iso(G - S) - \sum_{x \in S} f(x)$ vertices of $Iso(G - S)$. Hence $|H| \leq |G| - d$, where H is a maximum f-star subgraph.

We next prove the inverse inequality $|H| \geq |G| - d$ for a maximum f-star subgraph H of G. Add $2d$ new vertices $\{v_i, u_i : 1 \leq i \leq d\}$ together with d new edges $\{v_i u_i : 1 \leq i \leq d\}$ to G. Then join every v_i to every vertex of G by new edges. Denote the resulting graph by G^*, and define a function $f^* : V(G^*) \to \{2, 3, 4, \ldots\}$ by

$$f^*(u_i) = f^*(v_i) = 2 \qquad \text{for all } 1 \leq i \leq d, \text{ and}$$
$$f^*(x) = f(x) \qquad \text{for all } x \in V(G).$$

Then for every subset Y of $V(G^*)$, we shall estimate $iso(G^* - Y)$. If $u_i \in Y$, then we can consider $(Y - u_i) \cup \{v_i\}$ instead of Y, and so we may assume $Y \cap \{u_1, \ldots, u_d\} = \emptyset$. If $Y \not\supseteq \{v_1, v_2, \ldots, v_d\}$, then

$$iso(G^* - Y) = \#\{u_i : v_i \in Y\} \leq |Y| \leq \sum_{x \in Y} f^*(x).$$

If $Y \supseteq \{v_1, v_2, \ldots, v_d\}$, then

$$Iso(G^* - Y) = Iso(G - Y \cap V(G)) \cup \{u_i : 1 \leq i \leq d\},$$

an so

$$iso(G^* - Y) = iso(G - Y \cap V(G)) + d$$

$$\leq \sum_{x \in Y \cap V(G)} f(x) + d + d \qquad \text{by (7.7)}$$

$$= \sum_{x \in Y} f^*(x).$$

Hence by the f-star factor theorem, G^* has an f^*-star factor F^*. Then $R = F^* - \{u_i, v_i : 1 \leq i \leq d\}$ is an f-star subgraph of G, and satisfies

$$|H| \geq |R| \geq |G^*| - 3d = |G| - d,$$

since $\deg_{F^*}(v_i) \leq 2$ for all $1 \leq i \leq d$. Consequently the theorem is proved. $\square$

From Theorems 7.2, 7.3 and 7.6, the following problem is naturally proposed.

Problem 7.10 ([128]). Let G be a simple graph and λ be a positive rational number. If

$$iso(G - S) \leq \lambda|S| \qquad \text{for all} \quad \emptyset \neq S \subset V(G),$$

what factor does G have?

Partial answers to this problem are given in the next theorem and Theorem 7.24, where the condition is sharp.

Theorem 7.11 (Kano, Lu and Yu, [128]). *If a simple graph G satisfies*

$$iso(G - S) \leq \frac{|S|}{2} \qquad \text{for all} \quad S \subseteq V(G),$$

then G has a $\{K_{1,2}, K_{1,3}, K_5\}$-factor. In particular, G has a $\{K_{1,2}, K_{1,3}, K_{1,4}\}$-factor.

Recently, the above theorem was generalized as follows.

Theorem 7.12 (Kano and Saito [134]). *Let $m \geq 2$ be an integer. If a simple graph G satisfies*

$$iso(G - S) \leq \frac{1}{m}|S| \qquad \text{for all} \quad S \subseteq V(G),$$

then G has a $\{K_{1,m}, K_{1,m+1}, \ldots, K_{1,2m}\}$-factor.

We next consider a component factor where each component is K_2 or a cycle of order at least four, that is, a cycle of order three is not allowed. A **triangular cactus** is a connected graph each of whose blocks is a triangle (see Fig. 7.7). For convenience, we regard K_1 as a special triangular cactus. A triangular cactus with order at least three is called a **big triangular cactus**. We denote by $TriCa(G)$ the set of triangular cacti of G, and by $\tau(G)$ the number of triangular cacti of G, in particular, $\tau(G) = |TriCa(G)|$.

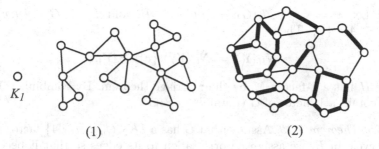

K_1

(1)   (2)

Fig. 7.7. (1) Two triangular cacti, K_1 and a big triangular cactus; (2) a $\{K_2, C_n : n \geq 4\}$-factor.

Theorem 7.13 (Cornúejols and Pulleyblank [58]). *A simple graph G has a $\{K_2, C_n : n \geq 4\}$-factor if and only if*

$$\tau(G - S) \leq |S| \qquad \text{for all} \quad S \subset V(G). \qquad (7.8)$$

We begin with some properties of a triangular cactus.

Lemma 7.14. *Let G be a big triangular cactus. Then*
(i) for every vertex v of G, $G - v$ has a 1-factor; and
(ii) for two distinct vertices x and y of G, G has an x-y path $P(x, y)$ of even length such that $G - V(P(x, y))$ has a 1-factor.

Proof. We first prove (i) by induction on $|G|$. If $G = C_3$, then the lemma follows immediately. Thus we may assume $|G| \geq 5$. Let w be a cut vertex of G. Then every component of $G-w$ is a graph obtained from a triangular cactus by removing one vertex w, and so it has a 1-factor by induction. If $w = v$, then the lemma holds. Hence we may assume that a component D of $G - w$ contains v. Then $\langle V(D) \cup \{w\}\rangle_G$ is a triangular cactus containing v. By the induction hypothesis, $\langle D \cup \{w\}\rangle_G - v$ has a 1-factor, and the other components of $G - w$ have 1-factors as stated above. Therefore $G - v$ has a 1-factor.

By (i), $G-x$ has a 1-factor M_x and $G-y$ has a 1-factor M_y. Then each component of $M_x \triangle M_y$ is either a x-y path of even length or an even cycle. Since G has no even cycle, $M_x \triangle M_y$ must be a x-y path P, and $G - V(P)$ is covered by $M_x \cap M_y$, which forms a 1-factor of $G - V(P)$. Therefore (ii) holds. $\square$

Lemma 7.15. *Let G be a bipartite graph with bipartition (A, B), and let $u \in A$ be a vertex. If $|A| = |B|$, $N_G(u) \neq \emptyset$, and*

$$|N_G(Y)| > |Y| \qquad \text{for all} \quad \emptyset \neq Y \subseteq A - \{u\},$$

then for any edge e incident with u, G has a 1-factor containing e.

Proof. Let $e = uv \in E(G)$, where $v \in B$, and $H = G - \{u, v\}$. Let $\emptyset \neq Y \subseteq A - \{u\}$. Then

$$|N_H(Y)| = |N_G(Y) \setminus \{v\}| \geq |Y|.$$

Hence H has a 1-factor M by the marriage theorem. By combining M and e, we obtain a 1-factor of G containing e. □

Proof of Theorem 7.13. Assume that G has a $\{K_2, C_n : n \geq 4\}$-factor F. For every cycle in F, we assign an orientation to its edges so that it becomes a directed cycle. Let S be any subset of $V(G)$. Since a triangular cactus has no $\{K_2, C_n : n \geq 4\}$-factor, for every triangular cactus D of $G - S$, there exists at least one vertex v in S that is adjacent to D by an edge of a K_2-component of F or by a directed edge of F from v to D. Thus

$$\tau(G - S) \leq \#\{\text{Edges or directed edges of } F \text{ adjacent from } S\}$$
$$\leq |S|.$$

Therefore (7.8) holds.

We shall prove the sufficiency by induction on the size $\|G\|$. Assume that (7.8) holds. We may assume that $|G| \geq 5$ and G is connected. We consider two cases.

Case 1. $\tau(G - S) = |S|$ *for some* $\emptyset \neq S \subset V(G)$.

Let $S \subset V(G)$ be a maximal subset such that $\tau(G - S) = |S|$. We first show that every component D of $G - S$ that is not a triangular cactus has a $\{K_2, C_n : n \geq 4\}$-factor. Let $\emptyset \neq X \subset V(D)$. Then

$$\tau(G - S) + \tau(D - X) = \tau(G - (S \cup X)) \leq |S \cup X| = |S| + |X|.$$

Hence $\tau(D - X) \leq |X|$. Thus D has a $\{K_2, C_n : n \geq 4\}$-factor by induction.

We define a bipartite graph B with bipartition $S \cup TriCa(G - S)$ as follows: a vertex $x \in S$ and an element $D \in TriCa(G - S)$ are joined by an edge of B if and only if x is adjacent to D in G. Let $\emptyset \neq Y \subset S$. If $|N_B(Y)| < |Y|$, then

$$\tau(G - (S - Y)) \geq |TriCa(G - S) - N_B(Y)| > \tau(G - S) - |Y|$$
$$= |S| - |Y| = |S - Y|,$$

which contradicts (7.8). Hence $|N_B(Y)| \geq |Y|$. Moreover, it is clear that $|N_B(S)| = \tau(G - S) = |S|$. Hence

$$|N_B(Y)| \geq |Y| \qquad \text{for all} \quad Y \subseteq S.$$

Therefore B has a 1-factor M by the marriage theorem.

By Lemma 7.14 (i), we can obtain a $\{K_2, C_n : n \geq 4\}$-factor of G from M and a $\{K_2, C_n : n \geq 4\}$-factor of every component of $G - S$ that is not a triangular cactus.

Case 2. $\tau(G - X) < |X|$ *for all* $\emptyset \neq X \subset V(G)$.

Let x and y be two adjacent vertices of G. Assume that

$$\tau(G - X) \leq |X| - 2 \qquad \text{for all} \quad \{x, y\} \subseteq X \subset V(G).$$

Then for every $Y \subset V(G) - \{x, y\}$,

$$\tau(G - \{x, y\} - Y) = \tau(G - (Y \cup \{x, y\}))$$
$$\leq |Y \cup \{x, y\}| - 2 = |Y|.$$

Hence by the induction hypothesis, $G - \{x, y\}$ has a $\{K_2, C_n : n \geq 4\}$-factor, and so does G by adding an edge xy to it. Therefore, there exists a vertex set Y_{xy} such that $\{x, y\} \subseteq Y_{xy}$ and $\tau(G - Y_{xy}) \geq |Y_{xy}| - 1$. By the condition of Case 2, we have $\tau(G - Y_{xy}) = |Y_{xy}| - 1$. Take such a minimal subset S, which depends on $\{x, y\}$. Then S satisfies

$$\{x, y\} \subseteq S, \qquad \tau(G - S) = |S| - 1 \qquad \text{and}$$
$$\tau(G - Y) \leq |Y| - 2 \qquad \text{for all} \quad \{x, y\} \subseteq Y \subset S. \tag{7.9}$$

Claim 1. *Every component of $G - S$ that is not a triangular cactus has a $\{K_2, C_n : n \geq 4\}$-factor.*

Let D be a component of $G - S$ that is not a triangular cactus. For every subset $X \subset V(D)$, we obtain

$$\tau(G - S) + \tau(D - X) = \tau(G - (S \cup X))$$
$$\leq |S \cup X| - 1,$$

which means $\tau(D - X) \leq |X|$ by (7.9). Hence by induction, D has a $\{K_2, C_n : n \geq 4\}$-factor.

Suppose $S \neq \{x, y\}$. Let B be the bipartite graph with bipartition $S \cup TriCa(G - S)$ defined in Case 1. Let $\emptyset \neq X \subset S - \{x, y\}$. If $|N_B(X)| \leq |X|$, then by $\{x, y\} \subseteq S - X \subset S$ and (7.9), we have

$$\tau(G - (S - X)) \geq |TriCa(G - S) - N_B(X)| \geq \tau(G - S) - |X|$$
$$= |S| - 1 - |X| = |S - X| - 1,$$

which contradicts (7.9). Hence $|N_B(X)| \geq |X| + 1$. If $N_B(S - \{x, y\}) \neq TriCa(G - S)$, then $\tau(G - \{x, y\}) \geq 1$, which implies $\tau(G - \{x, y\}) = 1$, and so $S = \{x, y\}$ by the choice of S, which contradicts $S \neq \{x, y\}$. Hence $N_B(S - \{x, y\}) = TriCa(G - S)$. Therefore, if $S \neq \{x, y\}$ then

$$|N_B(X)| \geq |X| + 1 \qquad \text{for all} \quad \emptyset \neq X \subseteq S - \{x, y\}. \tag{7.10}$$

Claim 2. *If $S \neq \{x, y\}$, then G has a $\{K_2, C_n : n \geq 4\}$-factor.*

Suppose $S \neq \{x, y\}$. If $N_B(x) = \emptyset$, then $\tau(G-(S-x)) = \tau(G-S) = |S-x|$, which contradicts the condition of Case 2. Hence $N_B(x) \neq \emptyset$, and similarly $N_B(y) \neq \emptyset$. Since $B - x$ and y satisfy the condition of Lemma 7.15 by (7.10), for any edge yD_y, $D_y \in TriCa(G-S)$, $B-x$ has a 1-factor M_x that contains yD_y. Similarly, for any edge xD_x, $D_x \in TriCa(G-S)$, $B-y$ has a 1-factor M_y that contains xD_x. If possible, we choose D_y and D_x so that they are distinct.

If $D_x = D_y$, then $N_B(x) = N_B(y) = \{D_x\}$ since we can arbitrarily choose D_y and D_x from $N_B(y)$ and $N_B(x)$. Thus

$$\tau(G - (S - \{x, y\})) \geq \tau(G - S) - 1 = |S| - 1 - 1 = |S - \{x, y\}|,$$

which contradicts the assumption of Case 2. Hence $D_x \neq D_y$.

Thus $M_x \bigtriangleup M_y$ contains an x-y path Q_B of even length and the vertices of $V(B) - V(Q_B)$ are covered by the edges of M_x not contained in Q_B (Fig. 7.8). For every two consecutive edges uD and Dv of Q_B, where $u, v \in S$ and $D \in TriCa(G - S)$, by Lemma 7.15, there exists a u-v path $P(u, v)$ in G which is of even length and contains two edges joining u to D and D to v such that $D \setminus V(P(u, v))$ has a 1-factor (see Fig. 7.8 x, D_x, a in G). Hence there exists a path Q_G in G of even length corresponding to Q_B such that Q_G together with a matching of G cover $S \cap V(Q_B)$ and the triangular cacti of $G - S$ corresponding to $V(Q_B) \cap TriCa(G - S)$.

For every edge zD of $M_x \setminus Q_B$, where $z \in S$ and $D \in TriCa(G-S)$, there exists a matching in G that covers $\{z\} \cup V(D)$ by Lemma 7.15 (see Fig. 7.8 c, D_2 in G). Since $Q_G + xy$ is a cycle of odd order, by Claim 1 we can obtain the desired $\{K_2, C_n : n \geq 4\}$-factor of G.

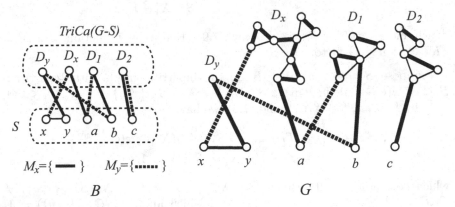

Fig. 7.8. Two matchings M_x and M_y of B; and an x-y path in G together with a matching that cover G.

By Claim 2, we may assume that $S = \{x, y\}$, that is, $\tau(G - \{x, y\}) = 1$ for every edge xy of G. In this case, G has the desired factor as shown in the following claim.

Claim 3. *If $\tau(G - \{x, y\}) = 1$ for every edge xy of G, then G has a $\{K_2, C_n : n \geq 4\}$-factor.*

Let $TriCa(G - \{x, y\}) = \{D_{xy}\}$. If D_{xy} contains two distinct vertices u, v which are adjacent to x and y, respectively, then by Lemma 7.14 $D_{xy} + xu + yv$ has an x-y path Q of even length such that $V(D_{xy}) \setminus V(Q)$ is covered by a matching. Since $Q + xy$ is an odd cycle of order at least five, and any other component of $G - \{x, y\}$ has a $\{K_2, C_n : n \geq 4\}$-factor by Claim 1, we can obtain the desired $\{K_2, C_n : n \geq 4\}$-factor of G. Therefore we may assume that

$$N_G(x) \cap V(D_{xy}) = N_G(y) \cap V(D_{xy}) = \{z\}.$$

In particular, $\langle V(D_{xy}) \cup \{x, y\} \rangle_G$ forms a triangular cactus for every edge xy.

For an edge $e = xy$ of G, let $f(e)$ denote the order of a largest component of $G - \{x, y\}$. An edge b is called **extremal** if

$$f(b) = \max\{f(e) : e \in E(G)\}.$$

Let $b = uv$ be an extremal edge of G. Assume that $G - \{u, v\}$ is not connected. Let D be a largest component of $G - \{u, v\}$. If $D = D_{uv}$, where D_{uv} is a triangular cactus of $G - \{u, v\}$, then $f(c) > f(b)$ for an edge $c = sw$, $s \in \{u, v\}$, $w \in V(G) - V(D) - \{u, v\}$, a contradiction. If $D \neq D_{uv}$, then $f(c) > f(b)$ for an edge $c = sw$, $s \in \{u, v\}$, $w \in V(D_{xy})$. Thus $G - \{u, v\}$ is connected, and so $G - \{u, v\} = D_{uv}$ is a triangular cactus. Hence $G = \langle D_{uv} \cup \{u, v\} \rangle_G$ is a triangular cactus. But this contradicts the fact that G is not a triangular cactus by taking $S = \emptyset$ in (7.8). Consequently the proof is complete. $\square$

For a set $\mathcal{Q}$ of odd integers greater than one, if we introduce $\mathcal{Q}$-critical graphs, then we can give a necessary and sufficient condition using $\mathcal{Q}$-critical graphs for a graph to have a $\{K_2, C_n : n \in \mathcal{Q}\}$-factor. A graph G is called a $\mathcal{Q}$-**critical graph** if G has no $\{K_2, C_n : n \in \mathcal{Q}\}$-factor but for every vertex v of G, $G - v$ has a $\{K_2, C_n : n \in \mathcal{Q}\}$-factor.

Theorem 7.16 (Cornuéjols and Pulleyblank [59]). *Let $\mathcal{Q}$ be a set of odd integers greater than one. Then a simple graph G has a $\{K_2, C_n : n \in \mathcal{Q}\}$-factor if and only if*

$$\#\{\mathcal{Q}\text{-critical components of } G - S\} \leq |S| \qquad \text{for all} \quad S \subset V(G). \quad (7.11)$$

Proof. We shall prove only the sufficiency since the necessity can be proved similarly as in Theorem 7.13. For convenience, we denote by $\tau_{\mathcal{Q}}(G - S)$ the number of $\mathcal{Q}$-critical components of $G - S$. Consider the Gallai-Edmonds decomposition $C(G) \cup A(G) \cup D(G)$ of G. Let $Comp(D(G))$ denote the set of components of $\langle D(G)\rangle_G$, and let $Comp^*(D(G))$ be the set of components of $\langle D(G)\rangle_G$ that have no $\{K_2, C_n : n \in \mathcal{Q}\}$-factor. Since every component in $Comp(D(G))$ is factor-critical, every component in $Comp^*(D(G))$ is $\mathcal{Q}$-critical.

Define a bipartite graph B with bipartition $A(G) \cup Comp(D(G))$ as follows: a vertex $x \in A(G)$ and a component $R \in Comp(D(G))$ are joined by an edge of B if and only if x and R are joined by an edge of G. Let $\emptyset \neq X \subseteq Comp^*(D(G))$. If $|N_B(X)| < |X|$, then

$$\tau_Q(G - N_B(X)) \geq |X| > |N_B(X)|.$$

This contradicts (7.11). Thus $|N_B(X)| \geq |X|$ for all $X \subseteq Comp^*(D(G))$, and so B has a matching that covers $Comp^*(D(G))$. By Theorem 2.21, B has a maximum matching M that covers $Comp^*(D(G))$. By the Galai-Edomonds Structure Theorem 2.47, M also covers $A(G)$.

For every edge xR of M, where $x \in A(G)$ and $R \in Comp(D(G))$, $\langle \{x\} \cup V(R)\rangle_G$ has a 1-factor. Moreover, $\langle C(G)\rangle_G$ has a 1-factor, and every component in $Comp(D(G)) - Comp^*(D(G))$ that is not covered by M has a $\{K_2, C_n : n \in \mathcal{Q}\}$-factor. Therefore, by combining these factors, we can get the desired $\{K_2, C_n : n \in \mathcal{Q}\}$-factor of G. $\square$

Notice that Theorem 7.13 can be proved based on Theorem 7.16 and some other additional properties and arguments ([182]; Section 6.4). The next theorem also can be proved by using Theorem 7.16, though it was proved in other way.

Theorem 7.17 (Hell and Kirkpatrick [100]). *A simple graph G has a $\{K_2, C_3\}$ -factor if and only if*

$$b(G - S) \leq |S| \qquad \text{for all} \quad S \subset V(G), \qquad (7.12)$$

where $b(G - S)$ denotes the number of components of $G - S$ such that D has no $\{K_2, C_3\}$-factor but $D - v$ has a K_2-factor for every $v \in V(D)$.

If a family $\mathcal{S}$ of connected graphs contains K_2, then K_2 often plays a very important role in $\mathcal{S}$-factor theorems. On the other hand, if $\mathcal{S}$ does not contain K_2, it often becomes very difficult to find a criterion for a graph to have an $\mathcal{S}$-factor. We will discuss this case later. Hong Wang found a criterion for a bipartite graph to have a $\{P_3, P_4, P_4\}$-factor, where $\{P_3, P_4, P_4\}$ does not contain K_2. This result is given in following theorem.

Theorem 7.18 (Wang [241]). *Let G be a bipartite simple graph. Then G has a $\{P_3, P_4, P_5\}$-factor if and only if*

$$iso(G - S) + k_2(G - S) \leq 2|S| \qquad \text{for all} \quad S \subset V(G), \qquad (7.13)$$

where $k_2(G-S)$ denotes the number of components of $G-S$ isomorphic to K_2.

Kaneko [111] found a criterion for a graph to have a $\{P_3, P_4, P_5\}$-factor, which we will now explain. For a factor-critical graph H with $V(H) = \{v_1, v_2, \ldots, v_n\}$, add new vertices $\{u_1, u_2, \ldots, u_n\}$ together with new edges $\{v_i u_i : 1 \leq i \leq n\}$ to H. Then the resulting graph is called a **sun**. Notice that K_2 is a sun since K_1 is factor-critical, and we regard K_1 also as a sun (Fig. 7.9). Moreover, a sun with order at least three has order at least six and is not a bipartite graph. We denote by $Sun(G)$ the set of sun components of G and by $sun(G) = |Sun(G)|$ the number of sun components of G. It is easy to see that every sun has no $\{P_3, P_4, P_5\}$-factor since its factor-critical subgraph does not have a 1-factor.

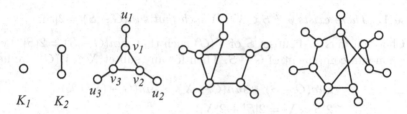

Fig. 7.9. A small sun K_1 and some other big suns.

Theorem 7.19 (Kaneko [111]). *Let G be a simple graph. Then G has a $\{P_3, P_4, P_5\}$-factor if and only if*

$$sun(G - S) \leq 2|S| \qquad for\ all \quad S \subset V(G). \qquad (7.14)$$

Kaneko proved this theorem by introducing a new concept named "trouble block" for $\{P_3, P_4, P_5\}$-factors, but its proof is involved and long. A shorter and simpler proof is given by Kano, Katona and Király [122]. Our proof depends on it. Recall that an edge incident with an end-vertex is called a **pendant edge**. We start with some properties of suns.

Lemma 7.20. *Let D be a sun of order at least six, and let $u_j v_j$ be a pendant edge of D. Then $D - \{u_j, v_j\}$ has a $\{P_4\}$-factor.*

Proof. Let H be the factor-critical subgraph of D. Then $v_j \in V(H)$, and $H - v_j$ has a 1-factor, and so the lemma follows immediately. □

Lemma 7.21. *Let D be a sun of order at least six, and u_j an end-vertex of D. Then $D - u_j$ has a $\{P_4, P_5\}$-factor.*

Proof. Let H be the factor-critical subgraph of D, and let $u_j v_j$ be an pendant edge of D. Let v_a be a vertex of H adjacent to v_j. Then $H - v_a$ has a 1-factor M. Let $v_j v_b$ be an edge of M incident with v_j. Then $G - u_j$ has a $\{P_4, P_5\}$-factor, which contains a path $(u_b, v_b, v_j, v_a, u_a)$ of order five. □

Proof of Theorem 7.19. Assume first G has a $\{P_3, P_4, P_5\}$-factor F. Let S be a vertex set of G. Then for every sun component D of $G - S$, an edge of F joins D to S since D does not have a $\{P_3, P_4, P_5\}$-factor, and for every vertex $x \in S$, at most two sun components of $G - S$ are joined to x by edges of F. Hence $sun(G - S) \le 2|S|$.

We prove the sufficiency by induction on the size $\|G\|$. Suppose that G satisfies (7.14). We may assume that G is connected since otherwise we can obtain a $\{P_3, P_4, P_5\}$-factor of G by applying the induction hypothesis to each component of G. It is also clear that we may assume $|G| \ge 5$.

We call K_1 a **small sun**, and we call a sun with order at least two a **big sun**. We consider the following three cases.

Case 1. *There exists $\emptyset \ne S \subset V(G)$ such that $sun(G - S) = 2|S|$.*

Choose a maximal subset S of $V(G)$ such that $sun(G - S) = 2|S|$. Let C be a non-sun component of $G - S$. Then for any subset $X \subset V(C)$, we have

$$sun(G - S) + sun(C - X) = sun(G - (S \cup X))$$
$$\le 2|S \cup X| = 2|S| + 2|X|.$$

Thus $sun(C - X) \le 2|X|$. Hence C has a $\{P_3, P_4, P_5\}$-factor by induction.

We define a bipartite graph B with bipartition $S \cup Sun(G - S)$, in which a vertex $v \in S$ and a sun component $D \in Sun(G - S)$ are joined by an edge of B if there exists an edge in G joining v to D. We shall show that

$$|N_B(X)| \ge 2|X| \qquad \text{for all } X \subseteq S. \tag{7.15}$$

Let $X \subset S$. Then $Sun(G - S) - N_B(X)$ is a set of sun components of $G - (S - X)$, and so

$$sun(G - S) - |N_B(X)| \le sun(G - (S - X))$$
$$\le 2|S - X| = 2|S| - 2|X|.$$

Hence $2|X| \le |N_B(X)|$, and (7.15) holds.

By the generalized marriage theorem (Theorem 2.10), B has a $\{K_{1,2}\}$-factor Q such that

$$\deg_Q(x) = 2 \qquad \text{for all } x \in S, \text{ and}$$
$$\deg_Q(D) = 1 \qquad \text{for all } D \in Sun(G - S).$$

By making use of this factor Q and by Lemma 7.20, we can obtain the desired $\{P_3, P_4, P_5\}$-factor of G (see Fig. 7.10).

Case 2. *$sun(G - X) < 2|X|$ for all $\emptyset \ne X \subset V(G)$, and there exists $\emptyset \ne S \subset V(G)$ such that $sun(G - S) = 2|S| - 1$.*

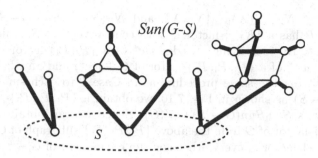

Fig. 7.10. A $\{P_3, P_4, P_5\}$-factor of G.

Take a maximal subset $S \subset V(G)$ such that $sun(G - S) = 2|S| - 1$. Let C be any non-sun component of $G - S$ and let $\emptyset \neq X \subset V(C)$. Then by the choice of S, we obtain

$$sun(G - S) + sun(C - X) = sun(G - (S \cup X))$$
$$\leq 2|S \cup X| - 2 = 2|S| - 1 + 2|X| - 1.$$

Thus

$$sun(C - X) \leq 2|X| - 1 \qquad \text{for all} \quad \emptyset \neq X \subset V(C). \tag{7.16}$$

Hence C has a $\{P_3, P_4, P_5\}$-factor by induction.

Claim 1. *If $G - S$ has a non-sun component, then G has the desired $\{P_3, P_4, P_5\}$-factor.*

Let C be a non-sun component of $G - S$, and let $v \in S$ and $w \in V(C)$ be vertices such that vw is an edge of G. Let $C + wv^*$ be a graph obtained from C by adding a new vertex v^* and a new edge wv^*. By (7.16), we have

$$sun(C + wv^* - Y) \leq 2|Y| \qquad \text{for all} \quad \emptyset \neq Y \subset V(C + wv^*) \tag{7.17}$$

since if $v^* \in Y$, then $sun(C+wv^*-Y) = sun(C-(Y-v^*))$, and if $v^* \notin Y$, then

$$sun(C + wv^* - Y) \leq sun(C - Y) + 1 \leq 2|Y| - 1 + 1.$$

If $C + wv^*$ itself is a sun, then by Lemma 7.20, $C + wv^*$ has a $\{P_2, P_4\}$-factor containing only one P_2, which can be chosen as wv^*. If $C + wv^*$ is not a sun, then by (7.17) and by the induction hypothesis, $C + wv^*$ has a $\{P_3, P_4, P_5\}$-factor, which contains a path $P = (\ldots, w, v^*)$.

Construct a bipartite graph B with bipartition $S \cup (Sun(G - S) \cup \{w^*\})$ as in Case 1, but add a new pendant edge vw^* to it, where v is the vertex of S given above. For any subset $\emptyset \neq X \subset S$, we have

$$sun(G - S) - |N_{B-w^*}(X)| \leq sun(G - (S - X))$$
$$\leq 2|S - X| - 1 = 2|S| - 1 - 2|X|.$$

Hence $2|X| \leq |N_{B-w^*}(X)| \leq |N_B(X)|$, and $|N_B(S)| = |Sun(G-S) \cup \{w^*\}| = 2|S|$. Thus B has a $\{K_{1,2}\}$-factor Q with central vertices S, which contains an edge vw^*. By identifying an edge wv^* of a $\{P_2, P_4\}$-factor of $C + wv^*$ (or an edge wv^* of a $\{P_3, P_4, P_5\}$-factor of $C + wv^*$) and an edge vw^* of Q, and by applying the same procedure as in Case 1 to each sun component of $Sun(G - S)$ as shown in Fig. 7.10, we obtain a $\{P_3, P_4, P_5\}$-subgraph of G that covers $S \cup Sun(G - S) \cup V(C)$. Hence we can obtain the desired $\{P_3, P_4, P_5\}$-factor of G from the above $\{P_3, P_4, P_5\}$-subgraph of G by adding a $\{P_3, P_4, P_5\}$-factor of every other non-sun component of $G - S$ to C.

By Claim 1, we may assume that every component of $G - S$ is a sun.

Claim 2. *If S has a vertex v that is adjacent to either no small sun component of $G - S$, or at least two small sun components of $G - S$, then G has the desired $\{P_3, P_4, P_5\}$-factor.*

Construct a bipartite graph B as in Case 1 with the additional pendant edge vw^*. Then B has a $\{K_{1,2}\}$-factor Q, which contains a component (D, v, w^*), $D \in Sun(G - S)$. Extend each $\{K_{1,2}\}$-component of Q except (D, v, w^*) to a $\{P_3, P_4, P_5\}$-subgraph of G as before (Fig. 7.10). If $|D| \geq 2$, then an edge Dv of Q can be extended to a $\{P_3, P_4, P_5\}$-factor of $D+vx$, where $x \in V(D)$ and $vx \in E(G)$, and thus G has a $\{P_3, P_4, P_5\}$-factor. Thus we may assume $D = \{x\}$, an isolated vertex of $G - S$. In this case $D + vx$ becomes a path $P_2 = (v, x)$ of order two. By our assumption, another small sun $\{z\}$ is also connected to v, and z is an end-vertex of a path of a $\{P_3, P_4, P_5\}$-subgraph of G. By adding an edge vz of G to connect $P_2 = (x, v)$ and the above path terminating at z, we can get the desired $\{P_3, P_4, P_5\}$-factor of G.

Claim 3. *If a small sun component $\{z\} \in Sun(G - S)$ is adjacent to at least two vertices of S in G, then G has the desired $\{P_3, P_4, P_5\}$-factor.*

Suppose that z is adjacent to two distinct vertices $u, v \in S$ in G. Construct a bipartite graph B as before with the addition of a pendant edge vw^* and the deletion of an edge vz (Fig. 7.11). Let $\emptyset \neq X \subset S$. If either $v \notin X$ or $u \in X$, then $N_B(X) = N_{B+zv}(X)$ and so $|N_B(X)| \geq 2|X|$ as in the proof of Claim 1. If $v \in X$ and $u \notin X$ (Fig. 7.11), then $w^* \in N_B(X)$ and it may occur that $\{z\} \notin Sun(G - (S - X))$ and $z \notin N_B(X)$. Thus we have

$$sun(G - S) - |N_B(X) - \{w^*\}| \leq sun(G - (S - X)) + 1$$
$$\leq 2|S - X| - 1 + 1 = 2|S| - 2|X|.$$

Hence $2|X| - 1 \leq |N_B(X) - \{w^*\}|$, and so

$$|N_B(X)| \geq 2|X| \quad \text{for all } X \subset S.$$

Thus B has a $\{K_{1,2}\}$-factor Q, which contains components (w^*, v, D_1) and (z, u', D_2), where $u' \in S - \{v\}$. If $|D_1| \geq 2$, then we obtain a $\{P_3, P_4, P_5\}$-factor of $D_1 + vx$, where $x \in V(D_1)$ and $vx \in E(G)$, and so G has the desired $\{P_3, P_4, P_5\}$-factor (Fig. 7.11). If $D_1 = \{x\}$, then by adding an edge zv to connect a path (v, x) and a path $(.., u', z)$ of a $\{P_3, P_4, P_5\}$-factor corresponding to (z, u', D_2) of Q, we can get the desired $\{P_3, P_4, P_5\}$-factor of G (Fig. 7.11). Therefore Claim 3 holds.

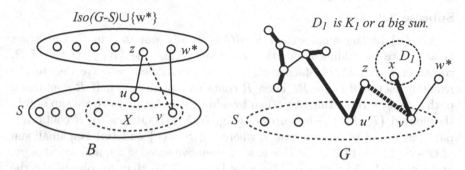

Fig. 7.11. A bipartite graph B not containing zv, and a $\{P_3, P_4, P_5\}$-factor of G corresponding to $Q - vw^*$.

By Claims 2 and 3, we may assume from now on that

(i) there are $|S|$ small suns and $|S| - 1$ big suns in $Sun(G - S)$; (ii) every small sun component of $G - S$ is an end-vertex of G, and (iii) every vertex of S is adjacent to exactly one end-vertex in G, which is a small sun component of $G - S$.

Claim 4. *If every vertex x of G with $\deg_G(x) \geq 2$ is adjacent to an end-vertex of G, then G has the desired $\{P_3, P_4, P_5\}$-factor.*

By the condition of Case 2, $sun(G - \{x\}) \leq 2|\{x\}| - 1 = 1$, and so every vertex x of G with $\deg_G(x) \geq 2$ is adjacent to exactly one end-vertex of G. Let U be the set of vertices x with $\deg_G(x) \geq 2$. If an induced subgraph $\langle U \rangle_G$ has a 1-factor, then G has a $\{P_4\}$-factor. Otherwise there exists a subset $X \subset U$ such that $\langle U \rangle_G - X$ has at least $|X| + 1$ factor-critical components by Theorem 2.28. Since every vertex of U is adjacent to exactly one end-vertex of G, $G - X$ contains $|X|$ isolated vertices and at least $|X| + 1$ big sun components, which correspond to the factor-critical components of $\langle U \rangle_G - X$. This contradicts (7.14). Therefore Claim 4 holds.

By Claim 4, we may assume that there exists a vertex z in G such that $\deg_G(z) \geq 2$ and $N_G(z)$ contains no end-vertices of G. By statement (iii) above, the vertex z is contained in a big sun component $D \in Sun(G - S)$. This means that $\langle D \rangle_G$ has an end-vertex y adjacent to $z \in D$, and then y is joined to a vertex $v \in S$ in G since $N_G(z)$ contains no end-vertices of G, and if D is a K_2-component then z is also connected to S.

Subcase 2.1 $|D| = 2$

Then $D = K_2$, $V(D) = \{x, y\}$, and both x and y are adjacent to S in G. Since y is adjacent to S, we can apply the same argument in the proof of Claim 3 to D, that is, D plays the same role as the vertex z. Thus we obtain the desired $\{P_3, P_4, P_5\}$-factor of G.

Subcase 2.2 $|D| \geq 3$

Let $\{x\}$ be the small sun of $Sun(G - S)$ adjacent to v in G. Construct B as before by adding a pendant edge vw^*. Take the $\{K_{1,2}\}$-factor of B, construct a $\{P_3, P_4, P_5\}$-factor of $G + vw^*$ and delete w^*. We denote the resulting factor of G by R. Then R contains a path (v, x). If R contains a path $(.., z, y)$ terminating at y, then by adding an edge yv to R, we can obtain the desired $\{P_3, P_4, P_5\}$-factor of G (Fig. 7.12 (1)). Otherwise, R contains a path (w, u, y, z) terminating at z, where $u \in S - \{v\}$ and $\{w\}$ is a small sun of $G - S$ (Fig. 7.12 (2)). In this case, we remove an edge zy, add an edge yv, and get a $\{P_4, P_5\}$-factor of $D - y$ by Lemma 7.21, then we obtain the the desired $\{P_3, P_4, P_5\}$-factor of G, which contains a path (w, u, y, v, x).

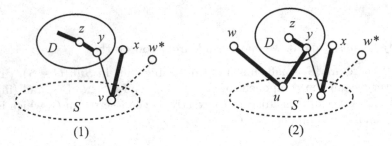

Fig. 7.12. (1) R contains a path $(.., z, y)$; (2) R contains a path (w, u, y, z).

Case 3. $sun(G - X) \leq 2|X| - 2$ for all $\emptyset \neq X \subset V(G)$.

If G has an end-vertex connected to a vertex v, then $sun(G - \{v\}) \geq 1$, which contradicts the assumption of this case. Thus G has no end-vertices, and so we can find an edge e for which $G - e$ is connected. For every subset $\emptyset \neq S \subset V(G - e)$, we have

$$sun((G - e) - S) \leq sun(G - S) + 2 \leq 2|S| - 2 + 2 = 2|S|.$$

Moreover, $G - e$ is not a sun because every sun of order at least six has at least three end-vertices, but $G - e$ has at most two end-vertices. Therefore, $G - e$ has a $\{P_3, P_4, P_5\}$-factor by the inductive hypothesis, and so does G.

Consequently, the proof is complete. □

The next theorems can be proved by using Theorem 7.19. The proofs of these theorems are left to the reader (Problem 7.2, 7.3).

Theorem 7.22 ([111]). *Let $r \geq 2$ be an integer. Then every r-regular simple graph has a $\{P_3, P_4, P_5\}$-factor.*

Theorem 7.23 ([111]). *Every 3-connected plane simple graph has $\{P_3, P_4, P_5\}$-factor.*

Theorem 7.24 ([128]). *If a simple graph G satisfies*

$$iso(G - S) \leq \frac{2}{3}|S| \quad \text{for all} \quad S \subset V(G),$$

then G has a $\{P_3, P_4, P_5\}$-factor.

A formula for the order of a maximum $\{P_3, P_4, P_5\}$-subgraph of a graph is given in the next theorem, and it uses two parameters S and T. Notice that the order of a maximum $\{P_3, P_4, P_5\}$-subgraph of a graph G is not generally equal to

$$|G| - \max_{S \subseteq V(G)} \{sun(G - S) - 2|S|\}.$$

This situation is different from other similar formulas. We omit its proof since it is much longer than the proofs of other similar formulas.

Theorem 7.25 (Kano, Katona, Király [122]). *Let G be a simple graph. Then the order of a maximum $\{P_3, P_4, P_5\}$-subgraph of G is*

$$|G| - \max_{T \subseteq S \subset V(G)} \{sun(G - S) - 2|S| + k_2(G - T) - 2|T|\}, \qquad (7.18)$$

where $k_2(G - T)$ denotes the number of K_2-components of $G - T$.

We introduce a new component factor, each of whose components is an induced subgraph. Namely, we call a subgraph H of a graph G a **strong S-subgraph** if every component of H is isomorphic to an element of S and is an induced subgraph of G. A spanning strong S-subgraph is called a **strong S-factor** (Fig. 7.13). So each component of a strong S-factor of G is an induced subgraph of G, which is different from other component factors. A strong $\{K_{1,i} : i \geq 1\}$-factor is briefly called a **strong star factor**.

Saito and Watanabe [216] and Kelmans [150] independently introduced a strong star factor and obtained the following theorem. In order to explain the theorem, we need some definitions. A graph G is called a **complete-cactus** if G is connected and every block of G is a complete graph. A complete-cactus is called an **odd complete-cactus** if every one of its blocks is a complete graph of odd order (Fig. 7.13). Note that K_1 is an odd complete-cactus.

Theorem 7.26 (Kelmans [150], Saito and Watanabe [216]). *A connected simple graph G has a strong star factor if and only if G is not an odd complete-cactus.*

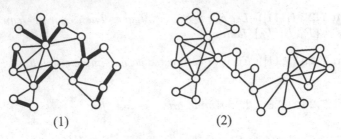

Fig. 7.13. (1) A strong $\{K_{1,1}, K_{1,2}, K_{1,3}\}$-factor; (2) An odd complete-cactus.

This theorem is generalized as follows by considering a strong $\{K_{1,1}, K_{1,2}, \ldots, K_{1,n}\}$-factor. For a graph, let $OddCa(G)$ denote the set of odd complete-cacti of G, and $oddca(G) = |OddCa(G)|$ denote the number of odd complete-cacti of G.

Theorem 7.27 (Egawa, Kano and Kelmans [71]). *Let $n \geq 2$ be an integer. Then a simple graph G has a strong $\{K_{1,1}, K_{1,2}, \ldots, K_{1,n}\}$-factor if and only if*

$$oddca(G - S) \leq n|S| \qquad \text{for all} \quad S \subset V(G). \tag{7.19}$$

We begin with basic properties of complete-cacti. For convenience, we briefly call a complete-cactus a **cactus** in the following proofs. Analogously, an odd complete-cactus is called an **odd cactus**. Every block of a cactus is a complete graph, and we call it an **odd block** or **even block** according to its order.

Lemma 7.28. *(i) A complete-cactus possessing exactly one even block has a 1-factor.*
(ii) A complete-cactus possessing precisely one or two even blocks has a strong $\{K_{1,1}, K_{1,2}\}$-factor.
(iii) If G is an odd complete-cactus, then for every vertex v of G, $G - v$ has a 1-factor .
(iv) An odd complete-cactus does not have a strong star factor.

Proof. We prove (i) by induction on $|G|$. Let D be the unique even block of G. Then D clearly has a 1-factor. Thus we may assume $G \neq D$. Clearly every component of $G - V(D)$ is a cactus with exactly one even block. Therefore, by the induction hypothesis, they have 1-factors, and hence G also has a 1-factor, and (i) holds.

Proof of (ii). By (i), we may assume that G has exactly two even blocks D_1 and D_2. Clearly there exists a cut vertex v of G such that $D_1 - v$ and $D_2 - v$ are contained in different components H_1 and H_2 of $G - v$, respectively, where if $v \notin V(D_i)$ then $D_i - v = D_i$. It is clear that every component of $G - v$ except H_1 and H_2 has a 1-factor by (i). If $v \in V(D_i)$, then let x_i be an arbitrary vertex of D_i adjacent to v. If $v \notin V(D_i)$, then let x_i be an arbitrary vertex of $H_i - V(D_i)$ adjacent to v. Then $\{v, x_1, x_2\}$ induces a $K_{1,2}$, and $H_i - x_i$ has a 1-factor by (i) since it is a cactus having exactly one even block. Therefore G has a strong $\{K_{1,1}, K_{1,2}\}$-factor.

Proof of (iii). If v is not a cut vertex, then $G - v$ is a cactus having exactly one even block, and so it has a 1-factor by (i). Otherwise, every component of $G - v$ satisfies condition (i), and so it has a 1-factor, and hence $G - v$ has a 1-factor.

We prove (iv) by induction on $|G|$. If G consists of exactly one block then the statement is true. Therefore, we may assume that G is not a complete graph. Then G has a cut vertex v. For every component C of $G - v$, the subgraph of G induced by $V(C) \cup \{v\}$ is an odd cactus and has no strong star factor by the induction hypothesis. On the other hand, if G has a strong star factor, then for at least one component D of $G - v$, the subgraph induced by $V(D) \cup \{v\}$ must have a strong star factor, which is a contradiction. Hence (iv) holds. $\square$

Lemma 7.29. *Let G be a bipartite simple graph with bipartition (A, B). Let $n \geq 1$ be an integer, and let $f : V(G) \to \{1, n\}$ a function such that $f(x) = n$ for all $x \in A$ and $f(y) = 1$ for all $y \in B$. Then G has a $(1, f)$-factor if and only if*

$$|N_G(X)| \geq |X| \qquad \text{for all} \quad X \subseteq A; \quad \text{and}$$
$$n|N_G(Y)| \geq |Y| \qquad \text{for all} \quad Y \subseteq B. \tag{7.20}$$

Proof. If G has a $(1, f)$-factor F, then (7.20) follows from

$$|N_G(X)| \geq |N_F(X)| \geq |X| \quad \text{and} \quad |N_G(Y)| \geq |N_F(Y)| \geq \frac{|Y|}{n}.$$

Conversely, assume that (7.20) holds. We may assume that G is connected since otherwise each component satisfies (7.20) and has a $(1, f)$-factor by induction, and hence G itself has a $(1, f)$-factor. For any subsets $X \subseteq A$ and $Y \subseteq B$, it follows from (7.20) that

$$\gamma^*(X, Y) = n|X| + \sum_{y \in Y} (\deg_{G-X}(y) - 1)$$
$$\geq n|X| - |S| \geq n|N_G(S)| - |S| \geq 0,$$
$$\text{where} \quad S = Iso(G - X) \cap Y \subseteq B, \ N_G(S) \subseteq X,$$
$$\gamma^*(Y, X) = |Y| + \sum_{x \in X} (\deg_{G-Y}(x) - 1)$$
$$\geq |Y| - |T| \geq |N_G(T)| - |T| \geq 0,$$
$$\text{where} \quad T = Iso(G - Y) \cap X \subseteq A, \ N_G(T) \subseteq Y.$$

Therefore by Theorem 4.3, G has the desired $(1, f)$-factor. $\square$

We are ready to prove Theorem 7.27.

Proof of Theorem 7.27. Suppose that G has a strong $\{K_{1,1}, \ldots, K_{1,n}\}$-factor F. Let $\emptyset \neq S \subset V(G)$. Since every odd cactus D of $G - S$ does not have a strong $\{K_{1,1}, \ldots, K_{1,n}\}$-factor by Lemma 7.28, F has an edge joining D to S. It is obvious that for every vertex $x \in S$, F has at most n edges joining x to odd cacti in $G - S$. Hence $oddca(G - S) \leq n|S|$.

We now prove the sufficiency by induction on $|G|$. Obviously we may assume that $|G| \geq 3$ and G is connected. By taking $S = \emptyset$, it follows that G is not an odd cactus. Assume that G has two adjacent vertices x and y such that

$$oddca(G - S) \leq n|S| - 2n \quad \text{for all} \quad \{x, y\} \subseteq S \subset V(G).$$

Then for every $T \subset V(G) - \{x, y\}$, we have

$$oddca(G - \{x, y\} - T) = oddca(G - (T \cup \{x, y\}))$$
$$\leq n|T \cup \{x, y\}| - 2n = n|T|.$$

Hence by the induction hypothesis, $G - \{x, y\}$ has a strong $\{K_{1,1}, \ldots, K_{1,n}\}$-factor, and so does G by adding xy to it. Therefore, we may assume that for every pair of adjacent vertices x and y of G, there exists a vertex set T_{xy} such that

$$\{x, y\} \subseteq T_{xy} \subset V(G), \quad \text{and}$$
$$oddca(G - T_{xy}) \geq n|T_{xy}| - 2n + 1 \geq 1. \tag{7.21}$$

Define

$$\beta = \min\{n|X| - oddca(G - X) : X \subset V(G) \text{ and } oddca(G - X) \geq 1\}.$$

By (7.19) and (7.21), we have $0 \leq \beta \leq 2n - 1$, and it immediately follows from the definition of β that

$$oddca(G - X) \leq n|X| - \beta \tag{7.22}$$

for every subset $X \subset V(G)$ with $oddca(G-X) \geq 1$. A proper subset $S \subset V(G)$ is said to be **tight** if

$$n|S| - oddca(G - S) = \beta \quad \text{and} \quad oddca(G - S) \geq 1.$$

Tight sets will play an essential role in our proof.

Claim 1. *Let S be a tight set of G. Then every component of $G - S$ which is not an odd cactus has a strong $\{K_{1,1}, \ldots, K_{1,n}\}$-factor.*

Let D be a component of $G - S$ that is not an odd cactus. For every subset $X \subset V(D)$, we obtain by (7.22) that

$$n|S \cup X| - \beta \geq oddca(G - (S \cup X)) = oddca(G - S) + oddca(D - X)$$
$$= n|S| - \beta + oddca(D - X),$$

which means $oddca(D - X) \leq n|X|$. Hence by induction, D has a strong $\{K_{1,1}, \ldots, K_{1,n}\}$-factor. Therefore Claim 1 follows.

We consider the following two cases.

Case 1. $oddca(G - S) \geq |S|$ *for some tight set S of G.*

We define a bipartite graph B with bipartition $S \cup OddCa(G - S)$ as follows: a vertex $x \in S$ and a component $C \in OddCa(G - S)$ are joined by an edge of B if and only if x is adjacent to C in G. We first show that

$$|N_B(X)| \geq |X| \qquad \text{and} \qquad n|N_B(Y)| \geq |Y|$$
$$\text{for all } X \subseteq S \text{ and } Y \subseteq OddCa(G - S). \tag{7.23}$$

Let $\emptyset \neq X \subseteq S$. If $|N_B(X)| < |X|$ and $oddca(G - (S - X)) \geq 1$, then

$$oddca(G - (S - X)) \geq oddca(G - S) - |N_B(X)|$$
$$> n|S| - \beta - |X| > n(|S - X|) - \beta,$$

which contradicts (7.22). Thus $|N_B(X)| \geq |X|$. If $oddca(G - (S - X)) = 0$, then it follows from the condition of Case 1 that

$$|N_B(X)| = |OddCa(G - S)| = oddca(G - S) \geq |S| \geq |X|.$$

Hence $|N_B(X)| \geq |X|$ for all $X \subseteq S$. Let $Y \subseteq OddCa(G - S)$. Then $N_B(Y) \subseteq S$, and by (7.19) we have

$$|Y| \leq oddca(G - N_B(Y)) \leq n|N_B(Y)|.$$

Therefore (7.23) holds.

By Lemma 7.29, the bipartite graph B has a $\{K_{1,1}, \ldots, K_{1,n}\}$-factor H, which is a $[1, n]$-factor with minimal edge set, and every component of $OddCa(G - S)$ has degree one in H. Consequently, by Lemma 7.28 (iii) and Claim 1, we can obtain a strong $\{K_{1,1}, \ldots, K_{1,n}\}$-factor of G from H.

Case 2. $oddca(G - S) < |S|$ *for every tight set S of G.*

Let S be a tight set of G. We shall first prove that $|S| = 2$ and $\beta = 2n - 1$. Recall that $0 \leq \beta \leq 2n - 1$. If $|S| = 1$, then $oddca(G - S) \geq 1$ as S is a tight set, which contradicts the assumption of this case. Hence $|S| \geq 2$. If $|S| \geq 3$, then since $n \geq 2$ we have

$$oddca(G - S) = n|S| - \beta \geq n|S| - 2n + 1$$
$$= (n - 2)(|S| - 2) + |S| - 3 + |S| \geq |S|,$$

which contradicts the assumption of this case. Hence $|S| = 2$. If $\beta \leq 2n - 2$, then

$$oddca(G - S) = n|S| - \beta = 2n - \beta \geq 2 = |S|,$$

and so $oddca(G - S) \geq |S|$, which contradicts the assumption of this case. Therefore $|S| = 2$ and $\beta = 2n - 1$.

Let x and y be two adjacent vertices of G. Then by (7.21), there exists a vertex subset T_{xy} of G such that

$$T_{xy} \supseteq \{x, y\} \quad \text{and} \quad \beta \leq n|T_{xy}| - oddca(G - T_{xy}) \leq 2n - 1 = \beta.$$

Therefore T_{xy} is a tight set of G, and so by the above observation, $T_{xy} = \{x, y\}$ and $oddca(G - \{x, y\}) = 1$. In particular, $\{x, y\}$ is a tight set, and if $G - \{x, y\}$ is connected, then $G - \{x, y\}$ is an odd cactus.

For an edge $a = uv$ of G, let $f(a)$ denote the order of a largest component of $G - \{u, v\}$. An edge e is called **extremal** if

$$f(e) = \max\{f(a) : a \in E(G)\}.$$

Let $e = xy$ be an extremal edge of G. We want to prove that $G_1 = G - \{x, y\}$ is connected. Suppose, to the contrary, that G_1 is not connected. Let D be a largest component of G_1. Let A denote the set of edges of G not incident to D and distinct from $e = xy$. Since G_1 is disconnected, the set A is not empty. Let $a = uv \in A$. Clearly if there is a vertex $z \in \{x, y\} \backslash \{u, v\}$ that is adjacent to D, then $f(a) > f(e)$, a contradiction. Since a is an arbitrary edge in A, every edge in A is incident to the same end of e, say x, and the other end y is not adjacent to D. Therefore y is an isolated vertex in $G - \{x\}$. Hence by (7.22), we have

$$1 \leq oddca(G - \{x\}) \leq n|\{x\}| - \beta = n - (2n - 1) = -n + 1 < 0.$$

This is a contradiction. Consequently, $G_1 = G - \{x, y\}$ is connected, and so it is an odd cactus.

Suppose that G_1 has a vertex v which is adjacent to exactly one of $\{x, y\}$. Then $G_1 - v = G - \{x, y, v\}$ has a 1-factor by Lemma 7.28, and thus G has a strong $\{K_{1,1}, K_{1,2}\}$-factor, which includes an induced $\{K_{1,2}\}$-subgraph $\langle\{x, y, v\}\rangle_G$. Hence we may assume that

$$Z = N_G(x) \cap V(G_1) = N_G(y) \cap V(G_1).$$

If $|Z| = 1$ or $Z = V(G_1)$, then G is an odd cactus, a contradiction. Therefore there is a matching xx', yy' in G with $\{x', y'\} \subset V(G_1)$. Suppose that x' and y' are not contained in the same block of G_1. Then every component of $G_2 = G_1 - \{x', y'\}$ is a cactus with at least one and at most two even blocks. By Lemma 7.28, G_2 has a strong $\{K_{1,1}, K_{1,2}\}$-factor, say F, and so F together with the matching xx', yy' forms a strong $\{K_{1,1}, K_{1,2}\}$-factor of G. Thus we may assume that Z is contained in a block B of G_1. Since $V(B) \cup \{x, y\}$ does not induce a complete graph because G is not an odd cactus, there exists a vertex $u \in V(B)$ such that $Z \subseteq V(B) - u$. Then $\{x, x', u\}$ includes a star

$K_{1,2}$. Obviously every component of $G_2 = G_1 - \{u, x', y'\}$ is a cactus with exactly one even block. By Lemma 7.28, G_2 has a strong $\{K_{1,1}, K_{1,2}\}$-factor F, and so F together with a star $\langle\{x, x', u\}\rangle_G$ and the edge yy' forms a strong $\{K_{1,1}, K_{1,2}\}$-factor of G. Consequently the proof is complete. □

A strong $\{K_{1,1}, \ldots, K_{1,n}\}$-subgraph H of a graph G said to be **maximum** if G has no strong $\{K_{1,1}, \ldots, K_{1,n}\}$-subgraph H' such that $|H'| > |H|$. Then a formula for the order of a maximum strong $\{K_{1,1}, \ldots, K_{1,n}\}$-subgraph is given in the next theorem.

Theorem 7.30. *Let $n \geq 2$ be an integer, and let G be a simple graph G. Then the order of a maximum strong $\{K_{1,1}, K_{1,2}, \ldots, K_{1,n}\}$-subgraph H is equal to*

$$|H| = |G| - \max_{X \subset V(G)} \{oddca(G - X) - n|X|\}. \tag{7.24}$$

Proof. We may assume that G is connected and $|G| \geq 4$. Let

$$d = \max_{X \subset V(G)} \{oddca(G - X) - n|X|\}.$$

Then $d \geq 0$ by considering the case $X = \emptyset$. Moreover, if $d = 0$, then (7.24) follows from Theorem 7.27. Hence we may assume $d \geq 1$. Let S be a subset $V(G)$ such that

$$oddca(G - S) - n|S| = d.$$

Then by considering a subgraph $\langle OddCa(G - S) \cup S \rangle_G$ and by Lemma 7.28, we have that every strong $\{K_{1,1}, \ldots, K_{1,n}\}$-subgraph of G cannot cover at least $oddca(G - S) - n|S|$ odd cacti of $OddCa(G - S)$. Hence $|H| \leq |G| - d$, where H is a maximum strong $\{K_{1,1}, \ldots, K_{1,n}\}$-subgraph of G.

We next prove the inverse inequality $|H| \geq |G| - d$ for a maximum strong $\{K_{1,1}, \ldots, K_{1,n}\}$-subgraph H of G. Add d new $K_{1,n-1}$ graphs $R_1, \ldots, R_d$ with centers $u_1, \ldots, u_d$ to G. Then join every $u_i, 1 \leq i \leq d$, to every vertex of G by a new edge. Denote the resulting graph by G^*. Let Y be a non-empty subset of $V(G^*)$. We may assume Y contains no end-vertices of $R_i, 1 \leq i \leq d$, when we estimate $oddca(G^* - Y)$. If $Y \not\supseteq \{u_1, \ldots, u_d\}$, then

$$oddca(G^* - Y) \leq (n - 1)|Y \cap \{u_1, \ldots, u_d\}| + 1 \leq n|Y|.$$

If $Y \supseteq \{u_1, \ldots, u_d\}$, then all the vertices of every $R_i - u_i$ become isolated vertices of $G^* - Y$, and so by the definition of d, we obtain

$$oddca(G^* - Y) \leq oddca(G - Y \cap V(G)) + (n - 1)d$$
$$\leq n|Y \cap V(G)| + d + (n - 1)d = n|Y|.$$

Hence by Theorem 7.27, G^* has a strong $\{K_{1,1}, \ldots, K_{1,n}\}$-factor F^*. Then $H = F^* - \bigcup_{1 \leq i \leq d} V(R_i)$ is a strong $\{K_{1,1}, \ldots, K_{1,n}\}$-subgraph of G, which covers at least $|G| - d$ vertices. Hence $|H| \geq |G| - d$. Consequently the theorem is proved. □

It is known that determining the order of a maximum P_3-subgraph of a graph is an NP-complete problem as shown in the next section. So by our experience, it seems to be impossible to find a good criterion for a graph to have a P_3-factor. However, there is a simple conjecture on P_3-factors of cubic graphs, which is given below. This conjecture is still open, but some results on the order of a P_3-subgraph in a connected cubic graph are obtained. We present one of them.

Conjecture 7.31 (Akiyama and Kano, [9]). Every 3-connected cubic simple graph of order $6n$ has a P_3-factor, where $n \geq 1$ is an integer (Fig. 7.14).

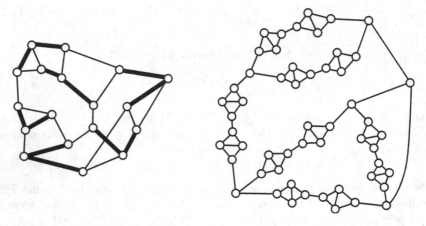

Fig. 7.14. A 3-connected cubic graph having a P_3-factor and a 2-connected cubic graph of order 54 having no P_3-factor.

There exist 2-connected cubic graphs of order $6n$ that have no P_3-factors. Such a cubic graph is shown in Fig. 7.14.

Theorem 7.32 (Kosowski and Żyliński [157]). *Every 2-connected cubic simple graph G with order at least 9 has a P_3-subgraph with order at least $9|G|/11$.*

Theorem 7.33 (Enomoto, Kaneko and Tuza [79]). *If a connected simple graph G with order $3n$ satisfies $\delta(G) \geq |G|/3$, then G has a P_3-factor.*

We now give a result on P_4-factors.

Theorem 7.34 (Akiyama and Kano [9]). *Let G be a 3-connected cubic simple graph of order $4p$. Then for any two edges of G, G has a P_4-factor containing them. In particular, G has a P_4-factor.*

Proof. Let e_1 and e_2 be two distinct edges of G. By Theorem 2.37, G has a 1-factor F_1 containing e_1. It may occur that F_1 also contains e_2. Denote by G^* the graph obtained from G by contracting all the edges of F_1 (see Fig. 7.15). Then G^* is a 4-regular multigraph with $V(G^*) = E(F_1)$ and $E(G^*) = E(G) - E(F_1)$. We now show that G^* is 3-connected. If $G^* - \{x, y\}$ is disconnected for some two vertices $x, y \in V(G^*)$, then $G - \{x, y\}$, $x, y \in E(G)$, is disconnected, which contradicts the connectedness of G because for a cubic graph, the edge connectivity is equal to the connectivity. Hence G^* is 3-connected. By Theorem 2.37, G^* has a 1-factor F_2 containing e_2 if $e_2 \notin E(F_1)$. Then $F_1 \cup F_2$ forms a P_4-factor of G containing both e_1 and e_2 (see Fig.7.15). □

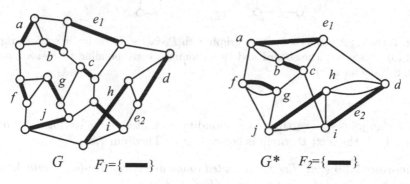

Fig. 7.15. A 3-connected cubic graph with 1-factor F_1; and a 3-edge connected 4-regular multiple graph G^* with 1-factor F_2.

Theorem 7.34 is best possible in the sense that there exists a 2-connected cubic bipartite graph of order $4n$ that has no P_4-factor ([9]).

We introduce a new graph called a **square graph** (Fig. 7.16). Let U be a set of lattice points $\{(x, y) : x, y \in \mathbf{Z}\}$ in the plane. Two points (x_1, y_1) and (x_2, y_2) in U are joined by an edge if and only if the distance between (x_1, y_1) and (x_2, y_2) is equal to one. Moreover, if the above graph G induced by U possesses the property that every edge is contained in a cycle of order four, then G is called a **square graph** (Fig. 7.16).

Theorem 7.35 (Akiyama and Kano [9]). *Every connected square graph of order $3p$ has a P_3-factor.*

The above theorem is equivalent to solving a tilling problem of a defective chessboard by triominos. A defective chessboard is said to be **tough** if its dual graph is a square graph, where the dual graph of a chessboard is a graph whose vertices correspond to the unit squares of the chessboard, and where

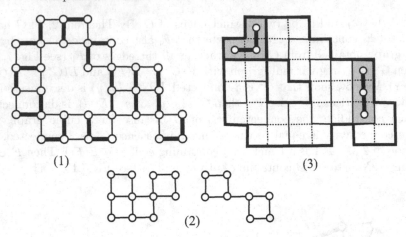

Fig. 7.16. (1) A connected square graph with P_3-factor; (2) two non square graphs; and (3) a defective chessboard and two triominos corresponding to P_3-component of the P_3-factor of (1).

each edge joins two vertices corresponding to unit squares having a common edge. Thus the next theorem is equivalent to Theorem 7.35.

Theorem 7.36 ([9]). *Every connected tough defective chessboard with $3n$ unit squares can be tiled with triominos (Fig. 7.16).*

7.2 Cycle factors and other component factors

In this section, we consider some other component factors. In particular, we will mainly deal with **cycle factors**, which are $\{C_n : n \geq 3\}$-factors. We first consider the complexity of some problems on component factors since it is believed by experience that if a problem on a component factor of a graph is NP-complete, then it is impossible to find a good criterion for a graph to have such a component factor.

The next theorem shows that many problems on K-factors are NP-complete.

Theorem 7.37 (Hell and Kirkpatrick [153]). *Let K be a connected simple graph of order at least three. Then the existence question for K-factor is NP-complete.*

Hell and Kirkpatrick also obtained the following theorem, which shows that if S does not contain K_2, then many S-factor problems are also NP-complete.

Theorem 7.38 (Hell and Kirkpatrick [153]). *Let G be a connected simple graph, let I be a set of integers greater than 2 (not necessary to be a finite* set), and let S be a family of connected graphs given below. Then the existence question for an S-factor is NP-complete.

(i) $S = \{K_n : n \in I\}$.

(ii) $S = \{K(1, 2n) : n \in I\}$.

(iii) $S = \{K(1, 2n + 1) : n \in I\}$.

(iv) $S = \{C_n : n \in I\}$, where $|I| < \infty$.

(v) $S =$ the set of trees of order m, where $m \geq 3$ is an integer.

(vi) $S =$ the set of all k-connected graph of order m, where $m \geq 3$ is an integer.

Since a 2-factor is nothing but a $\{C_n : n \geq 3\}$-factor, the next theorem is a restatement of Theorem 3.5.

Theorem 7.39 (Bäbler [22] and Petersen [209]). *Every 2-edge connected r-regular simple graph has a cycle factor.*

Notice that if we consider a 2-edge connected r-regular general graph, then it also has a 2-factor, but this 2-factor may contain a subgraph consisting of two vertices and two multiple edges joining them, and a loop. Thus if we restrict ourselves to a simple graph, a 2-factor is equal to a cycle factor. By Theorem 7.39, every 2-connected cubic simple graph has a cycle factor. This result can be strengthened to the following two theorems.

Theorem 7.40 (Kawarabayashi, Matsuda, Oda and Ota [149]). *Every 2-connected cubic simple graph has a $\{C_n : n \geq 4\}$-factor.*

Proof. Let G be a 2-connected cubic simple graph. We shall prove the theorem by induction on $|G|$. If $|G| = 4$, then $G = K_4$ and has the desired $\{C_n : n \geq 4\}$-factor. Assume $|G| \geq 6$. By Theorem 7.39, G has a cycle factor F. It is clear that $G - F = G - E(F)$ is a 1-factor of G.

We may assume that F contains a component D isomorphic to C_3 since otherwise F is the desired $\{C_n : n \geq 4\}$-factor. Let $V(D) = \{a, b, c\}$. Then since $G - F$ is a 1-factor of G, $G - F$ contains three edges ax, by, cz, where x, y and z are three distinct vertices (Fig. 7.17). By contracting D into one vertex, say v, we obtain a cubic simple graph G^* (Fig. 7.17). If v is a cut vertex of G^*, then a, b or c is a cut vertex of G, which contradicts the assumption. Hence G^* is 2-connected. Since the order of G^* is $|G| - 2$, by the induction hypothesis, G^* has a $\{C_n : n \geq 4\}$-factor F^*. Let C be a cycle of F^* passing through v. Without loss of generality, we may assume that C passes through (x, v, y) in this order. Then by replacing the cycle C by $C - xv - vy + xa + ac + cb + by$, we can obtain the desired $\{C_n : n \geq 4\}$-factor of G from F^* (Fig. 7.17). □

A cycle factor of a bipartite graph consists of cycles of order at least four. The next theorem shows that this cycle factor can also be strengthened.

Theorem 7.41 (Kano, Lee and Suzuki [127]). *Every connected cubic bipartite simple graph has a $\{C_n : n \geq 6\}$-factor.*

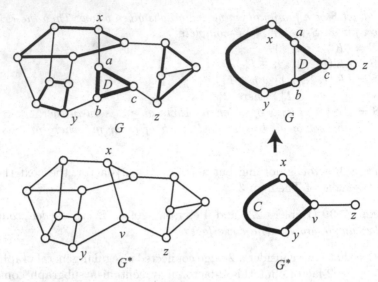

Fig. 7.17. A 2-connected cubic graph G with a 2-factor F; a 2-connected cubic graph G^* and its cycle C passing through (x, v, y); and a cycle of G obtained from C.

In order to prove the above theorem, we need the following easy lemma.

Lemma 7.42. *Let $r \geq 2$ be an integer. Then every connected r-regular bipartite multigraph is 2-edge connected.*

Proof. Let G be a connected r-regular bipartite multigraph with bipartition (A, B). Assume that G has a bridge $e = ab$, $a \in A$ and $b \in B$. Let D be a component of $G - e$ containing a. Then D does not contain b, and we have

$$r|V(D) \cap A| - 1 = \sum_{x \in V(D) \cap A} \deg_D(x)$$

$$= \sum_{y \in V(D) \cap B} \deg_D(y) = r|V(D) \cap B|.$$

This is a contradiction. Hence G is 2-edge connected.   □

We are ready to prove Theorem 7.41.

Proof of Theorem 7.41. Let G be a connected cubic bipartite simple graph. By Lemma 7.42, G is 2-edge connected, which implies G is 2-connected since G is a cubic graph. We prove the theorem by induction on $|G|$. There exists only one connected cubic bipartite graph of order six, which is $K_{3,3}$, and it has a $\{C_6\}$-factor. Thus we may assume $|G| \geq 8$.

By Theorem 7.39, G has a $\{C_n : n \geq 4\}$-factor F. We may assume that F contains a component D isomorphic to C_4 since otherwise F is the desired $\{C_n : n \geq 6\}$-factor. Let $V(D) = \{a, b, c, d\}$, and as, bt, cu, dw be the edges of $G - F = G - E(F)$ (see Fig. 7.18).

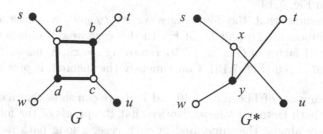

Fig. 7.18. Cubic graphs G and H; Bold lines are edges of D.

Since $G - F$ is a 1-factor of G, $\{as, bt, cu, dw\}$ is a set of independent edges, and so s, t, u, w are all distinct vertices of G. Let G^* be the graph obtained from G by removing the four vertices a, b, c, d and their incident edges, and by adding two new vertices x and y together with five new edges sx, ux, ty, wy, xy (see Fig. 7.19).

Then G^* is a connected cubic bipartite graph, and $|G^*| = |G| - 2$. Hence G^* has a $\{C_n : n \geq 6\}$-factor F^* by induction. We shall obtain the desired $\{C_n : n \geq 6\}$-factor of G from F^* by considering the following two cases.

Case 1. *A component of F^* contains the edge xy.*

In this case, without loss generality, we may assume that a component D of F^* contains xy, sx and yw by symmetry. Then $F^* - \{sx, xy, yw\} + \{sa, ab, bc, cd, dw\}$ is the desired $\{C_n | n \geq 6\}$-factor of G.

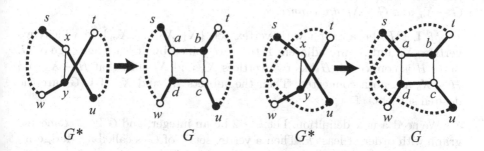

Fig. 7.19. Cubic bipartite graphs G and G^*, and their $\{C_n | n \geq 6\}$-factors.

Case 2. *No component of F^* contains the edge xy.*

In this case, F^* contains the four edges sx, xu, ty, yw. We first assume that these four edges are contained in the same component D of F^*. By symmetry, we may assume that a cycle D passes through s, x, u and then t, y, w (see Fig. 7.19). Then we can obtain the desired $\{C_n : n \geq 6\}$-factor from F^* by removing the edges sx, xu, ty, yw and by adding the edges sa, ab, bt, uc, cd, dw as shown in Fig. 7.19.

Next assume that the four edges sx, xu, ty, yw are contained in two distinct components D_1 and D_2 of F^*. In this case we can obtain the desired $\{C_n : n \geq 6\}$-factor of G from F^* by removing sx, xu, ty, yw and by adding sa, ab, bt, wd, dc, cu (Fig. 7.19). Consequently the theorem is proved. □

By making use of Theorems 7.40 and 7.41, we can show that a cubic graph has a long path factor as follows. Notice that the proofs of the following two theorems are almost the same, and in each case, a long path factor can be obtained from a long cycle factor given in the above two theorems.

Theorem 7.43 ([149]). *Every 2-connected cubic simple graph has a $\{P_n : n \geq 6\}$-factor.*

Theorem 7.44 ([127]). *Every connected cubic bipartite simple graph has a $\{P_n : n \geq 8\}$-factor.*

The next theorem gives a result on general component factors.

Theorem 7.45 (Egawa, Enomoto and Saito [67]). *Let K be a connected simple graph of order at least two, and p and n be integers such that $2 \leq p < n$. Let G be a connected simple graph of order $n|K|$. If every connected induced subgraph of G with order $p|K|$ has a K-factor, then G has a K-factor.*

In order to prove the above theorem, we need some lemmas.

Lemma 7.46. *Let G be a connected simple graph and $V(G) = X_1 \cup X_2 \cup \cdots \cup X_m$ be a partition of $V(G)$, where $m \geq 2$. Suppose that $\langle X_i \rangle_G$ is connected for all i. Then there exist two distinct elements X_a and X_b such that both $G - X_a$ and $G - X_b$ are connected.*

Proof. Let H be a graph with vertex set $\{X_1, X_2, \ldots, X_m\}$ in which two vertices X_i and X_j are adjacent if they are joined by at least one edge of G. Since H is connected, H has two vertices X_a and X_b such that $H - X_a$ and $H - X_b$ are both connected. Then the subsets X_a and X_b of $V(G)$ are the desired subsets. □

We need a new definition. Let $k \geq 2$ be an integer, and G be a connected graph with order at least k. Then a vertex set Y of G is called a k-**ledge** if
  (i) both $\langle Y \rangle_G$ and $G - Y$ are connected, and
  (ii) $|Y| \geq k$.
A k-ledge Y of G is called **minimal** if no proper subset of Y is a k-ledge. If G is a connected graph of order at least $k + 1$, then G has a vertex x such that $G - x$ is connected, and thus G has a k-ledge $V(G) - \{x\}$.

Lemma 7.47. *Let G be a connected simple graph and Y be a minimal k-ledge of G. Let $X = V(G) - Y$. Suppose $|Y| > k$. Then*
(i) $N_G(X) \cap Y \neq \emptyset$ and every vertex of $N_G(X) \cap Y$ is a cut vertex of $\langle Y \rangle_G$; and
(ii) there exists a vertex $v \in N_G(X) \cap Y$ such that a component D of $\langle Y \rangle_G - v$ satisfies $N_G(X) \cap Y \subseteq V(D) \cup \{v\}$.

Proof. Since G is connected, $N_G(X) \cap Y \neq \emptyset$. Assume that a vertex $y \in N_G(X) \cap Y$ is not a cut vertex of $\langle Y \rangle_G$. Then $Y - y$ is also a k-ledge, which contradicts the minimality of Y. Hence y is a cut vertex of $\langle Y \rangle_G$, and (i) holds.

We next prove (ii). Choose vertices $v, y \in N_G(X) \cap Y$ so that

$$dist_{\langle Y \rangle}(v, y) = \max\{dist_{\langle Y \rangle}(y_1, y_2) : y_1, y_2 \in N_G(X) \cap Y\},$$

where $dist_{\langle Y \rangle}(x, x')$ denotes the distance between x and x' in $\langle Y \rangle_G$. If $v = y$, then $N_G(X) \cap Y = \{v\}$ and so any component D of $Y - v$ satisfies the condition in (ii). If $v \neq y$, then let D be the component of $\langle Y \rangle_G - v$ containing y. Assume $N_G(X) \cap Y \not\subseteq V(D) \cup \{v\}$. Then there exists $z \in (N_G(X) \cap Y) - (V(D) \cup \{v\})$. However, since z and y belong to distinct components of $\langle Y \rangle_G - v$, we have

$$dist_{\langle Y \rangle}(y, z) = dist_{\langle Y \rangle}(y, v) + dist_{\langle Y \rangle}(v, z) > dist_{\langle Y \rangle}(v, y),$$

which contradicts the choice of the pair (v, y). Therefore $N_G(X) \cap Y \subseteq V(D) \cup \{v\}$, and statement (ii) holds. □

Proof of Theorem 7.45   We first prove the theorem when $p = n - 1$. Let $k = |K|$. Then $|G| \geq 3k$ as $n \geq 3$. We assume that if $G - S$ is connected for $S \subset V(G)$ with $|S| = k$, then $G - S$ has a K-factor.

Let Y be a minimal k-ledge, and let $X = V(G) - Y$. We consider the following two cases.

Case 1. $|Y| = k$.

Since $\langle X \rangle_G = G - Y$ is connected, $G - Y$ has a K-factor F. Let $D_1, D_2, \cdots, D_r$ be the components of F. Then every D_i is isomorphic to K. By Lemma 7.46, there exist two components D_a and D_b such that both $G - Y - V(D_a)$ and $G - Y - V(D_b)$ are connected.

We claim that either $G - V(D_a)$ or $G - V(D_b)$ is connected, which implies that G has a K-factor. Assume that neither $G - V(D_a)$ nor $G - V(D_b)$ is connected. Since $G - Y - V(D_a)$ and $\langle Y \rangle_G$ are connected, but $G - V(D_a)$ is not connected, we have $N_G(Y) \cap X \subseteq V(D_a)$. Similarly, by applying the above argument to D_b, we have $N_G(Y) \cap X \subseteq V(D_b)$. Therefore

$$N_G(Y) \cap X \subseteq V(D_a) \cap V(D_b) = \emptyset.$$

This contradicts the assumption that G is connected.

Case 2. $|Y| \geq k + 1$.

By Lemma 7.47, every vertex of $N_G(X) \cap Y$ is a cut vertex of $\langle Y \rangle_G$. Moreover, there exist a vertex $v \in N_G(X) \cap Y$ and a component D of $\langle Y \rangle_G - v$ such that $N_G(X) \cap Y \subseteq V(D) \cup \{v\}$. Let $\{D = Y_0, Y_1, Y_2, \cdots, Y_m\}$ be the set of components of $\langle Y \rangle_G - v$ (Fig. 7.20).

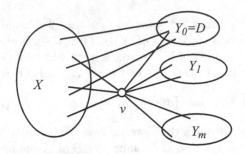

Fig. 7.20. $\langle X \rangle_G$ is connected, and $\langle Y \rangle_G - v = Y_0 \cup Y_1 \cup \cdots \cup Y_m$.

Assume that $|Y_c| \geq k$ for some $0 \leq c \leq m$. Then since $G - Y_c$ and $\langle Y_c \rangle_G$ are connected, Y_c is a k-ledge of G. This contradicts the minimality of Y. Hence $|Y_i| < k$ for all $0 \leq i \leq m$.

There exists an integer t such that

$$|Y_0 \cup Y_1 \cup \cdots \cup Y_t| < k \quad \text{and} \quad |Y_0 \cup Y_1 \cup \cdots \cup Y_t \cup Y_{t+1}| \geq k + 1.$$

Let Z be a vertex set of Y_{t+1} such that $Y_{t+1} \cup \{v\} - Z$ is connected and $|Z| = k - |Y_0 \cup \cdots \cup Y_t|$. Let $S = T_0 \cup \cdots \cup T_t \cup Z$. Then $|S| = k$ and $G - S$ is connected. Therefore, $G - S$ has a K-factor F. Let $C_1, C_2, \cdots, C_s$ be the components of F. Then every C_i is isomorphic to K and $s \geq 2$ since $|G| \geq 3k$.

If $V(C_j) \cap (V(Y_{t+1}) - Z) \neq \emptyset$, then $v \in C_j$ since $|Y_{t+1}| < k$ and $N_G(Y_{t+1}) - V(Y_{t+1}) = \{v\}$. This implies $(V(Y_{t+1}) \cup \{v\}) - Z \subseteq V(C_c)$ for some $0 \leq c \leq m$. Since $v \in C_c$ and $|Y_j| < k$ for every $t + 2 \leq j \leq m$, it follows that $V(Y_{t+2} \cup \cdots \cup Y_m) \subset V(C_c)$.

By Lemma 7.47, there exist two components C_a and C_b such that both $(G - S) - V(C_a)$ and $(G - S) - V(C_b)$ are connected. We may assume $b \neq c$. Then $V(C_b) \subseteq X$ and $G - V(C_b)$ is connected since $Y_1 \cup \cdots \cup Y_m \cup \{v\}$ are contained in $V(G) - V(C_b)$ and $G - S - V(C_b)$ is connected. Since $|C_b| = k$, $G - C_b$ has a K-factor F' by the assumption of the theorem. Thus $F' \cup C_b$ is the desired K-factor of G.

We now consider the case $n - p \geq 2$. We use induction on $n - p$. Let S be any vertex set of G such that $|S| = k$ and $G - S$ is connected. Let H be any induced connected subgraph of $G - S$ of order $p|K|$. Then H is also an induced subgraph of G, and thus H has a K-factor by the assumption of the theorem. Hence $G - S$ has a K-factor by the inductive hypothesis. Therefore G has a K-factor as shown above. Consequently the proof is complete. $\square$

Remark: Some other results on component factors are found in [13], [96], [69], [148], [254], and others. Most of them show that if the minimum degree of a graph G is large (for example, $\delta(G) \geq |G|/2$), then G has some component factor.

Problems

7.1. Prove Theorem 7.8 in the same way as the proofs of the sufficiency (1) and (2) of Theorem 7.6.

7.2. Prove Theorem 7.22 by using Theorem 7.19.

7.3. Prove Theorem 7.24 by using Theorem 7.19.

7.4. Prove Theorem 7.26 without using Theorem 7.27 .

8

Spanning Trees

A spanning tree can be considered as a special connected factor, and often appears in many branches of combinatorics. In this section we give a variety of sufficient conditions for a connected graph to have a spanning tree possessing a certain prescribed property. For example, in Section 8.2, we consider spanning trees with maximum degree at most k, and in Section 8.3, we deal with spanning trees having at most k leaves. We begin with some basic results on spanning trees, including minimum spanning trees in a weighted graph.

8.1 Preliminaries and minimum spanning trees

Let G be a connected graph without loops and T be a spanning tree of G. We often regard T as an edge set of G, and write

$$\overline{T} = E(G) - T.$$

An edge set S of G is called a **cutset** of G if $G - S$ is disconnected. Thus a **minimal cutset** S satisfies that $G - S$ is disconnected but $G - X$ is connected for every proper subset X of S. For any edge e of T, $T - e$ consists of two components, say X and Y. Then the set of edges of G joining X to Y forms a **minimal cutset**, denoted by $Cut(\overline{T} + e)$ (Fig. 8.1). It is easy to see that $Cut(\overline{T} + e)$ is the unique minimal cutset contained in $\overline{T} + e$. On the other hand, for any edge $b \in \overline{T}$, $T + b$ contains the unique cycle $Cyc(T + b)$ (Fig. 8.1). Note that if G is a multigraph, then it may occur that $Cyc(T + b)$ consists of two edges: b and an edge of T joining the two end-points of b.

We will use the following lemma on spanning trees without explicitly referring to it; its proof is also omitted.

Lemma 8.1. *Let G be a connected multigraph and T be a subgraph of G. Then the following four statements are equivalent.*
(i) T is a spanning tree of G.

J. Akiyama and M. Kano, *Factors and Factorizations of Graphs*,
Lecture Notes in Mathematics 2031, DOI 10.1007/978-3-642-21919-1_8,
© Springer-Verlag Berlin Heidelberg 2011

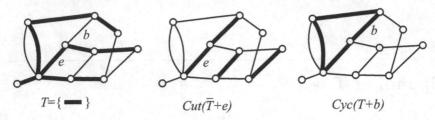

$$T=\{ \blacksquare \} \qquad\qquad Cut(\overline{T}+e) \qquad\qquad Cyc(T+b)$$

Fig. 8.1. A spanning tree T, $Cut(\overline{T}+e)$ and $Cyc(T+b)$ for $e \in T$ and $b \in \overline{T}$.

(ii) T *is a connected spanning subgraph of* G *and* $||T|| = |G| - 1$.
(iii) T *contains no cycle and* $||T|| = |G| - 1$.
(iv) $\overline{T}$ *contains no cutset and* $||\overline{T}|| = ||G|| - |G| + 1$.

We now give some basic results on spanning trees.

Lemma 8.2. *Let* G *be a connected multigraph and* T *be a spanning tree of* G. *Then for two edges* $e \in T$ *and* $b \in \overline{T}$, *the following three statements are equivalent.*
(i) $T - e + b$ *is a spanning tree of* G.
(ii) $b \in Cut(\overline{T} + e)$.
(iii) $e \in Cyc(T + b)$.

Proof. (ii) $\Rightarrow$ (i) Suppose $b \in Cut(\overline{T} + e)$. Since b joins two components of $T - e$, $T - e + b$ is connected, which implies that $T - e + b$ is a spanning tree of G.
(i) $\Rightarrow$ (ii) Assume $b \notin Cut(\overline{T} + e)$. Then b joins two vertices of the same component of $T - e$, and so $T - e + b$ is not connected. Hence $T - e + b$ is not a spanning tree.
(iii) $\Rightarrow$ (i) Suppose $e \in Cyc(T + b)$. Since $T + b$ contains the unique cycle $Cyc(T + b)$, $T + b - e$ has no cycle, and thus it is a spanning tree of G.
(i) $\Rightarrow$ (iii) Assume $e \notin Cyc(T + b)$. Then $T + b - e$ still contains a cycle $Cyc(T + b)$, and so it is not a tree. □

Lemma 8.3. *Let* G *be a connected multigraph and* Q *and* R *be two distinct spanning trees of* G. *Then for any edge* $e \in R \setminus Q$,

$$Cyc(Q + e) \cap Cut(\overline{R} + e) - e \neq \emptyset.$$

In particular, there exists an edge $b \in Q \setminus R$ *such that both* $R - e + b$ *and* $Q - b + e$ *are spanning trees of* G.

Proof. Assume that $e = xy$, where x and y are the end-points of e. Then $Cyc(Q + e) - e$ is the unique path in Q connecting x and y, and so it must intersect $Cut(\overline{R} + e)$ since x and y are contained in distinct components of $R - e$ (Fig. 8.2). Hence the first part holds.

Take an edge $b \in Cyc(Q + e) \cap Cut(\overline{R} + e) - e$. Then by Lemma 8.2, both $Q - b + e$ and $R - e + b$ are spanning trees of G. □

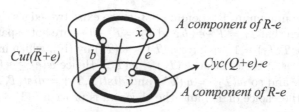

The figure labels:
Cut($\overline{R}$+e), A component of R-e, x, b, e, Cyc(Q+e)-e, y, A component of R-e

Fig. 8.2. $b \in Cyc(Q+e) \cap Cut(\overline{R}+e) - e$.

For two spanning trees Q and R, the distance $dist(Q, R)$ between Q and R is defined by

$$dist(Q, R) = |Q \setminus R| = |R \setminus Q|, \quad \text{where} \quad Q, R \subset E(G).$$

It is easy to see that this distance satisfies the following distance conditions for any spanning trees X, Y, Z:
(i) $dist(X, Y) = 0$ if and only if $X = Y$,
(ii) $dist(X, Y) = dist(Y, X)$,   and
(iii) $dist(X, Y) + dist(Y, Z) \geq dist(X, Z)$.

A graph is called a **weighted graph** if every edge e has a weight $w(e)$, which is a positive real number. Let G be a connected weighted multigraph. For a spanning tree T of G, the weight $w(T)$ of T is defined as the sum of weights of its edges, that is,

$$w(T) = \sum_{e \in T} w(e).$$

If the weight of a spanning tree T is minimum among all the spanning trees of G, then T is called a **minimum spanning tree** of G. Analogously, a **maximum spanning tree** can be defined. Notice that there might be many minimum spanning trees in G. In general, if $W_1 < W_2 < \cdots < W_m$ are the distinct weights of all the spanning trees of G, then a spanning tree T with weight W_k is called a k**-th minimum spanning tree** of G.

Theorem 8.4. *Let G be a weighted connected multigraph and T be a spanning tree of G. Then the following statements are equivalent:*
(i) T is a minimum spanning tree of G.
(ii) For every $b \in \overline{T}$ and every $e \in Cir(T + b)$, $w(e) \leq w(b)$.
(iii) For every $e \in T$ and every $b \in Cut(\overline{T} + e)$, $w(e) \leq w(b)$.
(iv) For every spanning tree R with $dist(T, R) = 1$, $w(T) \leq w(R)$.

Proof. By Lemma 8.2, (ii) and (iii) are equivalent. By Lemma 8.3, (iii) and (iv) are equivalent. Hence it suffices to show that (iv) implies (i) since (i) trivially implies (iv). Assume that T is a spanning tree that satisfies (iv), but T is not a minimum spanning tree. Choose a minimum spanning tree T^* so that

$dist(T, T^*)$ is minimum. By Lemma 8.3, there exist two edges $e \in T \setminus T^*$ and $b \in T^* \setminus T$ such that $T_1 = T - e + b$ and $T_2 = T^* - b + e$ are both spanning trees of G. Since $dist(T, T_1) = 1$, we have $w(T) \leq w(T_1)$ by (iv), which implies $w(e) \leq w(b)$. Moreover, it is clear that $w(T_2) \geq w(T^*)$. Hence $w(e) \geq w(b)$. Therefore $w(e) = w(b)$, and so $w(T_2) = w(T^*)$ and $dist(T, T_2) = dist(T, T^*) - 1$. This contradicts the choice of T^*, and the theorem is proved.   □

Let $\Omega = \Omega(G)$ denote the set of all the minimum spanning trees of a connected weighted multigraph G. We define the graph $\Gamma(\Omega, G)$ with vertex set Ω as follows: two vertices T_1 and T_2 are joined by an edge if and only if $dist(T_1, T_2) = 1$ in G. Then this graph is connected as the following theorem shows.

Theorem 8.5. $\Gamma(\Omega, G)$ *is connected.*

Proof. Assume that $\Gamma(\Omega, G)$ is disconnected. Choose a pair of minimum spanning trees Q and R of G contained in distinct components of $\Gamma(\Omega, G)$ such that $dist(Q, R)$ is minimum among all such pairs of spanning trees. Then $dist(Q, R) \geq 2$. By Lemma 8.3, there exists two edges $e \in Q \setminus R$ and $b \in R \setminus Q$ such that both $Q - e + b$ and $R - b + e$ are spanning trees of G. Since both Q and R are minimum spanning trees of G, we have $w(e) = w(b)$ since $w(Q) \leq w(Q - e + b)$ and $w(R) \leq w(R - b + e)$. Hence $R - b + e$ is a minimum spanning tree and contained in the same component containing R. But $dist(Q, R - b + e) = dist(Q, R) - 1$, which contradicts the choice of Q and R. Hence the theorem is proved.   □

By the above theorems, we can easily find a minimum spanning tree of a connected weighted graph G, and also obtain all the minimum spanning trees of G. There are enormous works on minimum and k-th minimum spanning trees, and the reader interested in this topic is referred to [250].

8.2 Spanning k-trees

Let G be a connected simple graph and $k \geq 2$ be an integer. A tree T is called a k-**tree** if $\deg_T(x) \leq k$ for all $x \in V(T)$, that is, the maximum degree of a k-tree is at most k. It is clear that a Hamiltonian path is nothing but a spanning 2-tree. Moreover, it is obvious that G has a spanning k-tree if and only if G has a connected $[1, k]$-factor. In this section, we investigate some sufficient conditions for a graph to have a spanning k-tree.

If a graph G has k independent vertices, define

$$\sigma_k(G) = \min_{S \subseteq V(G)} \left\{ \sum_{x \in S} \deg_G(x) \; : \; S \text{ is an independent set of } k \text{ vertices} \right\}.$$

If G has no k independent vertices, then define $\sigma_k(G) = \infty$. In particular, if G is not a complete graph, then

$$\sigma_2(G) = \min_{x,y}\{\deg_G(x) + \deg_G(y) : \text{two nonadjacent vertices } x \text{ and } y\}.$$

For a path P in G, a vertex of P that is not an end-vertex of P is called an **inner vertex** of P. We begin with Ore's Theorem, which gives a sufficient condition using $\sigma_2(G)$ for a graph to have a Hamiltonian path.

Theorem 8.6 (Ore [206]). *Let G be a connected simple graph. If $\sigma_2(G) \geq |G| - 1$, then G has a Hamiltonian path.*

Proof. Assume that G does not have a Hamiltonian path. Let P be a longest path in G, and let a and b be the end-vertices of P. We assign an orientation from a to b in P. Then for every inner vertex y of P, we can define the successor y^+ and the predecessor y^- of y. It is clear that $N_G(a) \cup N_G(b) \subseteq V(P)$ and $V(P) \neq V(G)$.

It is easy to see that $\langle V(P)\rangle_G$ has no Hamiltonian cycle. In particular, a and b are not adjacent in G. For every vertex $x \in N_G(a) - \{a^+\}$, x^- and b are not joined by an edge of G since otherwise $\langle V(P)\rangle_G$ contains a Hamiltonian cycle (Fig. 8.3).

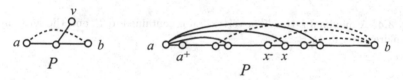

Fig. 8.3. A longest path P; broken lines present non-existence of such edges.

Let $X = \{x^- : x \in N_G(a)\}$, which contains a. Then $X \subset V(P)$, $N_G(b) \cap X = \emptyset$ and $b \notin X \cup N_G(b)$, and thus we have

$$\deg_G(b) = |N_G(b)| \leq |P| - |\{b\}| - |X|.$$

Hence

$$\sigma_2(G) \leq \deg_G(a) + \deg_G(b)$$
$$\leq |N_G(a)| + |P| - 1 - |X| = |P| - 1 < |G| - 1.$$

This is contrary to the assumption $\sigma_2(G) \geq |G| - 1$. Therefore the theorem is proved. $\square$

A generalization of Ore's theorem to a spanning k-tree was obtained by Win as follows.

Theorem 8.7 (Win [246]). *Let G be a connected simple graph and $k \geq 2$ be an integer. If $\sigma_k(G) \geq |G| - 1$, then G has a spanning k-tree.*

Proof. By Theorem 8.6, we may assume $k \geq 3$. Suppose that G satisfies the condition on $\sigma_k(G)$ but has no spanning k-tree. Let T be a maximal k-tree of G. Then G has no k-tree T' such that $V(T) \subset V(T')$. Since T is not a spanning tree, G has a vertex v not contained in T and adjacent to a vertex w of T. If $\deg_T(w) < k$, then $T + vw$ is a k-tree with vertex set $V(T) \cup \{v\}$, a contradiction. Hence $\deg_T(w) = k$. Let $D_1, D_2, \ldots, D_k$ be the components of $T - w$, and let u_i be the vertex in D_i adjacent to w and x_i be a leaf of T contained in D_i. It may happen that $u_j = x_j$ for some j's (Fig. 8.4).

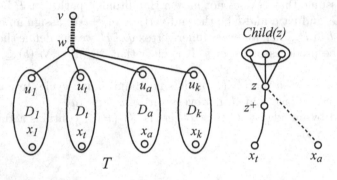

Fig. 8.4. A maximal k-tree T, a vertex v not contained in T, and the set $Child(z)$ of children of z in D_t.

Claim $\{x_1, x_2, \ldots, x_k\}$ *is an independent set of G, and $N_G(x_i) \subseteq V(T)$ for all $1 \leq i \leq k$.*

By the maximality of T, we easily see that $N_G(x_i) \subseteq V(T)$ for all i. If x_i and x_j are adjacent in G, then $T + x_i x_j - w u_i + wv$ is a k-tree of G, a contradiction. Hence $\{x_1, x_2, \ldots, x_k\}$ is an independent set of G. Notice that if $\sigma_k(G) = \infty$, then G has no k independent vertices, and so the theorem holds by Claim. Thus we may assume that $\sigma_k(G)$ is finite.

We consider an arbitrary D_t, $1 \leq t \leq k$. Choose a vertex x_a from $\{x_1, \ldots, x_k\} - \{x_t\}$ so that

$$|N_G(x_a) \cap D_t| = \max_{i \neq t} |N_G(x_i) \cap D_t|,$$

where $N_G(x_i) \cap D_t$ denotes $N_G(x_i) \cap V(D_t)$. Then

$$\deg_T(z) = k \qquad \text{for all} \quad z \in N_G(x_a) \cap D_t. \tag{8.1}$$

Because if $\deg_T(z) < k$ for some $z \in N_G(x_a) \cap D_t$, then $T + zx_a - wu_t + wv$ is a k-tree of G, which is contrary to the maximality of T.

We regard D_t as a rooted tree with root x_t. For a vertex x of D_t, the set of children of x is denoted by $Child(x)$. Let $z \in N_G(x_a) \cap D_t$. Then no child z_1 of z is adjacent to x_t in G since otherwise

$$T - z_1 z + z_1 x_t + z x_a - w u_t + w v$$

is a k-tree of G, a contradiction. Hence $Child(r)$, $r \in N_G(x_a) \cap D_t$, and $N_G(x_t) \cap D_t$ are pairwise disjoint. Therefore, by Claim, (8.1) and by the choice of x_a, we obtain

$$|D_t| \geq |N_G(x_t) \cap D_t| + \sum_{r \in N_G(x_a) \cap D_t} |Child(r)| + |\{x_t\}|$$

$$= |N_G(x_t) \cap D_1| + (k-1)|N_G(x_a) \cap D_t| + 1,$$

$$\geq \sum_{j=1}^{k} |N_G(x_j) \cap D_t| + 1.$$

Hence it follows from Claim and the above inequality that

$$\sum_{j=1}^{k} \deg_G(x_j)$$

$$\leq \sum_{j=1}^{k} \left(\sum_{i=1}^{k} |N_G(x_j) \cap D_i| + |\{w\}| \right)$$

$$\leq \sum_{i=1}^{k} \left(\sum_{j=1}^{k} |N_G(x_j) \cap D_i| \right) + k$$

$$\leq \sum_{i=1}^{k} (|D_i| - 1) + k = |T| - 1 \leq |G| - 2.$$

On the other hand, it follows from the assumption of the theorem that

$$\sum_{j=1}^{k} \deg_G(x_j) \geq \sigma_k(G) \geq |G| - 1.$$

This is a contradiction. Therefore the theorem is proved. $\square$

Corollary 8.8. *Let G be a connected simple graph and $k \geq 2$ be an integer. If*

$$\delta(G) \geq \frac{|G| - 1}{k},$$

then G has a spanning k-tree.

Proof. The corollary follows immediately from Theorem 8.7 and $\sigma_k(G) \geq k \cdot \delta(G) \geq |G| - 1$. $\square$

Win also found the following Posa-type criterion for the existence of spanning k-trees, which shows that a graph G with $\delta(G) < (|G| - 1)/k$ still has a spanning k-tree if the number of vertices with small degree is small.

Theorem 8.9 (Win [246]). *Let G be a connected simple graph and $k \geq 3$ be an integer. If G satisfies the following two conditions, then G has a spanning k-tree.*

(i) For every integer $1 \leq m < (|G| - 2)/k$, the number of vertices x with $\deg_G(x) \leq m$ is less than m.

(ii) The number of vertices x with $\deg_G(x) \leq (|G| - 2)/k$ is less than or equal to $(|G| - 2|)/k$.

Proof. Assume that G satisfies the above conditions but has no spanning k-tree. A k-tree T of G is said to be **maximum** if G has no k-tree T' such that $|T'| > |T|$. Choose a maximum k-tree T of G in the following way:

(a)   The number of leaves of T is minimum among all maximum k-trees of G;

(b) $\sum_{x \in Leaf(T)} \deg_G(x)$ is maximum subject to condition (a), where $Leaf(T)$ denotes the set of leaves of T.

There exists a vertex $v \in V(G) - V(T)$ that is adjacent to a vertex w of T. It is immediate that $\deg_T(w) = k$. Let $D_1, D_2, \ldots, D_k$ be the components of $T - w$, and let u_i be the vertex in D_i adjacent to w and x_i a leaf of T contained in D_i. It may happen that $u_j = x_j$ for some j's. Since we can apply the same argument as in the proof of Theorem 8.7 to T, we obtain that $\{x_1, x_2, \ldots, x_k\}$ is an independent set of G and

$$\sum_{i=1}^{k} \deg_G(x_i) \leq |G| - 2. \tag{8.2}$$

For any fixed x_i, and for every vertex y of G adjacent to x_i in G but not in T, let y^* denote the vertex that is adjacent to y and lies on the path connecting x_i and y in T (Fig. 8.5). Then the next claim holds.

Claim 1. $\deg_T(y^*) = 2$ *and* $\deg_G(y^*) \leq \deg_G(x_i)$ *for all* $y \in N_G(x_i) - N_T(x_i)$.

If $\deg_T(y^*) \geq 3$, then $T + yx_i - yy^*$ is a maximum k-tree with a smaller number of leaves than T, which contradicts the criterion in (a). Hence $\deg_T(y^*) = 2$. If $\deg_G(y^*) > \deg_G(x_i)$, then $R = T + yx_i - yy^*$ is a maximum k-tree such that

$$\sum_{x \in Leaf(R)} \deg_G(x) > \sum_{x \in Leaf(T)} \deg_G(x).$$

This contradicts the criterion in (b). Hence $\deg_G(y^*) \leq \deg_G(x_i)$, and so Claim 1 follows.

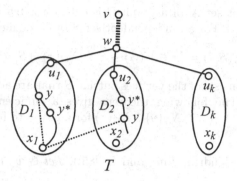

Fig. 8.5. Two vertices y^* such that $y \in N_G(x_1) - N_T(x_1)$

For every x_i, let us define

$$N^*[x_i] = \{y^* \ : \ y \in N_G(x_i) - N_T(x_i)\} \cup \{x_i\}.$$

If y_1 and y_2 are two distinct vertices in $N_G(x_i) - N_T(x_i)$, then $y_1^* \neq y_2^*$ since $\deg_T(y_1^*) = \deg_T(y_2^*) = 2$.

Claim 2. $\deg_G(x_i) = |N^*[x_i]| \geq (|G| - 2)/k$ *for every* $1 \leq i \leq k$.

It is immediate that $\deg_G(x_i) = |N^*[x_i]|$. Suppose $\deg_G(x_j) = |N^*[x_j]| < (|G| - 2)/k$ for some j. Then by the condition (i) in the theorem with $m = \deg_G(x_j)$ and by Claim 1, we have

$$|N^*[x_j]| \leq \#\{x \in V(G) \ : \ \deg_G(x) \leq \deg_G(x_j)\}$$
$$< \deg_G(x_j) = |N^*[x_j]|.$$

This is a contradiction. Therefore $\deg_G(x_i) = |N^*[x_i]| \geq (|G| - 2)/k$ for all i.

If $\deg_G(x_j) = |N^*[x_j]| > (|G| - 2)/k$ for some j, then by Claim 2, we have

$$\sum_{i=1}^{k} \deg_G(x_i) > k \cdot \frac{|G| - 2}{k} = |G| - 2.$$

This contradicts (8.2). Therefore

$$\deg_G(x_i) = |N^*[x_i]| = \frac{|G| - 2}{k} \qquad \text{for every} \quad 1 \leq i \leq k.$$

By Claim 1 and the condition (ii) of the theorem, we obtain

$$\left| \bigcup_{i=1}^{k} N^*[x_i] \right| \leq \#\left\{x \in V(T) \ : \ \deg_G(x) \leq \frac{|G| - 2}{k}\right\} \leq \frac{|G| - 2}{k}.$$

This implies $N^*[x_1] = N^*[x_2] = \cdots = N^*[x_k]$. In particular, $x_1 \in N^*[x_1] = N^*[x_2]$. But this contradicts Claim 1 since $\deg_T(x_1) = 1$. Consequently the theorem is proved. $\square$

An independent set X of a graph G is called a k-**frame** if $|X| = k$ and $G - X$ is connected. For an independent set S of G, define

$$N_G(S; i) = \{x \in V(G) - S : |N_G(x) \cap S| = i\}.$$

Hence $N_G(S; i)$ consists of the vertices outside S that are adjacent to precisely i vertices of S. The following theorem gives a sufficient condition using $N_G(\{x, y, z\}; 3) = N_G(x) \cap N_G(y) \cap N_G(z)$ for a graph to have a Hamiltonian path.

Theorem 8.10 (Flandrin, Jung and Li [85]). *Let G be a connected simple graph. If*

$$\deg_G(x) + \deg_G(y) + \deg_G(z) - |N_G(\{x, y, z\}; 3)| \geq |G| - 1$$

for all independent sets $\{x, y, z\}$ of G, then G has a Hamiltonian path.

An extension of the above theorem to spanning k-trees was obtained by Kyaw, which is stronger than Theorem 8.7 and given in the following theorem.

Theorem 8.11 (Kyaw [162]). *Let G be a connected simple graph and $k \geq 3$ be an integer. If for every $(k + 1)$-frame S of G,*

$$\sum_{x \in S} \deg_G(x) + \sum_{i=2}^{k+1} (k - i)|N_G(S; i)| \geq |G| - 1, \qquad (8.3)$$

then G has a spanning k-tree.

Another well-known sufficient condition for a graph to have a Hamiltonian path is given in the following theorem, which was proved by Chvátal and Erdös in 1972. We consider this type of sufficient condition for a graph to have a spanning k-tree.

Theorem 8.12 (Chvátal and Erdös [55]). *If a simple graph G satisfies $\alpha(G) \leq \kappa(G) + 1$, then G has a Hamiltonian path.*

Proof. If $\alpha(G) = 1$, then G is a complete graph, and has a Hamiltonian path. Hence we may assume $\alpha(G) \geq 2$, which implies $\kappa(G) \geq 1$. In particular, G is connected. Let $k = \kappa(G) \geq 1$. Assume that G has no Hamiltonian path. Let P be a longest path in G, and v a vertex not contained in P. Then $|P| \geq 2k + 1$ (by Problem 1.2). We assign an orientation to P, and so for every inner vertex y of P, we can define the successor y^+ and the predecessor y^- of y. By Theorem 1.6, there exists a $(v, V(P))$-fan $\{Q_1, Q_2, \ldots, Q_k\}$ of size k such that every Q_i connects v to a vertex x_i of P (Fig. 8.6).

It is clear that no x_i is an end-vertex of P, that is, every x_i is an inner vertex of P. Let y be the initial end-vertex of the oriented path P.

Claim $\{v, y, x_1^+, x_2^+, \ldots, x_k^+\}$ *is an independent set of G.*

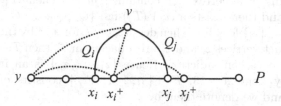

Fig. 8.6. A longest path P, and broken lines denote the non-existence of edges.

It is clear that v and x_i^+ are not adjacent since P is a longest path in G. Moreover, it is easy to see that v and y are not adjacent, and y and x_i^+ are not adjacent in G. If two distinct vertices x_i^+ and x_j^+ are joined by an edge of G, then there is a path $P + Q_i + Q_j + x_i^+ x_j^+ - x_i x_i^+ - x_j x_j^+$, which is longer than P, and so this is a contradiction (Fig. 8.6). Hence the claim holds.

By the claim, we have $\alpha(G) \geq k+2 = \kappa(G)+2$, a contradiction. Therefore G has a Hamiltonian path. $\square$

The above theorem is generalized to spanning k-trees in n-connected graphs as Theorem 8.14 below. In order to prove the theorem, we need the next lemma.

Lemma 8.13. *Let T be a tree and X be a set of independent vertices of T. Then $T - X$ has $\sum_{x \in X}(\deg_T(x) - 1) + 1$ components.*

Proof. We prove the lemma by induction on $|X|$. If $X = \{v\}$, then $T - v$ has $\deg_T(v)$ components, and so the lemma holds. Assume $|X| \geq 2$. Then there exists a vertex $y \in X$ such that one component of $T - y$, say D, contains all the vertices in $X - y$. By the inductive hypothesis, $D - (X - y)$ has $\sum_{x \in X-y}(\deg_D(x) - 1) + 1$ components. Since $\deg_T(x) = \deg_D(x)$ for all $x \in X - y$, the number of components of $T - X$ is

$$\deg_T(y) - 1 + \sum_{x \in X-y}(\deg_T(x) - 1) + 1 = \sum_{x \in X}(\deg_T(x) - 1) + 1.$$

Therefore the lemma is proved. $\square$

Theorem 8.14 (Neumann-Lara and Rivera-Campo [193]). *Let $k \geq 2$ and $n \geq 2$ be integers. If an n-connected simple graph G satisfies $\alpha(G) \leq (k-1)n + 1$, then G has a spanning k-tree.*

Proof. Assume first $k = 2$. Then G satisfies $\alpha(G) \leq n+1 \leq \kappa(G)+1$, and so by Theorem 8.11, G has a Hamiltonian path P, which is the desired spanning 2-tree. Hence we may assume $k \geq 3$.

Suppose that G does not have a spanning k-tree. Let T be a maximal k-tree of G and v a vertex not contained in T. Then $|T| \geq 2n + 1$ by Problem 1.2, and there exists a $(v, V(T))$-fan $\{Q_1, Q_2, \ldots, Q_n\}$ of size n. Let $V(Q_i) \cap V(T) = \{z_i\}$ for all i. Then $\deg_T(z_i) = k$ for all i by the maximality of T. If two distinct vertices z_i and z_j are adjacent in T, then $T - z_i z_j + Q_i + Q_j$ is a k-tree of G, a contradiction. Hence $\{z_1, \ldots, z_n\}$ is an independent set of T. Thus by Lemma 8.13, $T - \{z_1, z_2, \ldots, z_n\}$ consists of $(k-1)n + 1$ components, and we denote them by

$$T - \{z_1, z_2, \ldots, z_n\} = D_1 \cup D_2 \cup \ldots \cup D_N, \qquad N = (k-1)n + 1.$$

Let w be a vertex of T that is not a leaf of T and not contained in $\{z_1, \ldots, z_n\}$. We regard T as a rooted tree with root w. If D_i contains a vertex with degree less than k in T, then choose such a vertex, say x_i. If every vertex of D_i has degree k in T, then there exists a vertex $z_s \in \{z_1, \ldots, z_n\}$ such that the parent of z_s in T is contained in D_i, and we choose the parent of z_s as x_i (see Fig. 8.7). Notice that for every vertex $z_i \in \{z_1, \ldots, z_n\}$, there exists exactly one component D_i that contains the parent of z_i, and so the x_i's are distinct.

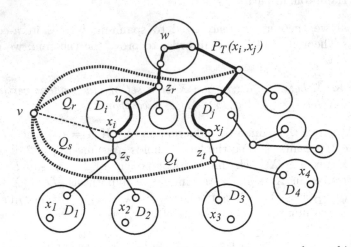

Fig. 8.7. Two components D_i and D_j with their vertices x_i and x_j, which are the parents of z_s and z_t, respectively.

We now show that $\{v, x_1, x_2, \ldots, x_N\}$ is an independent set of G. Namely, we prove the following two claims.

Claim 1. v and x_i are not adjacent in G for all $1 \leq i \leq N$.

Assume that v and x_i are joined by an edge of G for some i. If $\deg_T(x_i) \leq k - 1$, then $T + x_i v$ is a k-tree of G, which contradicts the maximality of T.

If $\deg_T(x_i) = k$, then there exists a vertex $z_s \in \{z_1, \ldots, z_n\}$ corresponding to D_i. Then $T + vx_i + Q_s - x_i z_s$ is a k-tree of G, a contradiction (see Fig. 8.7).

Claim 2. x_i and x_j are not adjacent in G for all $1 \leq i < j \leq N$.

Assume that x_i and x_j are adjacent in G for some $i \neq j$. Let $P_T(x_i, x_j)$ be the path in T connecting x_i and x_j. Then there exists $z_r \in \{z_1, \ldots, z_n\}$ in $P_T(x_i, x_j)$ that is adjacent to a vertex u of D_i. If $\deg_T(x_i) \leq k - 1$ and $\deg_T(x_j) \leq k - 1$, then $T + x_i x_j + Q_r - z_r u$ is a k-tree of G, a contradiction. Thus may assume that $\deg_T(x_i) = k$. Then every vertex of D_i has degree k in T, and there exists $z_s \in \{z_1, \ldots, z_n\}$ which is adjacent to x_i in T. If $\deg_T(x_j) \leq k - 1$, then

$$T + x_i x_j + Q_s + Q_r - x_i z_s - z_r u$$

is a k-tree, a contradiction (Fig. 8.7). Hence we may assume that $\deg_T(x_j) = k$. Then there exists $z_t \in \{z_1, \ldots, z_n\}$ which are adjacent to x_j in T. Then

$$T + x_i x_j + Q_s + Q_t + Q_r - x_s x_i - x_t x_j - z_r u$$

is a k-tree of G since $k \geq 3$, a contradiction. Therefore the claim follows.

By Claims 1 and 2, we obtain $\alpha(G) \geq N + 1 = (k - 1)n + 2$. This contradicts the assumption of the theorem, and the theorem is proved. □

The next theorem, Theorem 8.16, gives a sufficient condition using toughness for a graph to have a spanning k-tree. The proof given here is due to Ellingham and Zha [76].

For a vertex set S of a simple graph G, a **bridge** of (S, G) is a connected subgraph of G which is either an edge joining two vertices of S and its end-points, or a component of $G - S$ together with the edges joining S to it and their end-points (see Fig. 8.8). The next lemma is used in the proof of Theorem 8.16.

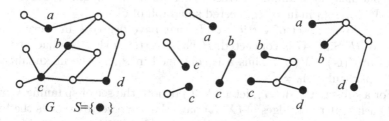

Fig. 8.8. The bridges of (S, G).

Lemma 8.15. *Let T be a tree and X a set of vertices of T that are not leaves. Then $T - X$ has at least $\sum_{x \in X}(\deg_T(x) - 2) + 2$ components.*

Proof. Consider an induced subgraph $\langle X \rangle_T$ that contains at most $|X| - 1$ edges. For every edge e of $\langle X \rangle_T$, we insert a new vertex $v(e)$ with degree 2 into T, and denote the resulting tree by T^*. Then X becomes an independent set of T^* and the degree of each vertex of X in T is equal to that in T^*. By Lemma 8.13, we obtain

$$\omega(T - X) = \omega(T^* - X) - ||\langle X \rangle_T||$$
$$\geq \sum_{x \in X} (\deg_{T^*}(x) - 1) + 1 - (|X| - 1)$$
$$= \sum_{x \in X} (\deg_T(x) - 2) + 2.$$

Hence the lemma holds. □

Theorem 8.16 (Win [248]). *Let $k \geq 3$ be an integer. If a connected simple graph G satisfies*

$$\omega(G - S) \leq (k - 2)|S| + 2 \qquad \text{for all} \quad S \subset V(G), \tag{8.4}$$

then G has a spanning k-tree. In particular, if $tough(G) \geq 1/(k - 2)$, then G has a spanning k-tree.

Proof. [76] We first show that the former implies the latter. Suppose $tough(G) \geq 1/(k - 2)$, and let $S \subset V(G)$. If $\omega(G - S) \geq 2$, then

$$\frac{|S|}{\omega(G - S)} \geq tough(G) \geq \frac{1}{k - 2}.$$

Thus $\omega(G - S) \leq (k - 2)|S|$. If $\omega(G - S) = 1$, then clearly $\omega(G - S) \leq (k - 2)|S| + 2$. Hence G has a spanning k-tree by (8.4).

We now prove the first statement. Suppose that G satisfies the condition (8.4) but has no spanning k-tree. Let T be a maximum k-tree of G and $H = \langle V(T) \rangle_G$ an induced connected subgraph of G.

Let A_0 be the set of vertices of H that have degree k in every spanning k-tree of H. Since G is connected, H has a vertex u that is adjacent to some vertex of $V(G) - V(H)$. Thus u is contained in A_0 by the maximality of T, and in particular $A_0 \neq \emptyset$.

For a subset $X \subset V(H)$, let $\Omega(X, T)$ denote the set of spanning k-trees R of H such that the bridges of (X, R) have the same vertex sets as the bridges of (X, T). Namely, for every bridge D_R of (X, R), there is a bridge B_T of (X, T) such that $V(D_R) = V(B_T)$. Then $\deg_{D_R}(x) = \deg_{B_T}(x) = 1$ for all $x \in X \cap V(D_R)$. By this fact and since T and R are trees, both $T - B_T + D_R$ and $R - D_R + B_T$ are spanning k-trees of H contained in $\Omega(X, T)$, where T, R, D_R and B_T are considered as edge sets. We are ready to prove the following claim.

Claim   *There exists a non-empty vertex set U in H such that every edge of G joining $V(G) - V(H)$ to $V(H)$ is incident with a vertex in U, $\omega(H - U) = \omega(T - U)$ and $\deg_T(x) = k$ for all $x \in U$. In particular, no edge of $E(H) - E(T)$ joins two distinct components of $T - U$.*

Let A_1 be the set of vertices of H that have degree k in every spanning k-tree in $\Omega(A_0, T)$. It is clear that $A_0 \subseteq A_1$. Assume that there is an edge $xy \in E(H)$ that joins two distinct components of $T - A_0$ (see Fig. 8.9). We shall show that at least one of vertices x and y is contained in A_1.

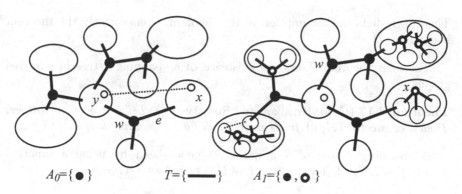

$A_0 = \{\bullet\}$ $T = \{\textbf{———}\}$ $A_1 = \{\bullet, \circ\}$

Fig. 8.9. A spanning tree T with A_0 and A_1.

Suppose that $x \notin A_1$ and $y \notin A_1$. Then there exist spanning trees T_x and T_y in $\Omega(A_0, T)$ such that $\deg_{T_x}(x) < k$ and $\deg_{T_y}(y) < y$ by the definition of A_1. Let D be the bridge of (A_0, T_x) containing y and B be the bridge of (A_0, T_y) containing y. Then $V(D) = V(B)$ and $T_z = T_x - D + B$ is a spanning k-tree of H contained in $\Omega(A_0, T)$, and satisfies

$$\deg_{T_z}(x) = \deg_{T_x}(x) < k \quad \text{and} \quad \deg_{T_z}(y) = \deg_{T_y}(y) < k.$$

Consider the path $P(x, y)$ in T_z connecting x and y. Note that $P(x, y)$ passes through a vertex $w \in A_0$, and let e be an edge of $P(x, y)$ incident with w (see Fig. 8.9). Then $T^* = T_z - e + xy$ is a spanning k-tree of H and satisfies $\deg_{T^*}(w) < k$, which contradicts $w \in A_0$. Therefore, at least one of x and y is contained in A_1. Namely, the edge xy does not connect two distinct components of $T - A_1$.

Define A_2 to be the set of vertices of H that have degree k in every spanning k-tree in $\Omega(A_1, T)$. Then $\Omega(A_1, T) \subseteq \Omega(A_0, T)$ since $A_0 \subseteq A_1$, which implies $A_1 \subseteq A_2$, and we can similarly show that for every edge of H joining two distinct components of $T - A_1$, at least one of its end-points is contained in A_2. Notice that such an edge joins two components of $T - A_1$ contained in a single bridge of (A_0, T) (see Fig. 8.9).

By repeating this procedure, we obtain a sequence

$$A_0 \subset A_1 \subset A_2 \subset \cdots \subset A_m = A_{m+1}.$$

Moreover, no edge of H joins two distinct components of $T - A_m$ since $A_{m+1} = A_m$. Therefore $U = A_m$ is the desired vertex subset, and the claim is proved.

Let U be the vertex subset of $V(H)$ obtained in the claim. Since every edge of G joining $V(G) - V(H)$ to H is incident with U, we have $\omega(G - U) > \omega(H - U)$. By the above claim and Lemma 8.15, we have

$$\omega(G - U) > \omega(H - U) = \omega(T - U) \geq (k-2)|U| + 2.$$

This contradicts the assumption in the theorem. Consequently the theorem is proved. $\square$

The next theorem shows the existence of a spanning k-tree in star-free graphs.

Theorem 8.17 (Caro, Krasikov and Roditty [42]). *Let $k \geq 3$ be an integer. Then a connected $\{K_{1,k}\}$-free simple graph has a spanning k-tree.*

A spanning k-tree of a graph G is generalized by using a function $f : V(G) \to \mathbb{Z}^+$ as follows: a tree T of G is called an f-**tree** if

$$\deg_T(x) \leq f(x) \qquad \text{for all} \quad x \in V(T).$$

The next theorem is a generalization of Theorem 8.16.

Theorem 8.18 (Ellingham, Nam and Voss [75]). *Let G be a connected simple graph, and let $f : V(G) \to \mathbb{Z}^+$ be a function. If*

$$\omega(G - S) \leq \sum_{x \in S}(f(x) - 2) + 2 \qquad \text{for all} \quad S \subset V(G),$$

then G has a spanning f-tree.

Other results on spanning f-trees can be found in Frank and Gyárfás [87]. We conclude this section with the following theorem, which guarantees the existence of a spanning tree satisfying another bounded degree condition.

Theorem 8.19 (Ellingham, Nam and Voss [75]). *Every m-edge connected simple graph has a spanning tree T such that*

$$\deg_T(x) \leq 2 + \left\lceil \frac{\deg_G(x)}{m} \right\rceil \qquad \text{for all} \quad x \in V(G).$$

8.3 Spanning k-ended tree

A Hamiltonian path is a spanning tree having exactly two leaves. From this point of view, some sufficient conditions for a graph to have a Hamiltonian path are modified to those for a spanning tree having at most k leaves. A tree having at most k leaves is called a k-**ended tree**, and we now turn our attention to **spanning k-ended trees** (Fig. 8.10). It it clear that if $s \leq t$ then a spanning s-ended tree is also a spanning t-ended tree. We first give a generalization of Ore's Theorem 8.6.

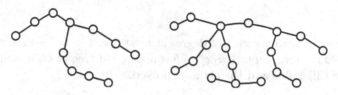

Fig. 8.10. A 3-ended tree and a 6-ended tree.

Theorem 8.20 (Las Vergnas [163], Broersma and Tuinstra [37]). *Let $k \geq 2$ be an integer and G be a connected simple graph. If $\sigma_2(G) \geq |G| - k + 1$, then G has a spanning k-ended tree.*

Proof. By Theorem 8.6, the theorem is true for $k = 2$, and so we may assume $k \geq 3$. Suppose that G does not have a spanning k-ended tree. Let T be a maximal k-ended tree of G. Since G is connected, there exist two vertices $v \in V(G) - V(T)$ and $w \in V(T)$ which are adjacent in G. It is immediate that T has precisely k leaves since otherwise $T + vw$ is a larger k-ended tree than T.

Let $U = \{u_1, u_2, \ldots, u_k\}$ be the set of k leaves of T. By the maximality of T, it is clear that $N_G(u_i) \subset V(T)$ for every $1 \leq i \leq k$. In particular, $N_G(v) \cap U = \emptyset$. We regard T as a rooted tree with root w, and for a vertex x of T, let x^+ denote the parent of x (Fig. 8.11).

Claim *(i) U is an independent set of G; (ii) the vertex v is not adjacent to any vertex of $N_T(w)$ in G; and (iii) if a vertex x of $V(T) - w$ is adjacent to v in G, then u_1 is not adjacent to x^+ in G (Fig. 8.11).*

Assume that two distinct vertices u_a and u_b of U are adjacent in G. The path $P_T(u_a, u_b)$ in T contains an edge e incident with a vertex of degree at least 3 in T since $k \geq 3$. Thus $T + u_a u_b - e + vw$ is a k-ended tree of G, which contradicts the maximality of T. Hence U is an independent set of G.

If v is adjacent to $y \in N_T(w)$ in G, then $T - wy + vy + vw$ is a k-ended tree, a contradiction (Fig. 8.11). Hence (ii) holds.

Assume that v is adjacent to $x \in V(T) - w$ in G, and u_1 is adjacent to x^+ in G. Then $T - xx^+ + vx + vw + u_1 x^+$ has an unique cycle, which contains an

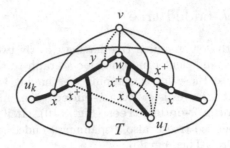

Fig. 8.11. A maximal k-ended tree T of a graph G.

edge b incident with a vertex of degree at least 3 in $T - xx^+ + vx + vw + u_1x^+$. Thus $T - xx^+ + vx + vw + u_1x^+ - b$ is a k-ended tree of G, a contradiction. Therefore (iii) holds and the claim is proved.

By the claim, we have

$$N_G(v) \subseteq (V(G) - V(T) - \{v\}) \cup (N_G(v) \cap V(T)),$$
$$N_G(u_1) \subseteq V(T) - U - \{x^+ : x \in N_G(v) \cap V(T) - \{w\}\}.$$

Hence by letting $m = |N_G(v) \cap V(T)| \geq 1$, we have

$$\deg_G(v) + \deg_G(u_1) \leq |G| - |T| - 1 + m + |T| - k - (m - 1) = |G| - k.$$

This contradicts the assumption $\sigma_2(G) \geq |G| - k + 1$. Consequently the theorem is proved. □

Theorem 8.12 says that every graph G satisfying $\alpha(G) \leq \kappa(G) + 1$ has a Hamiltonian path. Las Vergnas conjectured the following theorem, which is a generalization of Theorem 8.12. This conjecture was proved by Win, who introduced a new proof technique called a k-*ended system*, and this proof technique seems to be useful for the proofs of other results on spanning k-ended trees.

Theorem 8.21 (Win [247]). *Let $k \geq 2$ be an integer and G be a connected simple graph. If $\alpha(G) \leq \kappa(G) + k - 1$, then G has a spanning k-ended tree.*

Win also showed that the following theorem holds when $k = 3$.

Theorem 8.22 (Win [247]). *Let G be a connected simple graph. If $\alpha(G) \leq \kappa(G) + 2$, then G has a spanning 3-ended tree that includes a longest path of G.*

Proof. We may assume that G does not have a Hamiltonian path. Let Q be a longest path of G, and a_Q and b_Q be the two end-vertices of Q. We orient the path Q from a_Q to b_Q, and so for every inner vertex x of $V(Q)$, the successor x^+ and the predecessor x^- of x are defined.

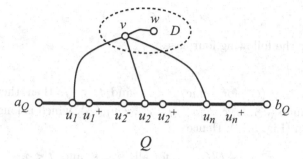

Fig. 8.12. A longest path Q of G and a component D of $G - V(Q)$.

Let $n = \kappa(G) \geq 1$. For any vertex $v \in V(G) - V(Q)$, by Theorem 1.6 there exists a $(v, V(Q))$-fan $\{P_1, P_2, \ldots, P_n\}$ with size n. Let $u_i = V(P_i) \cap V(Q)$ for every i. Then all u_i's are distinct, and we may assume that $u_1, u_2, \ldots, u_n$ lie on Q in this order. Let

$$U = \{u_1, \ldots, u_n\}, \quad U^+ = \{u_i^+ : u_i \in U\}, \quad U^- = \{u_i^- : u_i \in U\}.$$

Then $U \cap U^+ = \emptyset$ and $U \cap U^- = \emptyset$ since Q is a longest path of G. Moreover, every edge of G joining v to Q can be chosen as a path P_i, that is,

at least n or all vertices of $N_G(v) \cap V(Q)$ are contained in U.   (8.5)

Since Q is a longest path of G, we can easily show that $a_Q, b_Q \notin U$, and the following two sets are independent in G.

$$\{v, a_Q\} \cup U^+ \quad \text{and} \quad \{v, b_Q\} \cup U^-.   (8.6)$$

Let D be a component of $G - V(Q)$ containing v, and let A be an independent set of D with $\alpha(D)$ vertices. Then for each vertex w of A, we can show that $\{w, a_Q\} \cup U^+$ is an independent set of G (Fig. 8.12), which implies that $A \cup \{a_Q\} \cup U^+$ is an independent set of G. Thus $\alpha(G) \geq n + 1 + \alpha(D)$. Since $\alpha(G) \leq n + 2$, it follows that $\alpha(D) = 1$, and thus D is a complete graph. Since the vertex v was arbitrarily chosen in $V(G) - V(Q)$, every component of $G - V(Q)$ is a complete graph. Hence, if we show that $G - V(Q)$ consists of exactly one component, then the theorem is proved.

Assume that $G - V(Q)$ has at least two components, and take two vertices x and y from two distinct components of $G - V(Q)$. We consider a $(x, V(Q))$-fan and a $(y, V(Q))$-fan, and define vertex sets $X = U(x)$ and $Y = U(y)$ of Q as above. Put

$$X = \{x_1, x_2, \ldots, x_n\} \quad \text{and} \quad Y = \{y_1, y_2, \ldots, y_n\}.$$

If $X^+ \cap Y = \emptyset$, then $X^+ \cap (N_G(y) \cap V(Q)) = \emptyset$ by (8.5), and so $\{x, y, a_Q\} \cup X^+$ is an independent set of G, which contradicts $\alpha(G) \leq n + 2$. Hence $X^+ \cap Y \neq \emptyset$. Choose a vertex

$$x_s^+ = y_t \in X^+ \cap Y,$$

and consider the following four cases.

Case 1. $1 < t \leq s$.

If $x_i^+ y_j^- \in E(G)$ for some $i < s$ and $t < j$, then there is a path passing through $(a_Q, x_i, x, x_s, x_i^+, y_j^-, x_s^+, y, y_j, b_Q)$, which is longer than Q, a contradiction (Fig. 8.13). Hence

$$x_i^+ y_j^- \notin E(G) \qquad \text{for all} \quad i < s \text{ and } t < j. \tag{8.7}$$

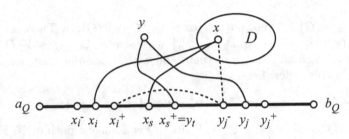

Fig. 8.13. $x_s^+ = y_t$, $i < s$, $t < j$ and $1 < t \leq s$.

If $xy_j^- \in E(G)$ for some $t < j$, then there is a path passing through $(a_Q, x_s, x, y_j^-, x_s^+, y, y_j, b_Q)$, a contradiction. Thus

$$xy_j^- \notin E(G) \qquad \text{for all} \quad t < j. \tag{8.8}$$

Similarly,

$$yx_i^+ \notin E(G) \qquad \text{for all} \quad i < s. \tag{8.9}$$

By (8.7), (8.8) and (8.9),

$$\{x_1^+, \ldots, x_{s-1}^+\} \cup \{y_{t+1}^-, \ldots, y_n^-\} \cup \{a_Q, b_Q, x, y\}$$

is an independent set of G. But, by the condition of this case, this implies

$$\alpha(G) \geq (s-1) + (n-t) + 4 \geq n + 3.$$

This contradicts the condition $\alpha(G) \leq n + 2$.

Case 2. $s < t$.

If $y_j^+ x_i^- \in E(G)$ for some $j < t$ and $s < i$, then there is a path passing through $(a_Q, y_j, y, x_s^+, x_i^-, y_j^+, x_s, x, x_i, b_Q)$. This is longer than Q, a contradiction. Hence .

$$y_j^+ x_i^- \notin E(G) \qquad \text{for} \quad j < t \text{ and } s < i. \tag{8.10}$$

If $xy_j^+ \notin E(G)$ for all $j < t$, then $\{x, a_Q, b_Q\} \cup \{x_i^- : s < i\} \cup \{y_j^+ : j < t\}$ is an independent set of G, which implies $\alpha(G) \geq 3 + (n - s) + (t - 1) \geq n + 3$ and contradicts $\alpha(G) \leq n + 2$. Therefore

$$xy_p^+ \in E(G) \qquad \text{for some} \quad p < t.$$

If $xy_j^+ \in E(G)$ for some $j < t$, $j \neq p$, then there is a path passing through $(a_Q, y_c, y, y_d, y_c^+, x, y_d^+, b_Q)$, where $\{c, d\} = \{j, p\}$ and $c < d$, a contradiction. Hence

$$xy_j^+ \notin E(G) \qquad \text{for all} \quad j < t, \ j \neq p. \tag{8.11}$$

If $yx_i^- \in E(G)$ for some $s < i$, then G has a longer path passing through $(a_Q, y_p, y, x_i^-, y_p^+, x, x_i, b_Q)$ than Q, a contradiction. Hence

$$yx_i^- \notin E(G) \qquad \text{for all} \quad s < i. \tag{8.12}$$

Therefore, by (8.10), (8.11) and (8.12),

$$\{x, y, a_Q, b_Q\} \cup \{x_i^- : s < i\} \cup \{y_j^+ : j < t, j \neq p\}$$

is an independent set of G, which implies $\alpha(G) \geq 4 + (n - s) + (t - 2) \geq n + 3$ and contradicts $\alpha(G) \leq n + 2$.

Case 3. $s = t = 1$.

If $xy_j^- \in E(G)$ for some $1 < j$, then there is a path passing through $(a_Q, x_1, x, y_j^-, x_1^+, y, y_j, b_Q)$, a contradiction. Hence $xy_i^- \notin E(G)$ for all $1 < i$, and so $\{a_Q, b_Q, x, y\} \cup (Y^- - \{y_1^-\})$ is an independent set of G containing $n + 3$ vertices. This is a contradiction.

Case 4. $s = t = n$.

By a similar argument as in Case 3, we can show that $\{a_Q, b_Q, x, y\} \cup (X^+ - \{x_n^+\})$ is an independent set, and derive a contradiction. Consequently the proof is complete. $\square$

We now introduce the concept of a k-ended system in order to prove Theorem 8.21. Let G be a simple graph and S be a set of paths, cycles and vertices of G such that no two distinct elements of S have a vertex in common. Define a function $f : S \to \{1, 2\}$ as follows: for each element X of S,

$$f(X) = \begin{cases} 2 & \text{if } X \text{ is a path of order at least three,} \\ 1 & \text{otherwise (i.e., if } X \text{ is } K_1, K_2 \text{ or a cycle).} \end{cases}$$

For the set S of G, let

$$\mathcal{S}_P = \{X \in \mathcal{S} : f(X) = 2\},$$
$$\mathcal{S}_C = \{X \in \mathcal{S} : f(X) = 1\}, \text{ and}$$
$$V(\mathcal{S}) = \bigcup_{X \in \mathcal{S}} V(X).$$

Moreover, $V(\mathcal{S}_P)$ and $V(\mathcal{S}_C)$ can be defined analogously, and $|\mathcal{S}|$, $|\mathcal{S}_P|$, $|\mathcal{S}_C|$ denote the number of elements in $\mathcal{S}$, $\mathcal{S}_P$ and $\mathcal{S}_C$, respectively. Then $\mathcal{S}$ is called a **k-ended system** if

$$\sum_{X \in \mathcal{S}} f(X) = 2|\mathcal{S}_P| + |\mathcal{S}_C| \le k.$$

If $V(\mathcal{S}) = V(G)$, then $\mathcal{S}$ is called a **spanning k-ended system** of G (see Fig. 8.14). The following lemma shows why a k-ended system is important for spanning k-ended trees.

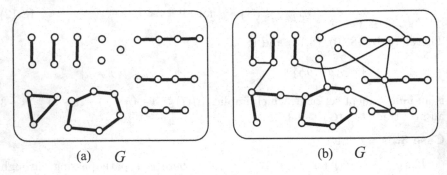

(a) G (b) G

Fig. 8.14. (a): A spanning 14-ended system $\mathcal{S}$ of G with $|\mathcal{S}_C| = 8$ and $|\mathcal{S}_P| = 3$; (b) A spanning tree of G with 14 leaves.

Lemma 8.23 ([247]). *Let $k \ge 3$ be an integer, and G be a connected simple graph. If G has a spanning k-ended system, then G has a spanning k-ended tree.*

Proof. Let $\mathcal{S}$ be a spanning k-ended system of G. We may assume that $\mathcal{S}$ does not contain a Hamiltonian cycle of G. Since G is connected, we can take a minimal set X of edges such that $\mathcal{S}$ together with X determines a connected spanning subgraph of G.

For each cycle $C \in \mathcal{S}$, there exists an edge $e_C \in X$ incident with a vertex v_C of C. Next, we delete one edge of C incident with v_C. By repeating this procedure for every cycle of $\mathcal{S}$, we can obtain a spanning tree with at most k leaves (see Fig. 8.14). $\square$

We call a k-ended system $\mathcal{S}$ of G a **maximal k-ended system** if there exists no k-ended system $\mathcal{S}'$ in G such that $V(\mathcal{S}) \subset V(\mathcal{S}')$. The following lemma expresses some nice properties of k-ended systems. Note that we say that **two distinct elements of S are connected by a path in $G - V(S)$** if there exists a path in G that connects the two elements of S and whose inner vertices are all contained in $V(G) - V(S)$, where a path may consist of only one edge of G.

Lemma 8.24. *Let $k \geq 3$ be an integer, and G be a connected simple graph. Suppose that G has no spanning k-ended system, and let S be a maximal k-ended system of G such that $|S_P|$ is maximum subject to the maximality of $V(S)$. Then the following statements hold.*

(i) No two elements of S_C are connected by a path in $G - V(S)$ (Figure. 8.15).
(ii) No element of S_C is connected to an end-vertex of an element of S_P by a path in $G - V(S)$.
(iii) No end-vertex of an element of S_P is connected to an end-vertex of another element of S_P by a path in $G - V(S)$.
(iv) No vertex u in $V(G) - V(S)$ is connected to two distinct elements of S_C by two internally disjoint paths Q_1 and Q_2 in $G - V(S)$ that satisfy $V(Q_1) \cap V(Q_2) = \{u\}$ (Figure. 8.15).

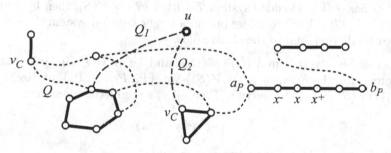

Fig. 8.15. For a maximal k-ended system, there exist no paths consisting of dot lines.

Proof. Assume that two elements D_1 and D_2 of S_C are joined by a path Q in $G - V(S)$. Then there is a path P with $V(P) = V(D_1) \cup V(D_2) \cup V(Q)$ in G, and $S - \{D_1, D_2\} + P$ is either a larger k-ended system of G than S or a maximal k-ended system with more S_P elements. This contradicts the choice of S. Hence the statement (i) holds. Statement (ii) and (iii) can be shown in the same way. Statement (iv) is an easy consequence of (i) since $Q_1 + Q_2$ is a path connecting two elements of S_C. $\square$

We are ready to prove Theorem 8.21.

Proof of Theorem 8.21 Suppose that G does not have a spanning k-ended tree. Then G does not have a spanning k-ended system by Lemma 8.23. Choose a maximal k-ended system $\mathcal{S}$ such that $|\mathcal{S}_P|$ is maximum subject to the maximality of $V(\mathcal{S})$.

For each path $P \in \mathcal{S}_P$, let a_P and b_P denote the two end-vertices of P. For each element $C \in \mathcal{S}_C$, let v_C be an arbitrarily chosen vertex of C. Then define

$$End(\mathcal{S}_P) = \bigcup_{P \in \mathcal{S}_P} \{a_P, b_P\},$$

$$End(\mathcal{S}_C) = \bigcup_{C \in \mathcal{S}_C} \{v_C\}, \text{ and}$$

$$End(\mathcal{S}) = End(\mathcal{S}_P) \cup End(\mathcal{S}_C).$$

For every path P in $\mathcal{S}_P$, we orient P from a_P to b_P, and for every inner vertex x of P, let x^+ and x^- denote the successor and the predecessor of x, respectively (Figure 8.15). For every cycle $C \in \mathcal{S}_C$, we arbitrarily orient C and for every vertex y of C, let y^+ and y^- denote the successor and the predecessor of y, respectively. The following claim is easy and we use it without referring to it explicitly.

Claim 1. G has no $(k-1)$-ended system $\mathcal{T}$ with $V(\mathcal{T}) = V(\mathcal{S})$.

If G has a $(k-1)$-ended system $\mathcal{T}$ with $V(\mathcal{T}) = V(\mathcal{S})$, then by adding a vertex in $V(G) - V(\mathcal{S})$ to $\mathcal{T}$, we obtain a larger k-ended system than $\mathcal{S}$. This is a contradiction, and so the claim holds.

Let w be a vertex in $V(G) - V(\mathcal{S})$, and let $n = \kappa(G) \geq 1$. Then by Theorem 1.6 there exists a $(w, V(\mathcal{S}))$-fan $\{P_1, P_2, \ldots, P_n\}$ of size n. Let $V(P_i) \cap V(\mathcal{S}) = \{u_i\}$ for every i. Then all u_i's are distinct, and let

$$U = \{u_1, u_2, \ldots, u_n\}.$$

By Lemma 8.24, the next two claims hold.

Claim 2. $End(\mathcal{S})$ is an independent set of G.

Claim 3. $U \cap End(\mathcal{S}_P) = \emptyset$, and at most one element of $\mathcal{S}_C$ contains vertices of U.

We first consider the following case.

Case 1. An element of $\mathcal{S}_C$ contains a vertex of U.

It is clear that for every K_1-component $\{x\}$ of $\mathcal{S}_C$, it follows that $N_G(x) \cap (V(G) - V(\mathcal{S})) = \emptyset$. Thus no K_1-component of $\mathcal{S}_C$ contains a vertex of U.

First assume that a K_2-component D of $\mathcal{S}_C$ contains a vertex, say u_a, of U (see Fig. 8.16). Let $V(D) = \{u_a, v_D\}$, where $v_D \in End(\mathcal{S}_C)$. Obviously v_D is not contained in U. We shall show that

$$End(\mathcal{S}) \cup \{u_i^+ : u_i \in U - u_a\} \cup \{w\}$$

is an independent set of G. By Claim 2, $End(\mathcal{S})$ is independent in G. It is clear that v_D and w are not adjacent in G. If $u_j^+ = b_P \in End(\mathcal{S})$, where $P \in \mathcal{S}_P$, then there exists a path Q passing through $(V(P) - b_P) \cup \{w\} \cup V(D)$, which implies that $\mathcal{S} - \{P, D\} + Q + \{b_P\}$ is a k-ended system, a contradiction. Hence $End(\mathcal{S}) \cap \{u_i^+ : u_i \in U - u_a\} = \emptyset$. If v_D and $u_i^+ \in V(P)$, where $P \in \mathcal{S}_P$, are adjacent in G, then there exists a path Q in G passing through $V(P) \cup V(D) \cup \{w\}$, and so $\mathcal{S} - \{P, D\} + Q$ is a $(k-1)$-end system, a contradiction. Hence v_D and $u_i^+, i \neq a$, are not adjacent in G. If $u_i^+ \in V(P)$ and $u_j^+ \in V(P)$, where $P \in \mathcal{S}_P$, are adjacent in G, then G has a path passing through $V(P) \cup \{w\}$, a contradiction. If $u_i^+ \in V(P)$ and $u_j^+ \in V(P')$ are adjacent in G, where $P, P' \in \mathcal{S}_P$ and $P \neq P'$, then there exist two disjoint paths in G that cover $V(P) \cup V(P') \cup \{w\}$ (see Fig. 8.16). This contradicts the maximality of $\mathcal{S}$. Therefore $End(\mathcal{S}) \cup \{u_i^+ : u_i \in U - u_a\} \cup \{w\}$ is an independent set of G with size $k + n$.

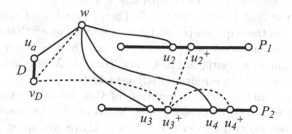

Fig. 8.16. No broken curves are edges of G.

Next assume that a cycle C of $\mathcal{S}_C$ contains at least one vertex of U. We shall show that

$$(End(\mathcal{S}) - v_C) \cup \{u_i^+ : u_i \in U\} \cup \{w\}$$

is an independent set of G with size $k + n$. If $u_i^+ \in V(C)$ and $u_j^+ \in V(C)$ are adjacent in G, then G has a cycle passing through $V(C) \cup \{w\}$, which contradicts the maximality of $\mathcal{S}$. If $u_i^+ \in V(C)$ and $u_j^+ \in V(P)$, where $P \in \mathcal{S}_P$, are adjacent in G, then there exists a path in G that passes through $V(P) \cup V(C) \cup \{w\}$, a contradiction. By the same argument given above, we can show that $u_i^+ \in V(P)$ and $u_j^+ \in V(P)$, where $P \in \mathcal{S}_P$, are not adjacent in G and that $u_i^+ \in V(P)$ and $u_j^+ \in V(P')$, where $P, P' \in \mathcal{S}_P$ and $P \neq P'$, are not adjacent in G. Therefore $(End(\mathcal{S}) - v_C) \cup \{u_i^+ : u_i \in U\} \cup \{w\}$ is an independent set of G with size $k + n$. This fact contradicts the assumption $\alpha(G) \leq k + n - 1$.

Case 2. *No element of $\mathcal{S}_C$ contains a vertex of U.*

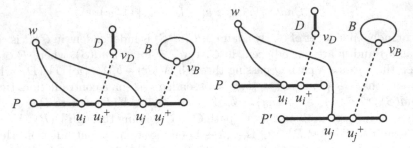

Fig. 8.17. A path Q of $\mathcal{S}_P$ contains U.

We shall show that

$$X = End(\mathcal{S}) \cup U^+ \cup \{w\}$$

contains an independent set of G with size $k + n$.

Suppose first that $End(\mathcal{S}) \cap U^+ = \emptyset$. Then $|X| = k+n+1$, and in order to show that X has the required property, it suffices to prove that (i) X does not contain four distinct vertices x_1, x_2, x_3, x_4 such that x_1 and x_2 are adjacent in G and x_3 and x_4 are adjacent in G, and that (ii) X does not contain three distinct vertices which are pairwise adjacent in G.

By Claim 2 and by the assumption of this case, we may assume that $End(\mathcal{S}) \cup \{w\}$ is an independent set of G. Moreover, by the previous argument, it is shown that no two vertices of U^+, which are contained in elements of $\mathcal{S}_P$, are adjacent in G. Therefore statement (ii) holds by these facts. Assume that $v_A \in End(\mathcal{S}_C)$ and $u_i^+ \in V(P)$ are adjacent in G and $v_B \in End(\mathcal{S}_C)$ and $u_j^+ \in V(P')$ are adjacent in G (see Fig. 8.17). If $P \neq P'$, then there exist three disjoint paths that cover $V(P) \cup V(P') \cup V(A) \cup V(B) \cup \{w\}$, a contradiction. If $P = P'$, then there exist two disjoint paths that cover $V(P) \cup V(A) \cup V(B) \cup \{w\}$, a contradiction. Therefore X includes an independent set of G with size $k + n$.

Next suppose that $End(\mathcal{S}) \cap U^+ \neq \emptyset$, which implies that there exists a path $P \in \mathcal{S}_P$ such that $b_P = u_a^+ \in End(P) \cap U^+$. If there exists another path $P' \in \mathcal{S}_P - P$ such that $b_{P'} = u_b^+ \in End(P') \cap U^+$, then there exists a path Q covering $(V(P) - b_P) \cup \{w\} \cup (V(P') - b_{P'})$. So $\mathcal{S} - \{P, P'\} + Q + \{b_P\} + \{b_{P'}\}$ is a k-ended system, a contradiction. Thus $U^+ \cap End(\mathcal{S}) = \{b_P\}$, and $|X| = k+n$. Assume that $v_D \in End(\mathcal{S}_C)$ and $u_i^+ \in V(P_1)$ are adjacent in G. Then there exist two disjoint paths that cover $(V(P) - b_P) \cup V(P_1) \cup V(D) \cup \{w\}$, which contradicts the maximality of $\mathcal{S}$. Therefore X is an independent set of G with size $k + n$. This contradicts the assumption $\alpha(G) \leq k + n - 1$. Consequently the proof is complete. $\square$

We now consider claw-free graphs, which do not contain $K_{1,3}$ as an induced subgraph. The next theorem gives a sufficient condition for a claw-free graph to have a spanning k-ended tree and this condition is much weaker than that of Theorem 8.20, as expected. We give a sketch of its proof, and show that the notion of a k-ended system is also useful for this theorem.

Theorem 8.25 ([126]). *Let $k \geq 2$ be an integer and G be a connected claw-free simple graph. If $\sigma_{k+1}(G) \geq |G| - k$, then G has a spanning k-ended tree.*

The graph constructed from two complete graphs K_m and K_n by identifying one vertex of K_m with one vertex of K_n is called a **double complete graph** and denoted by $DC(m, n)$, where $m, n \geq 2$. The common vertex of K_m and K_n is called the **center**, and the other vertices are called **non-central vertices** (Figure 8.18). Note that the order of $DC(m, n)$ is $m + n - 1$, and the path P_3 of order three is a double complete graph $DC(2, 2)$.

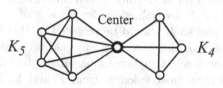

Fig. 8.18. The double complete graph $DC(m, n)$, whose order is $m + n - 1$.

Let G be a connected claw-free graph, and $\mathcal{F}$ be a set of paths, cycles, double complete graphs and vertices of G such that no two distinct elements of $\mathcal{F}$ have a vertex in common. We call $\mathcal{F}$ an **extended system** of G. For an extended system $\mathcal{F}$, we define a mapping f from $\mathcal{F}$ to $\{1, 2\}$ as follows. For every element $X \in \mathcal{F}$,

$$f(X) = \begin{cases} 2 & \text{if } X \text{ is a path of order at least four,} \\ 1 & \text{otherwise (i.e., if } X \text{ is } K_1, K_2, \text{ a cycle or} \\ & \text{a double complete graph}). \end{cases}$$

Let

$$\mathcal{F}_P = \{X \in \mathcal{F} : f(X) = 2\}, \quad \mathcal{F}_C = \{X \in \mathcal{F} : f(X) = 1\},$$

and define

$$f(\mathcal{F}) = \sum_{X \in \mathcal{F}} f(X) = 2|\mathcal{F}_P| + |\mathcal{F}_C|.$$

An extended system $\mathcal{F}$ is called a k-**extended system** if $f(\mathcal{F}) \leq k$. The following lemma is easy but is an important observation nonetheless.

Lemma 8.26. *Let G be a claw-free simple graph and D be an induced double complete subgraph of G. If a vertex $v \in V(G) - V(D)$ is adjacent to the center of D, then v is also adjacent to a non-central vertex of D.*

Proof. Let D_1 and D_2 be the two complete graphs of D, let y be the center of D and let $x_i \in D_i - \{y\}$ $(i = 1, 2)$. Since D is an induced subgraph of G, $x_1 x_2 \notin E(G)$. Since $\{x_1, x_2, v\} \subseteq N_G(y)$ and G is claw-free, v must be adjacent to at least one of x_1 and x_2. $\square$

The next lemma shows a relationship between a k-extended system and a spanning k-ended tree in a claw-free graph.

Lemma 8.27. *Let $k \geq 2$ be an integer and G be a connected claw-free simple graph. If G has a spanning k-extended system, then G has a spanning k-ended tree.*

Proof. Choose a spanning k-extended system $\mathcal{F}$ so that the number of double complete graphs is as small as possible. Then every double complete graph of $\mathcal{F}$ is an induced subgraph of G since if two non-central vertices of a double complete graph D of $\mathcal{F}$ are joined by an edge e of G, then $D + e$ has a Hamiltonian cycle, and so D should be replaced by this Hamiltonian cycle.

Since G is connected, there exists a minimal set Y of edges such that $\mathcal{F}$ together with Y forms a connected spanning subgraph of G. We shall construct a spanning tree with at most k leaves from $\mathcal{F}$ and Y. By Lemma 8.26, we can choose Y so that no edge in Y is incident with the center of a double complete graph.

For any double complete graph D of $\mathcal{F}$, there exists an edge $e_D \in Y$ incident with a vertex v_D of D, where v_D is not the center of D. Thus D has a Hamiltonian path starting at v_D, and we replace D by this Hamiltonian path.

For any cycle C of $\mathcal{F}$, there exists an edge $e_C \in Y$ incident with C. For a vertex v_C of C incident with e_C, delete one edge of C incident with v_C. By repeating the above procedure for every double complete graph and every cycle of $\mathcal{F}$, we obtain a spanning tree T. By the construction, for each $X \in \mathcal{F}$, the number of leaves of T contained in X is at most $f(X)$. Hence T has at most k leaves. $\square$

By making use of Lemma 8.27, we can prove Theorem 8.25, where a $(k+1)$-extended system $\mathcal{F}$ is chosen as follows.

Suppose that G has no spanning k-extended system. Take a maximal $(k+1)$-extended system $\mathcal{F}$ so that

(F1) $\sum_{X \in \mathcal{F}_P} |X|$ is as large as possible,
(F2) The number of cycles in $\mathcal{F}_C$ is as large as possible subject to (F1),
(F3) $\sum_{P \in \mathcal{F}_P} \left(\deg_{\langle V(P) \rangle_G}(a_P) + \deg_{\langle V(P) \rangle_G}(b_P) \right)$ is as small as possible, subject to (F1) and (F2), where a_P and b_P are the end-vertices of P.

Then $\mathcal{F}$ possesses the following nice properties as an k-ended system. By making use of this lemma, Theorem 8.25 is proved, but we omit these proofs.

Lemma 8.28. *(i)   Every double complete graph D of $\mathcal{F}_C$ is an induced subgraph of G.*
(ii) No two components of $\mathcal{F}_C$ are connected by an edge of G.
(iii) No end-vertex of a path in $\mathcal{F}_P$ is connected to a component of $\mathcal{F}_C$ by an edge of G.
(iv) No two end-vertices of two distinct paths or of the same path in $\mathcal{F}_P$ are joined by an edge of G.

By generalizing Theorem 8.10, the following theorem is obtained.

Theorem 8.29 (Kano and Kyaw [125]). *Let $k \geq 2$ be an integer, and G be a connected simple graph. If*

$$\deg_G(x) + \deg_G(y) + \deg_G(z) - |N_G(\{x, y, z\}; 3)| \geq |G| - k + 1$$

for all 3-frame $\{x, y, z\}$ of G, then G has a spanning k-ended tree.

For a tree T, let $Leaf(T)$ denote the set of leaves of T. A graph G is said to be k-**leaf-connected** if for every subset S of $V(G)$ with $|S| = k$, G has a spanning tree T with $Leaf(T) = S$. Gurgel and Wakabayashi [95] obtained the following theorem about k-leaf-connected graphs. This result can be considered as a generalization of Ore's theorem [207], which says that if $\sigma_2(G) \geq |G| + 1$, then for any two distinct vertices x and y of G, G has a Hamiltonian path connecting x and y.

Theorem 8.30 (Gurgel and Wakabayashi [95]). *Let $k \geq 2$ be an integer, and G be a connected simple graph of order at least $k + 1$. If*

$$\sigma_2(G) \geq |G| + k - 1,$$

then G is k-leaf-connected.

The above theorem is an easy consequence of Theorem 8.31 below. By Theorem 8.31, a **closure concept** for a k-leaf-connected graph can be defined. Namely, the **closure** of a graph G is the graph obtained from G by recursively joining pairs of nonadjacent vertices whose degree sum is at least $|G| + k - 1$ until no such pair remains. Furthermore, G is k-leaf-connected if and only if the closure of G is k-leaf-connected. It is obvious that the closure of a graph in Theorem 8.30 becomes a complete graph, which is clearly k-leaf-connected, and thus G itself is k-leaf-connected.

Theorem 8.31 ([95]). *Let $k \geq 2$ be an integer, and G be a connected simple graph of order at least $k + 1$. Assume that two nonadjacent vertices u and v of G satisfy*

$$\deg_G(u) + \deg_G(v) \geq |G| + k - 1. \tag{8.13}$$

Then G is k-leaf-connected if and only if $G + uv$ is k-leaf-connected.

Proof. It is sufficient to show that if $G + uv$ is k-leaf-connected, then G is also k-leaf-connected.

Let $S \subset V(G)$ with $|S| = k$. Suppose that $G + uv$ has a spanning tree T with $Leaf(T) = S$. It suffices to show that G also has a spanning tree with leaf set S. If $uv \notin E(T)$, then T is the desired spanning tree of G, and so we may assume $uv \in E(T)$. Let T_u and T_v be the components of $T - uv$ that contain u and v, respectively.

First assume that neither u nor v is a leaf of T. We regard T_u and T_v as rooted trees with root u and v, respectively (Fig. 8.19). Hence for every vertex $x \in V(T_u) \setminus S$ and $y \in V(T_v) \setminus S$, their children x^- in T_u and y^- in T_v are well-defined.

Hereafter we assume that G has no spanning tree T' with $Leaf(T') = S$ in order to derive a contradiction. Let

$$N_G(v) \cap (V(T_u) \setminus S) = \{x_1, x_2, \cdots, x_r\} \not\ni u, \qquad r \geq 0,$$
$$N_G(u) \cap (V(T_v) \setminus S) = \{y_1, y_2, \cdots, y_s\} \not\ni v, \qquad s \geq 0. \quad \text{(Fig. 8.19)}$$

Note that if $N_G(v) \cap (V(T_u) \setminus S) = \emptyset$ then $r = 0$, and the same situation holds for s.

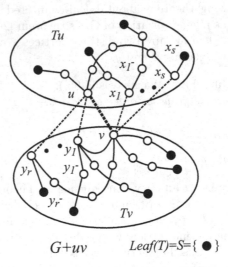

$$G+uv \qquad Leaf(T){=}S{=}\{\,\bullet\,\}$$

Fig. 8.19. $G + uv$ has a spanning tree T with $Leaf(T) = S$ that contains uv.

If u and x_i^- are adjacent in G for some i, then G has a spanning tree $T - uv - x_i x_i^- + vx_i + ux_i^-$ with leaf set S, a contradiction. Hence the following holds.

$$ux_i^- \notin E(G) \qquad \text{for every} \quad 1 \leq i \leq r, \quad \text{and}$$
$$vy_j^- \notin E(G) \qquad \text{for every} \quad 1 \leq j \leq s. \tag{8.14}$$

Then by (8.14) and $u \notin N_G(v) \cap V(T_u)$, we obtain

$$
\begin{aligned}
\deg_G(u) + \deg_G(v) &= |N_G(u) \cap V(T_u)| + |N_G(u) \cap V(T_v)| \\
&\quad + |N_G(v) \cap V(T_v)| + |N_G(v) \cap V(T_u)| \\
&\leq |V(T_u)| - r - 1 + s + |N_G(u) \cap V(T_v) \cap S| \\
&\quad + |V(T_v)| - s - 1 + r + |N_G(v) \cap V(T_u) \cap S| \\
&\leq |G| + |S| - 2 = |G| + k - 2.
\end{aligned}
$$

This contradicts the degree sum condition in the theorem.

Next assume that u or v is a leaf of T. Since T contains the edge uv and $|T| = |G| \geq k + 1$, we may assume that u is a leaf of T, but not v. Then $T_u = \{u\} \subset S$ and we regard T_v as a rooted tree with root v. For a vertex $x \in N_G(u) \cap (V(T_v) \setminus S)$, if v and x^- are adjacent in G, then $T - uv - xx^- + ux + vx^-$ is a spanning tree with leaf set S, a contradiction. Hence if u and $x \in V(G) - (S \cup \{v\})$ are adjacent in G, then v and x^- are not adjacent in G. Thus we have

$$
\begin{aligned}
\deg_G(u) + \deg_G(v) &= |N_G(u)| + |N_G(v)| \\
&\leq |N_G(u) \cap S| + |N_G(u) \setminus S| \\
&\quad + |V(G) - \{u, v\}| - |N_G(u) \setminus S| \\
&\leq |G| - 2 + |S| = |G| + k - 2.
\end{aligned}
$$

This is a contradiction. Consequently the proof is complete. □

The next theorem gives a sufficient condition for a $(k+1)$-connected graph to be k-leaf-connected.

Theorem 8.32 (Egawa, Matsuda, Yamashita and Yoshimoto [72]). *Let $k \geq 2$ be an integer, and G be a $(k+1)$-connected simple graph. If*

$$
\sigma_2(G) \geq |G| + 1,
$$

then G is k-leaf-connected.

As a corollary of Theorem 8.32, we can obtain the following theorem.

Theorem 8.33 ([72]). *Let k and s be integers such that $2 \leq k$ and $0 \leq s \leq k$. Assume that a $(s+1)$-connected simple graph G satisfies*

$$
\sigma_2(G) \geq |G| - k + 1 + s.
$$

Then for every subset $S \subset V(G)$ with $|S| = s$, G has a spanning tree T such that $S \subseteq Leaf(T)$ and $|Leaf(T)| \leq k$.

Proof. Construct the join $H = G + K_{k-s}$, where every vertex of G and every vertex of K_{k-s} are joined by edges of H. Then H satisfies the condition in Theorem 8.32, and thus H has a spanning tree T with $Leaf(T) = S \cup V(K_{k-s})$. Therefore $T - V(K_{k-s})$ is the desired spanning tree of G. □

8.4 Spanning trees with miscellaneous properties

We have studied spanning k-trees and spanning k-ended trees. We now consider spanning trees possessing some other properties. Let T be a tree. The **leaf degree of a vertex** v in T is defined to be the number of leaves adjacent to v, and the **leaf degree of** T is the maximum leaf degree among all vertices of T. The **leaf distance** of T is defined to be the minimum of distances between any two leaves of T (Fig. 8.20). Notice that a tree with leaf degree one has leaf distance at least three. We begin with a necessary and sufficient condition for a graph to have a spanning tree with bounded leaf degree.

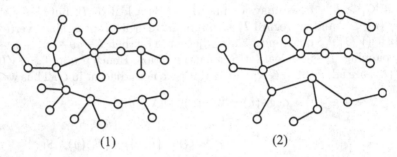

(1)   (2)

Fig. 8.20. (1) A tree with leaf degree 2; and (2) a tree with leaf distance 4.

Theorem 8.34 (Kaneko [110]). *Let G be a connected simple graph and $m \geq 1$ be an integer. Then G has a spanning tree with leaf degree at most m if and only if*

$$iso(G - S) < (m + 1)|S| \qquad \text{for all} \quad \emptyset \neq S \subset V(G). \tag{8.15}$$

Proof. For a vertex v of a tree T, let

$$leafd_T(v) = \text{ the leaf degree of } v \text{ in } T.$$

Assume that G has a spanning tree T with leaf degree at most m. Suppose, to the contrary, that there exists a vertex set S such that $iso(G-S) \geq (m+1)|S|$. Let

$$W = \{x \in Iso(G - S) : \deg_T(x) \geq 2\},$$

where $Iso(G - S)$ denotes the set of isolated vertices of $G - S$. Consider an induced subgraph $\langle W \cup S \rangle_T$. Since the leaf degree of T is at most m and $iso(G - S) \geq (m + 1)|S|$, it follows that W is not an empty set and

$$m|S| \geq \#\{x \in Iso(G - S) : \deg_T(x) = 1\}$$
$$= |Iso(G - S) - W| = iso(G - S) - |W|$$
$$\geq (m + 1)|S| - |W|,$$

which implies $|W| \geq |S|$. Since $|W| \geq |S|$ and $\deg_T(x) \geq 2$ for all $x \in W$, $\langle W \cup S \rangle_T$ contains at least $|W| + |S|$ edges, which implies that $\langle W \cup S \rangle_T$ has a cycle. This contradicts the fact that T is a tree. Hence $iso(G-S) < (m+1)|S|$ for all $\emptyset \neq S \subset V(G)$.

We next prove the sufficiency. We need some definitions. A graph R is called a **triangle-tree** if

  (i) R is connected and every cycle of R, if there are any, has order 3.
  (ii) No two cycles have a vertex in common.

In particular, a tree and a C_3 are special triangle-trees.

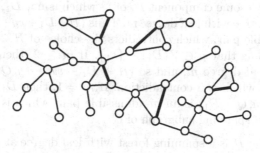

Fig. 8.21. A triangle-tree R with leaf degree one such that every non-leaf vertex x with $leafd_R(x) = 0$ is contained in a cycle.

A pair (H, Q) of two subgraphs of G is called m-**admissible** if the following five conditions hold.

  (A1) $V(H) \cap V(Q) = \emptyset$ and $V(H) \cup V(Q) = V(G)$.
  (A2) Each component of Q is a star $K(1, m+1)$.
  (A3) Each component of H has order at least two.
  (A4) If $m \geq 2$, then every component of H is a tree with leaf degree at most m.
  (A5) If $m = 1$, then every component of H is a triangle-tree with leaf degree at most one such that every non-leaf vertex x with $leafd_H(x) = 0$ is contained in a cycle of H (Fig. 8.21).

We first show the existence of an m-admissible pair. Since

$$iso(G - S) < (m + 1) \qquad \text{for all} \quad \emptyset \neq S \subset V(G),$$

G has a $\{K(1,1), K(1,2), \ldots, K(1, m+1)\}$-factor F by Theorem 7.6. Let Q be the set of components isomorphic to $K(1, m+1)$. Then $(F - Q, Q)$ is an m-admissible pair.

Choose an m-admissible pair (H, Q) so that $|V(H)|$ is as large as possible. We shall show that H is a spanning subgraph of G, that is, $Q = \emptyset$. Assume

$Q \neq \emptyset$. Let $D_1, D_2, \ldots, D_k$ be the components of Q. Then every D_i is isomorphic to $K(1, m + 1)$. Let s_i be the center of D_i for every i, and let $S = \{s_1, s_2, \ldots, s_k\}$. Then by (8.15), we have

$$iso(Q - S) = (m + 1)k \qquad \text{and} \qquad iso(G - S) < (m + 1)k.$$

Hence there exists an edge xy in G joining a vertex $x \in V(Q) - S$ to a vertex y of $H \cup (Q - S)$. We may assume $x \in D_1$. Then $x \neq s_1 \in S$.

If y belongs to some component X of H, then $X + xy + D_1$ is a tree with leaf degree at most m (when $m \geq 2$) or a triangle-tree with leaf degree at most one (when $m = 1$). Thus $(H + xy + D_1, Q - D_1)$ is an m-admissible pair, which contradicts the maximality of $|V(H)|$.

If y belongs to some component D_i of Q which is not D_1, then $y \neq s_i$ and $D_1 + xy + D_i$ is a tree with leaf degree m. Thus $(H + D_1 + xy + D_i, Q - D_1 - D_i)$ is an m-admissible pair, which contradicts the choice of H.

Finally, assume that $y \in V(D_1) - \{s_1\}$. If $m \geq 2$, then $D_1 - s_1x + xy$ is a tree with leaf degree m, and so $(H + D_1 - sx + xy, Q - D_1)$ is an m-admissible pair, which is a contradiction. If $m = 1$, then $D_1 + xy$ is C_3, and so $(H + D_1 + xy, Q - D_1)$ is an m-admissible pair, which is a contradiction. Therefore H is a spanning subgraph of G.

If $m \geq 2$, then H is a spanning forest with leaf degree at most m. We add some edges of G to H so that the resulting subgraph H^* is a spanning tree of G. It is clear that H^* is the desired spanning tree with leaf degree at most m. If $m = 1$, then by adding some edges of G to H, we obtain a connected spanning subgraph H'. Thus by removing one suitable edge from each cycle of H', we obtain the desired spanning tree of G with leaf degree one. For example, by removing bold edges from the triangle-tree R in Fig. 8.21, we obtain a tree with leaf degree one.   □

The next theorem shows that a graph having a certain property on isolated vertices contains a spanning tree with constrained leaf distance. Though we omit its proof, a proof technique using β, (which can be seen, for example, in the proof of Theorem 7.6), is adapted to it.

Theorem 8.35 (Kaneko, Kano and Suzuki [112]). *Let G be a connected simple graph of order at least five. If*

$$iso(G - S) < |S| \qquad \text{for all} \quad \emptyset \neq S \subseteq V(G), \qquad (8.16)$$

then G has a spanning tree with leaf distance at least four (Fig. 8.20).

For a spanning tree with constrained leaf distance, Kaneko made the following conjecture. Notice that the conjecture is true for $d = 3$ and 4 by Theorem 8.34 with $m = 1$ and Theorem 8.35, but other cases are unsolved.

Conjecture 8.36 (Kaneko [110]). Let $d \geq 3$ be an integer and G be a connected simple graph of order at least $d + 1$. If

$$iso(G - S) < \frac{2|S|}{d - 2} \qquad \text{for all} \quad \emptyset \neq S \subseteq V(G),$$

then G has a spanning tree with leaf distance at least d.

The next theorem is used in the proof of Theorem 8.35, but is of interest in its own right.

Theorem 8.37 (Frank and Gyárfás [87], Kaneko and Yoshimoto [113]). *Let G be a connected bipartite simple graph with bipartition $A \cup B$, and $f : A \to \{2, 3, 4, \ldots\}$ be a function. Then G has a spanning tree T such that*

$$\deg_T(x) \geq f(x) \qquad \text{for all} \quad x \in A, \tag{8.17}$$

if and only if

$$|N_G(S)| \geq \sum_{x \in S} f(x) \ - |S| + 1 \qquad \text{for all} \quad \emptyset \neq S \subseteq A. \tag{8.18}$$

In particular, if $|N_G(S)| \geq |S| + 1$ for all $\emptyset \neq S \subseteq A$, then G has a spanning tree whose leaves are all contained in B.

Proof. The last statement follows from (8.18) by setting $f(x) = 2$ for all $x \in A$. We now prove the main part. Assume that there exists a spanning tree T satisfying (8.17). Then for any subset $S \subseteq A$, S is independent in T, and so by Lemma 8.13, $T - S$ has $\sum_{x \in S}(\deg_T(x) - 1) \ + 1$ components. Hence

$$\sum_{x \in S} f(x) \ - |S| + 1$$
$$\leq \sum_{x \in S} \deg_T(x) - |S| + 1$$
$$= \omega(T - S)$$
$$= \text{the number of edges of } T \text{ joining } S$$
$$\text{to the components of } T - S$$
$$\leq |N_T(S)| \leq |N_G(S)|.$$

Hence the necessity is proved.

We now prove the sufficiency. By the generalized marriage theorem (Theorem 2.10) and by

$$N_G(S) \geq \sum_{x \in S}(f(x) - 1) \qquad \text{for all} \quad \emptyset \neq S \subseteq A,$$

G has a factor F such that for every vertex x of A, F contains a star component with center x and $\deg_F(x) = f(x) - 1$. In particular, the leaves of F is contained in B.

We now show that by adding some edges of G to F, we can obtain a forest H such that $\deg_H(x) = f(x)$ for all $x \in A$. If such a forest exists, then by adding suitable edges of G to H so that the resulting subgraph is connected, we can obtain the desired spanning tree.

We prove the existence of the above forest H by induction on the number of vertices x with degree $f(x) - 1$. Assume that R is a forest that is obtained from F by adding some edges and satisfies $\deg_R(x) \leq f(x)$ for all $x \in A$. Let

$$S = \{x \in A : \deg_R(x) = f(x) - 1\}.$$

Then by (8.18),

$$\sum_{x \in S}(f(x) - 1) < |N_G(S)|,$$

which implies that there exists an edge e that joins a vertex in S to a vertex of $V(R) \cap B$ contained in the other component of R. Hence $R + e$ is also a forest and the number of vertices x of degree $f(x) - 1$ in $R + e$ is less than that of R. Therefore, we can obtain the desired forest by induction, and the proof is complete. □

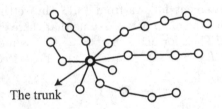

The trunk

Fig. 8.22. A spider with trunk.

Finally, we consider spanning trees called spiders. In a tree T, a vertex with degree at least three is called a **branch**. A tree having at most one branch is called a **spider**. Namely, a spider is a path or a tree having precisely one branch (see Fig. 8.22). The unique branch of a spider, if there are any, is called its **trunk**.

Theorem 8.38 (Gargano, Hammar, Hell, Stacho and Vaccaro [94]). *If a connected simple graph G satisfies*

$$\delta(G) \geq \frac{|G| - 1}{3},$$

then G has a spanning spider.

Proof. If G has a Hamiltonian path, we are done, and so we may assume that G does not have a Hamiltonian path. Let

$$P = (v_0, v_1, \ldots, v_t)$$

be a longest path in G. We assign an orientation in P from v_0 to v_t, and for a vertex v_i, its successor v_i^+ and the predecessor v_i^- are well-defined, if they exist. Since P is a longest path, neither v_0 nor v_t is adjacent to $V(G) - V(P)$. For a vertex $y \in V(G)$, let

$$N_P(y) = N_G(y) \cap V(P),$$

and if $v_t \notin N_P(y)$ or $v_0 \notin N_P(y)$, then denote

$$N_P^+(y) = \{v_i^+ | v_i \in N_P(y)\} \quad \text{and} \quad N_P^-(y) = \{v_i^- | v_i \in N_P(y)\}.$$

Claim 1. *For the longest path P, the following statements hold.*
(i) $N_G(v_0) \cup N_G(v_t) \subseteq V(P)$, in particular, $N_G(v_0) = N_P(v_0)$ and $N_G(v_t) = N_P(v_t)$.
(ii) G has no cycle Q with $V(Q) = V(P)$.
(iii) $N_P(x) \cap N_P^-(v_0) = N_P(x) \cap N_P^+(v_t) = \emptyset$ for all $x \in V(G) - V(P)$.
(iv) $N_P(v_0) \cap N_P^+(v_t) = \emptyset$ and $\{v_0, v_t\} \cup (N_P^-(v_0) \cap N_P^+(v_t))$ is an independent set of G.
(v) $N_P^-(x)$ is an independent set of G for every $x \in V(G) - V(P)$.
(vi) If $N_P^-(v_0) \cap N_P^+(v_t) \neq \emptyset$, then
    (a) no two consecutive vertices of P both have neighbors in $V(G) - V(P)$.
    (b) $V(G) - V(P)$ is an independent set of G.

Since P is a longest path of G, (i), (ii) and (iii) follow immediately (Fig. 8.23 (1) for (iii)). We now prove (iv). If $N_G(v_0) \cap N_G^+(v_t) \neq \emptyset$, then G has a cycle Q with $|Q| = |P|$, which contradicts (ii) (Fig. 8.23 (2)). Hence $N_G(v_0) \cap N_G^+(v_t) = \emptyset$. By symmetry, $N_G(v_t) \cap N_G^-(v_0) = \emptyset$. If two vertices $v_i, v_j \in N_G^-(v_0) \cap N_G^+(v_t)$ are adjacent in G, then G has a cycle Q with $V(Q) = V(P)$ (Fig. 8.23 (3)). Hence $\{v_0, v_t\} \cup (N_G^-(v_0) \cap N_G^+(v_t))$ is an independent set, and thus (iv) holds.

If two vertices v_i and v_j in $N_P^-(x)$ are adjacent in G, we can find a longer path than P (Fig. 8.23 (4)), a contradiction. Hence (v) holds.

Let $v_k \in N^-(v_0) \cap N^+(v_t)$ (Fig. 8.23 (5)). If two consecutive vertices of P both have neighbors in $V(G) - V(P)$, then G has a longer path than P or a cycle Q with $|Q| = |P|$, each of which does not pass through v_k (see Fig. 8.23 (5) with v_a, v_{a+1} or v_b, v_{b+1}). This is a contradiction. Hence (a) of (vi) holds. If two vertices of $V(G) - V(P)$ are adjacent, then we can find a path R that passes these two vertices and terminates at a vertex of P. Then we can find a path with length $|P| + |R| - 1$, which does not pass through v_k (Fig. 8.23 (5)). This is a contradiction, and thus (vi) is proved.

Claim 2. *If $N_P^-(v_0) \cap N_P^+(v_t) = \emptyset$, then G has a spanning spider.*

Assume $N_P^-(v_0) \cap N_P^+(v_t) = \emptyset$. Let x be any vertex in $V(G) - V(P)$. By Claim 1 (iii) and $N_P^-(v_0) \cap N_P^+(v_t) = \emptyset$, it follows that $N_P(x)$, $N_P^-(v_0)$ and $N_P^+(v_t)$ are disjoint sets of $V(P)$. Thus we have

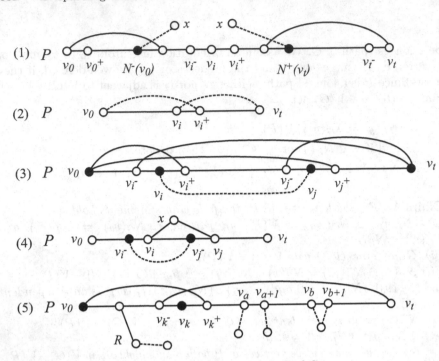

Fig. 8.23. A longest path P of G.

$$\deg_G(x) + \deg_G(v_0) + \deg_G(v_t)$$
$$\leq |N_G(x) \setminus (V(P) \cup \{x\})| + |N_P(x)| + |N_P^-(v_0)| + |N_P^+(v_t)|$$
$$\leq |V(G)| - |V(P)| - 1 + |V(P)| = |G| - 1.$$

Since $\deg_G(x) + \deg_G(v_0) + \deg_G(v_t) \geq 3\delta(G) \geq |G| - 1$ by the condition of the theorem, we obtain

$$V(P) = N_P(x) \cup N_G^-(v_0) \cup N_G^+(v_t). \qquad \text{(disjoint union)}$$

Hence x is adjacent to all the vertices in $V(P) - (N_G^-(v_0) \cup N_G^+(v_t))$, which is not an empty set. Therefore for a vertex $v_k \in V(P) - (N_G^-(v_0) \cup N_G^+(v_t))$, it follows that $N_G(v_k) \supseteq V(G) - V(P)$. Thus we can obtain a spanning spider with trunk v_k, and Claim 2 is proved.

By Claim 2, we may assume that

$$N_P^-(v_0) \cap N_P^+(v_t) \neq \emptyset \quad \text{and Claim 1 (vi) holds.}$$

Claim 3. *If $|V(G) - V(P)| \leq 2$, then G has a spanning spider.*

If $|V(G) - V(P)| = 1$, then G has a spanning spider immediately. Assume $|V(G) - V(P)| = 2$, and let $V(G) - V(P) = \{x_1, x_2\}$. If $N_P(x_1) \cap N_P(x_2) \neq \emptyset$, then clearly G has a spanning spider, and so we may assume $N_P(x_1) \cap N_P(x_2) = \emptyset$. By Claim 1 (vi), we have $N_P(x_1) \cap N_{\bar{P}}(x_2) = \emptyset$, and so

$$N_P(x_1) \subseteq V(P) - (N_P(x_2) \cup N_{\bar{P}}(x_2)).$$

By $N_P(x_2) \cap N_{\bar{P}}(x_2) = \emptyset$, and since x_1 and x_2 are not adjacent by Claim 1 (vi), we obtain

$$
\begin{aligned}
\deg_G(x_1) &= |N_P(x_1)| \\
&\leq |P| - |N_P(x_2)| - |N_{\bar{P}}(x_2)| \\
&\leq |G| - 2 - 2\delta(G) \leq |G| - 2 - \frac{2(|G| - 1)}{3} \\
&= \frac{|G| - 4}{3}.
\end{aligned}
$$

This contradicts $\delta(G) \geq (|G| - 1)/3$. Hence Claim 3 is proved.

By Claim 3, we may assume $|V(G) - V(P)| \geq 3$. Let

$$X = V(G) - V(P), \qquad \text{where} \quad |X| \geq 3.$$

Claim 4. *For each vertex $x \in X$, $N_{\bar{P}}(x) \cup X$ is an independent set of G, and $V(P) - N_{\bar{P}}(x)$ contains at least one vertex v^* that satisfies*

$$|N_G(v^*) \cap (N_{\bar{P}}(x) \cup X)| \geq \frac{|G| - 1}{6} + \frac{3|X| - 1}{4}.$$

By Claim 1 (vi), X is independent and no edge of G joins X to $N_{\bar{P}}(x)$. By Claim 1 (v), $N_{\bar{P}}(x)$ is independent, and so $N_{\bar{P}}(x) \cup X$ is an independent set.

Let $|G| = n$. Take a vertex $x \in X$, and enumerate the number of edges joining $N_{\bar{P}}(x) \cup X$ and $V(P) - N_{\bar{P}}(x)$. Then the number of those edges is at least

$$\delta(G)|N_{\bar{P}}(x) \cup X| \geq \frac{n-1}{3} \cdot |N_{\bar{P}}(x) \cup X|.$$

Hence there exists a vertex $v^* \in V(P) - N_{\bar{P}}(x)$ that is adjacent to at least the following number of vertices in $N_{\bar{P}}(x) \cup X$.

$$\frac{\frac{n-1}{3}\cdot|N_P^-(x)\cup X|}{|V(P)-N_P^-(x)|}=\frac{\frac{n-1}{3}(|N_P^-(x)|+|X|)}{n-|X|-|N_P^-(x)|}$$

$$\geq\frac{\frac{n-1}{3}(\frac{n-1}{3}+|X|)}{n-|X|-\frac{n-1}{3}}\qquad\left(\text{by }|N_P^-(x)|\geq\delta(G)\geq\frac{n-1}{3}\right)$$

$$\geq\frac{(n-1)(n-1+3|X|)}{6n-9|X|+3}$$

$$\geq\frac{(n-1)(n-\frac{3}{2}|X|+\frac{1}{2}+\frac{9|X|-3}{2})}{6(n-\frac{3}{2}|X|+\frac{1}{2})}$$

$$\geq\frac{n-1}{6}+\frac{3|X|-1}{4}.$$

$$\left(\text{by }(n-1)(9|X|-3)\geq 3(n-\tfrac{3}{2}|X|+\tfrac{1}{2})(3|X|-1)\right)$$

Let x^* be a vertex of X, and v^* a vertex in $V(P)-N_P^-(x^*)$ obtained in Claim 4. We construct a spider S, which is a subgraph of G, from P as follows:

(i) Add all the edges joining v^* to $N_G(v^*)\cap X$.
(ii) If v_{i-1} is adjacent to v^* and v_i is adjacent to x^*, then remove the edge $v_{i-1}v_i$ and add $v_{i-1}v^*$.
(iii) If v^* is adjacent to a vertex v_j lying between v^* and v_t, then remove v^*v^{*+} (Fig. 8.24). Note that v^{*+} is not adjacent to x^* by the choice of v^*.

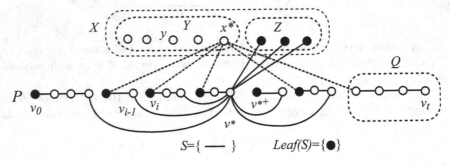

Fig. 8.24. Assume here x^* and v^* are not adjacent. A spider S with trunk v^*, which does not include $Q\cup Y$.

Let

$$Z=X\cap N_G(v^*)\qquad\text{and}\qquad Y=X-Z,$$

and let $Leaf(S)$ denote the set of leaves of S (Fig. 8.24).

Claim 5. *There exists a matching between Y and $Leaf(S)\cap V(P)$ that covers all Y.*

Fig. 8.24 illustrates the case where x^* and v^* are not adjacent. We will also make this assumption here. As we explain below, the other case can be proved similarly.

Let Q denote the tail of P that is not included in the spider S (Fig. 8.24). Let

$$\Delta = |N_G(v^*) \cap (N_{\bar{P}}(x^*) \cup X)| \geq \frac{n-1}{6} + \frac{3|X|-1}{4}. \quad \text{(by Claim 4)} \quad (8.19)$$

The number of leaves in S is $\Delta + 1$, and since X is independent by Claim 1, and since x^* is not adjacent to v_0 or v^{*+}, we have

$$\deg_{Leaf(S)}(x^*) = |N_G(x^*) \cap Leaf(S)| \geq \Delta - |Z| - 2, \quad (8.20)$$

where $\deg_{Leaf(S)}(x^*)$ is defined by the above equation. Since $|P| \geq 2\delta(G)+1 \geq (2n+1)/3$, $|X| = |G| - |P| \leq (n-1)/3$ and

$$\Delta - |Z| - 2 \geq |Y| \quad \Leftrightarrow \quad \Delta \geq |X| + 2 \quad \Leftarrow \quad \frac{2n}{3} - \frac{29}{3} \geq |X|, \quad \text{(by (8.19))}$$

we have $\Delta - |Z| - 2 \geq |Y|$. Since x^* is adjacent to $(Leaf(S) \cap V(P)) - \{v_0, v^{*+}\}$, which contains at least $|Y|$ vertices, it suffices to show that there exists a matching between $Y - x^*$ and $Leaf(S) \cap V(P)$.

Let $Int(S) = (V(S) - Leaf(S)) \cup V(Q)$, which is the set of vertices of P without leaves of S. Let $y \in Y - x^*$. Then

$$\deg_G(y) = |N_P(y) \cap Int(S)| + |N_P(y) \cap Leaf(S)|$$
$$= \deg_{Int(S)}(y) + \deg_{Leaf(S)}(y).$$

There is no edge between $\{x^*, y\}$ and $\{v_0, v_t\}$, and for any two consecutive vertices v_j, v_{j+1} of $V(P) - \{v_0, v_t\}$, there exists at most one edge between $\{x^*, y\}$ and $\{v_j, v_{j+1}\}$ by Claim 1 (vi). Thus

$$\deg_{Int(S)}(y) + \deg_{Leaf(S)}(x^*) \leq \frac{|P| - 2}{2}. \quad (8.21)$$

Hence

$$\deg_{Leaf(S)}(y) = \deg_G(y) - \deg_{Int(S)}(y)$$
$$\geq \deg_G(y) + \deg_{Leaf(S)}(x^*) - (|P| - 2)/2.$$

Since $\deg_{Leaf(S)}(x^*) \geq \Delta - |Z| - 2$ by (8.20) and $\delta(G) \geq (n-1)/3$, we obtain

$$\deg_{Leaf(S)}(y) \geq \frac{n-1}{3} + (\Delta - |Z| - 2) - \frac{n - |X| - 2}{2}.$$

By using (8.19), we obtain

$$\deg_{Leaf(S)}(y) \geq \frac{n-1}{3} + \frac{n-1}{6} + \frac{3|X|-1}{4} - |Z| - 2 - \frac{n-|X|-2}{2}$$
$$= |X| - |Z| + (|X|-7)/4$$
$$\geq |Y| - 1. \qquad \text{(by } |X| \geq 3\text{)}$$

Therefore, each vertex of $Y - x^*$ is adjacent to at least $|Y|-1$ leaves of $S \cap V(P)$, and thus there exists a matching between $Y - x^*$ and $Leaf(S) \cap V(P)$ covering $Y - x^*$. Therefore Claim 5 holds.

It is easy to construct a spanning spider of G from S by adding a matching between Y and $Leasf(S) \cap V(P)$ and by adding an edge joining x^* to Q and one more edge joining x^* to $Leaf(S) \cap V(P)$. Consequently the proof is complete. □

In [94], the authors consider a maximal path possessing the properties given in Claim 1, which plays a similar role as a longest path in G, and propose a polynomial time algorithm for finding a spanning spider in a connected graph G with $\delta(G) \geq (|G| - 1)/3$. Moreover, they show that a connected graph G with $\sigma_3(G) \geq |G| - 1$ has a spanning spider, though its proof is difficult to follow. It is not known whether the condition of Theorem 8.33 is sharp or not. In particular, Yamashita [251] gives the following problem.

Problem 8.39. Is it true that a connected simple graph G with $\delta(G) \geq (|G| - 1)/4$ has a spanning spider? If it is true, then the condition $\delta(G) \geq (|G| - 1)/4$ is sharp. If it is not true, find a minimum $\delta(G)$ that guarantees the existence of a spanning spider in G.

The interested reader is referred to a current survey [208] by Ozeki and Yamashita for more results on spanning trees.

A spanning tree can be generalized to an n-edge connected (g, f)-factor, which is an n-edge connected spanning subgraph F satisfying $g(x) \leq \deg_F(x) \leq f(x)$ for all $x \in V(F)$, and much research has been done on this topic. We refer the reader to the survey [161] by Kouider and Vestergaard and the book [254] by Yu and Liu. We conclude this section with the following theorem, which is one of the strongest results on this topic and guarantees the existence of many such connected factors in a graph G with $\sigma_2(G) \geq |G|$.

Theorem 8.40 (Egawa [65]). *Let $k \geq 2$ be an integer and G be a connected simple graph. If $|G| \geq 44(k - 1)$,*

$$\delta(G) \geq 4k - 2 \qquad and \qquad \sigma_2(G) \geq |G|,$$

then G has k edge-disjoint Hamiltonian cycles. Moreover, if $|G| \geq 8(2k - 2)^2$,

$$\delta(G) \geq 2k + 1 \qquad and \qquad \sigma_2(G) \geq |G|,$$

then G has k edge-disjoint Hamiltonian cycles.

References

1. J. Akiyama, Three Developing Topics in Graph Theory, (1980) unpublished book.
2. J. Akiyama, Factorizations and linear arboricity of graphs, Doctoral thesis, Tokyo University of Science (1982).
3. J. Akiyama, D. Avis and H. Era, On a $\{1,2\}$-factor of a graph, *TRU Math.* **16** (1980) 97–102.
4. J. Akiyama and V. Chvátal, A short proof of the linear arboricity for cubic graphs, *Bull Liber. Arts & Sci. NMS (Nippon Medical School)* **2** (1981) 1–3.
5. J. Akiyama, G. Exoo and F. Harary, Covering and packing in graphs III: Cyclic and acyclic invariants, *Math. Slovaca* **30** (1980) 405–417.
6. J. Akiyama and P. Frankl, On the size of graphs with complete factors, *J. Graph Theory* **9** (1985) 197–201.
7. J. Akiyama and M. Kano, Factors and factorizations of graphs — a survey, *J. Graph Theory* **9** (1985) 1–42.
8. J. Akiyama and M. Kano, Almost-regular factorization of graphs, *J. Graph Theory* **9** (1985) 123–128
9. J. Akiyama and M. Kano, Path factors of a graph, *Graphs and applications (Boulder, Colo., 1982)*, (1985) 1–21, Wiley-Intersci. Publ., Wiley, New York, 1985.
10. J. Akiyama and M. Kano, 1-factors of triangle graphs, *Number theory and combinatorics, Japan 1984 (Tokyo, Okayama and Kyoto, 1984)* (1985) 21–35, World Sci. Publishing, Singapore.
11. N. Alon, C. McDiarmid and B. Reed, Star arboricity, *Combinatorica* **12** (1992) 375–380.
12. N. Alon, The linear arboricity of graphs, *Israel J. Combin.* **62** (1988) 311–325.
13. N. Alon and R. Yuster, H-factors in dense graphs, *J. Combin. Theory Ser. B* **66** (1996) 260–282.
14. A. Amahashi, On factors with all degrees odd, *Graphs Combin.* **1** (1985) 1–6.
15. A. Amahashi and M. Kano, On factors with given components, *Discrete math.*, **42** (1983) 1–6.
16. I. Anderson, Perfect matchings of a graph, *J. Combin. Theory Ser. B* **10** (1971) 183–186.
17. I. Anderson, Sufficient conditions for matching, *Proc. Edinburgh Math. Soc.* **18** (1973) 129–136.

J. Akiyama and M. Kano, *Factors and Factorizations of Graphs*,
Lecture Notes in Mathematics 2031, DOI 10.1007/978-3-642-21919-1,
© Springer-Verlag Berlin Heidelberg 2011

18. K. Ando, Y. Egawa, A. Kaneko, K. Kawarabayashi and H. Matsuda, Path factors in claw-free graphs, *Discrete Math.* **243** (2002) 195–200.

19. R.P. Anstee, Simplified existence theorems for (g, f)-factors, *Discrete Appl. Math.* **27** (1990) 29–38.

20. R.P. Anstee, Matching theory: fractional to integral, *New Zealand J. Math.* bf 21 (1992) 17–32.

21. K. Appel, W. Haken and J. Koch, Every planar map is four-colorable, *Illinois J. Math.* **21** (1977) 429–567.

22. F. Bäbler, Über die Zerlegung regulärer Streckenkomplexe ungerader Ordnung, *Comment. Math. Helvetici* **10** (1938) 275-287.

23. D. Bauer, H. J. Broersma and H.J. Veldman, Not every 2-tough graph is hamiltonian, *Discrete Appl. Math.* **99** (2000) 317–321.

24. H.B. Belck, Reguläre Faktoren von Graphen, *J. Reine Angew. Math.* **188** (1950) 228–252.

25. C. Berge, Two theorems in graph theory, *Proc. Nat. Acad. Sci. U.S.A.* **43** (1957) 842–844.

26. C. Berge, Sur le couplage maximum d'un graphe, *C.R. Acad. Sci. Paris Ser.I Math.* **247**, (1958) 258-259.

27. C. Berge, *Graphs and Hypergraphs*, North-Holland, (1973)

28. C. Berge, Regulariable graphs, *Ann. Discrete Math.* **3** (1978) 11–19.

29. J.C. Bermond and M. Las Vergnas, Regular factors in nearly regular graphs, *Discrete Math.* **50** (1984) 9–13.

30. E. Bertram and P. Horák, Decomposing 4-regular graphs into triangle-free 2-factors, *SIAM J. Discrete Math.* **10** (1997) 309–317.

31. N.L. Biggs, E.K. Lloyd and R.J. Wilson, *Graph Theory 1736–1936*, Oxford University Press, Oxford (1976)

32. B. Bollobás, *Extremal Graph Theory*, Academic Press, (1978).

33. B. Bollobás, A. Saito and N.C. Wormald, Regular factors of regular graphs, *J. Graph Theory* **9** (1985) 97–103.

34. J.A. Bondy and U.S.R. Murty, *Graph Theory with Applications*, The MacMillian Press (1976).

35. H.J. Broersma, J. van den Heuvel and H.J. Veldman, A generalization of Ore's theorem involving neighborhood unions, *Discrete Math.* **122** (1993) 37–49.

36. H.J. Broersma, M. Kriesell and Z. Ryjáček, On factors of 4-connected claw-free graphs, *J. Graph Theory* **37** (2001) 125–136.

37. H. Broersma and H. Tuinstra, Independence trees and Hamilton cycles, *J. Graph Theory* **29** (1998) 227–237.

38. Maocheng Cai, $[a, b]$-factorizations of graphs, *J. Graph Theory* **15** (1991) 283–301.

39. Maocheng Cai, On some factor theorems of graphs, *Discrete Math.* **98** (1991) 223-229.

40. Maocheng Cai, Yanjun Li and M. Kano, A $[k, k + 1]$-factor containing a given Hamiltonian cycle, *Electron. J. Combin.* **6** (1999) R4.

41. K. Cameron, Induced matchings, *Discrete App. Math.* **24** (1989) 97–102.

42. Y. Caro, I. Krasikov and Y. Roditty, Spanning trees and some edge reconstructible graphs, *Ars Combin.* **20** (1985) A 109–118.

43. G. Chartrand and L. Lesniak, *Graphs & Digraphs* (third edition), Chapman & Hall, London, 1996.

44. Boliong Chen, M. Matsumoto, Jianfang Wang, Zhongfu Zhang, Jianxun Zhang, A short proof of Nash-Williams' theorem for the arboricity of a graph. *Graphs Combin.* **10** (1994) 27–28.

45. Ciping Chen, A note on matching extension (Chinese), *J. Beijing Agri. Engin. Univ.* **10** (1990) 293–297.

46. Ciping Chen, Binding number and minimum degree for [a, b]-factors, *J. Sys. Sci. and Math. Scis.* **6** (1993) 179–185.

47. Ciping Chen, Binding number and toughness for matching extension, *Discrete Math.* **146** (1995) 303–306.

48. Ciping Chen, Y. Egawa and M. Kano, Star factors with given properties, *Ars Combin.* **28** (1989) 65–70.

49. Ciping Chen and Guizhen Liu, Toughness of Graphs and [a, b]-factors with prescribed properties, *J. Combin. Math. Combin. Comput.* **12** (1992) 215–221.

50. Ciping Chen and Jianfang Wang, Factors in graphs with odd-cycle property, *Discrete Math.* **112** (1993) 29–40.

51. William Y.C. Chen, Maximum (g, f)-factors of a general graph, *Discrete Math.* **91** (1991) 1–7.

52. A.G. Chetwynd and A.J.W. Hilton, Star multigraphs with three vertices of maximum degree, *Math. Proc. Cambridge Philos. Soc.* **100** (1986) 303–317.

53. S.A. Choudum and M.S. Paulraj, Regular factors in $K_{1,3}$-free graphs, *J. Graph Theory* **15** (1991), 259–265.

54. V. Chvátal, Tough graphs and Hamiltonian circuits, *Discrete Math.* **5** (1973) 215–228.

55. V. Chvátal and P. Erdös, A note on hamiltonian circuits, *Discrete Math.* **2** (1972) 111–113.

56. C. Colbourn and J. Dinitz (Editors), *Handbook of Combinatorial Designs*, Discrete Mathematics and its Applications (Boca Raton), CRC Press, 2006.

57. G. Cornuéjols, General factors of graphs, *J. Combin. Theory Ser. B* **45** (1988) 185–198.

58. G. Cornuéjols and W. Pulleyblank, Perfect triangle-free 2-matchings, *Combinatorial optimization, II (Proc. Conf., Univ. East Anglia, Norwich, 1979). Math. Programming Stud.* **13** (1980), 1–7.

59. G. Cornuéjols and W. Pulleyblank, Critical graphs, matchings and tours or a hierarchy of relaxations for the travelling salesman problem, *Combinatorica* **3** (1983) 35–52.

60. Yuting Cui and M. Kano, Some results on odd factors of graphs, *J. Graph Theory* **12** (1988) 327–333.

61. G.A. Dirac, In abstrakten Graphen vorhandene vollständige 4-Graphen und ihre Unterteilungen, *Math. Nachr.* **22** (1960) 61–85.

62. J. Edmonds, Paths, trees and flowers, *Canada. J. Math.* **17** (1965) 449–467.

63. J. Edmonds and D.R. Fulkerson, Transversals and matroid partition, *J. Res. Nat. Bur. Standards Sect. B* **69B** (1965) 147–153.

64. Y. Egawa, Era's conjecture on [k, k + 1]-factorizations of regular graphs, *Ars Combin.* **21** (1986) 217–220.

65. Y. Egawa, Edge-disjoint Hamiltonian cycles in graphs of Ore type, *SUT J. Math.* **29** (1993) 15–50.

66. Y. Egawa and H. Enomoto, Sufficient conditions for the existence of k-factors, *Recent Studies in Graph Theory* (V.R.Kull, Ed) (1989) 97–103.

67. Y. Egawa, H. Enomoto and A. Saito, On component factors, *Graphs Combin.* **2** (1986) 223–225.

68. Y. Egawa, H. Enomoto and A. Saito, Factors and induced subgraphs, *Discrete Math.* **68** (1988) 179–189.

69. Y. Egawa, S. Fujita and K. Ota, $K_{1,3}$-factors in graphs, *Discrete Math.* **308** (2008) 5965–5973.

70. Y. Egawa and M. Kano, Sufficient conditions for graphs to have (g, f)-factors, *Discrete Math.* **151** (1996) 87–90.

71. Y. Egawa, M. Kano and A. Kelmans, Star partitions of graphs, *J. Graph Theory* **25** (1997) 185–190.

72. Y. Egawa, H. Matsuda, T. Yamashita and K. Yoshimoto, On a spanning tree with sepcified leaves, *Graphs Combin.* **24** (2008), pp. 13–18.

73. Y. Egawa and K. Ota, Regular factors in $K_{1,n}$-free graphs, *J. Graph Theory* **15** (1991) 337–344.

74. E. Egerváry, On combinatorial peoperties of matrices (Hungarian), *Mat. Lapok* **38** (1931) 16–28.

75. M. N. Ellingham, Y. Nam and H-J. Voss, Connected (g, f)-factors, *J. Graph Theory* **39** (2002) 62–75.

76. M.N. Ellingham and Xiaoya Zha, Toughness, Trees and Walks, *J. Graph Theory* **33** (2000) 125–137.

77. H. Enomoto and M. Hagita, Toughness and the existence of k-factors IV, *Discrete Math.* **216** (2000) 111–120.

78. H. Enomoto, B. Jackson, P. Katerinis and A. Saito, Toughness and the existence of k-factors, *J. Graph Theory* **9** (1985) 87–95.

79. H. Enomoto, A. Kaneko and Zs. Tuza, P_3-factors and covering cycles in graphs of minimum degree $n/3$, *Combinatorics (Eger, 1987)*, 213–220, Colloq. Math. Soc. János Bolyai, 52, North-Holland, Amsterdam, (1988).

80. H. Enomoto, K. Ota and M. Kano, A sufficient condition for a bipartite graph to have a k-factor, *J. Graph Theory* **12** (1988) 141–151.

81. H. Era, A note on semi-regular factorizations of regular graphs, *Proc. Fac. Sci. Tokai Univ.* **19** (1984) 7–13.

82. H. Era, Semiregular factorizations of regular graphs, *Graphs and applications (Boulder, Colo., 1982)*, 101–116, Wiley-Intersci. Publ., (1985).

83. G. Fan and A. Raspaud, Fulkersonfs conjecture and circuit covers. *J. Comb. Theory Ser. B*, **61** (1994) 133–138.

84. O. Favaron, On k-factor-critical graphs, *Discuss. Math. Graph Theory* **16** (1996) 41–51.

85. E. Flandrin, H. A. Jung and H. Li, Hamiltonism, degree sum and neighborhood intersections, *Discrete Math.* **90** (1991), 41–52.

86. J. Folkman and D.R. Fulkerson, Flows in infinite graghs, *J. Combin. Theory* **8** (1970) 30–44.

87. A. Frank and E. Gyárfás, How to orient the edges of a graph?, *Colloq. Math. Soc. Janos Bolyai* **18** (1976) 353–364.

88. S. Fujita, Vertex-disjoint $K_{1,t}$'s in graphs, *Ars Combin.* **64** (2002) 211–223.

89. D.R. Fulkerson, Blocking and anti-blocking pairs of polyhedra, *Math. Programming* bf 1 (1971) 168–194.

90. D.R. Fulkerson, A.J. Hoffman and M.H. McAndrew, Some properties of graphs with multiple edges, *Canadian J. Math.* **17** (1965) 166–177.

91. T. Gallai, On factorisation of graphs, *Acta Math. Acad. Sci. Hung* **1** (1950), 133–153.

92. T. Gallai, Neuer Beweis eines Tutte'schen Satzes, *Magyar Tud. Akad. Mat. Kutató Int. Közl* **8** (1963) 135–139.

93. T. Gallai, König Dénes (1884–1944), *Mat. Lapok* **15** (1964) 277–293 (Hungarian).

94. L. Gargano, M. Hammar, P. Hell, L. Stacho, and U. Vaccaro, Spanning spiders and light-splitting switches, *Discrete Math.* **285** (2004) 83–95.

95. M.A. Gurgel and Y. Wakabayashi, On k-leaf-connected graphs, *J. Combin. Theory Ser. B* **41** (1986) 1–16.

96. A. Hajnal and E. Szemerédi, Proof of a conjecture of P. Erdős, *Combinatorial theory and its applications, II (Proc. Colloq., Balatonfüred, 1969)*, pp. 601–623. North-Holland, Amsterdam, 1970.

97. P.R. Halmos and H.E. Vaughan, The marriage problem, *Amer. J. Math.* **72** (1950) 214–215.

98. P. Hall, On representatives of subsets, *J. London Math. Soc.* **10** (1935) 26–30.

99. K. Heinrich, P. Hell, D. Kirkpatrick and Guizhen Liu, A simple existence ciriterion for $(g < f)$-factor, *Discrete Math.* **85**(1990) 313–317.

100. P. Hell and D.G. Kirkpatrick, Packings by cliques and by finite families of graphs, *Discrete Math.* **49** (1984) 45–59.

101. G.R.T. Hendry, Maximum graphs with a unique k-factor, *J. Combin. Theory Ser. B* **37** (1984) 53–63.

102. G. Hetyei [see L. Lovász, On the structure of factorizable graphs, *Acta Math. Acad. Sci. Hungar.* **23** (1972) 179–195; ibid. 23 (1972), 465–478; (Corollary 1.6)]

103. G. Hetyei, A new proof of a factorization theorem, *Acta Acad. Paedagog. Civitate Pécs Ser.6 Math. Phys. Chem. Tech.* **16** (1972) 3–6 (Hungarian).

104. A. Hilton, Factorizations of regular graphs of high degree, *J. Graph Theory* **9** (1985) 193–196.

105. A. Hoffmann and L. Volkmann, On unique k-factors and unique $[1, k]$-factors in graphs, *Discrete Math.* **278** (2004) 127–138.

106. A. Hoffmann and L. Volkmann, On regular factors in regular graphs with small radius, *Electron. J. Combin.* **11** (2004) R7.

107. T. Iida and T. Nishimura, An Ore-type condition for the existence of k-factors in graphs, *Graphs Combin.* **7** (1991) 353–361.

108. A. Joentgen and L. Volkmann, Factors of locally almost regular graphs, *Bull. London Math. Soc.* **23** (1991) 121–122.

109. P. Johann, On the structure of graphs with a unique k-factor, *J. Graph Theory* **35** (2000) 227–243.

110. A. Kaneko, Spanning trees with constraints on the leaf degree, *Discrete Applied Math.*, **115** (2001) 73–76.

111. A. Kaneko, A necessary and sufficient condition for the existence of a path factor every component of which is a path of length at least two, *J. Combin. Theory Ser. B* **88** (2003) 195–218.

112. A. Kaneko, M. Kano and K. Suzuki, Spanning trees with leaf distance at least four, *J. Graph Theory*, **55** (2007) 83–90.

113. A. Kaneko and K. Yoshimoto, On spanning trees with restricted degrees, *Inform. Process. Lett.* **73** (2000) 163–165.

114. M. Kano, Graph factors with given properties, *Graph theory, Singapore 1983*, Springer Lecture Notes in Math. **1073** (1984) 161–168.

115. M. Kano, $[a, b]$-factorization of a graph, *J. Graph Theory* **9** (1985) 129–146.

116. M. Kano, $[a, b]$-factorizations of nearly bipartite graphs, *Graph Theory with Applications to Algorithms and Computer Science, Kalamazoo, Michigan, 1984*, Wiley-Interscience Publication, Wiley, New York, (1985) 471–474.

342   References

117. M. Kano, Factors of regular graphs, *J. Combin. Theory, Series B* **41** (1986) 27–36.
118. M. Kano, Sufficient conditions for a graph to have factors, *Discrete Math.* **80** (1990) 159–165.
119. M. Kano, A sufficient condition for a graph to have [a, b]-factors, *Graphs Combin.* **6** (1990) 245–251.
120. M. Kano and G.Y. Katona, Odd subgraphs and matchings, *Discrete Math.* **250** (2002) 265–272.
121. M. Kano and G.Y. Katona, Structure theorem and algorithm on (1, f)-odd subgraphs, *Discrete Math.*, **307** (2007) 1404–1417.
122. M. Kano, G.Y. Katona and Z. Király, Packing paths of length at least two, *Discrete Math.* **283** (2004) 129–135.
123. M. Kano, G.Y. Katona and J. Szabó, Elementary graphs with respect to f-parity factors, *Graphs Combin.* **25** (2009) 717-726.
124. M. Kano and H. Kishimoto, Spanning k-trees of n-connected graphs, preprint.
125. M. Kano and A. Kyaw, A note on leaf-constrained spanning trees in a graph, preprint.
126. M. Kano, A. Kyaw, H. Matsuda, K. Ozeki, A. Saito and T. Yamashita, Spanning trees with a bounded number of leaves in a claw-free graph, preprint.
127. M. Kano, C. Lee and K. Suzuki, Path factors and cycle factors of cubic bipartite graphs, *The Discussiones Mathematicae Graph Theory*, **28** (2008) 551–556.
128. M. Kano, Hongliang Lu and Qinglin Yu, Component factors with large components in graphs, *Applied Mathematics Letters*, **23** (2010) 385–389.
129. M. Kano and H. Matsuda, Some results on (1, f)-odd factors, Combinatorics, Graph Theory, and Algorithms, Vol.II (1999) 527–533. (New Issues Press; Kalamazoo, Michigan, ISBN 0-932826-74-8).
130. M. Kano and H. Matsuda, Partial parity (g, f)-factors and subgraphs covering given vertex subsets, *Graphs Combin.*, **17** (2001) 501–510.
131. M. Kano and S. Poljak, Graphs with the Balas-Uhry property, *J. Graph Theory* **14** (1990) 623–628.
132. M. Kano and A. Saito, [a, b]-factors of graphs, *Discrete Math.* **47** (1983) 113–116.
133. M. Kano and A. Saito, A short proof of Lovasz's factor theorem, *Memories of Akashi Technological College* **26** (1984) 167–170.
134. M. Kano and A. Saito, Star Factors with Large Components, preprint.
135. M. Kano and N. Tokushige, Binding numbers and f-factors of graphs, *J. Combinatorial Theory, Series B* **54** (1992) 213–221.
136. M. Kano and Qinglin Yu, Pan-fractional property in regular graphs, *Electron. J. Combin.* **12** (2005) (1) N23.
137. P. Katerinis, Some conditions for the existence of f-factors, *J. Graph Theory* **9** (1985) 513–521.
138. P. Katerinis, A Chvátal-Erdös condition for an r-factor in a graph, *Ars Combinat.* **20-B** (1985) 185–191.
139. P. Katerinis, Minimum degree of a graph and the existence of k-factors, *Proc. Indian. Acad. Sci. Math. Sci.* **94** (1985) 123–127.
140. P. Katerinis, Two sufficient conditions for a 2-factor in a bipartite graph, *J. Graph Theory* **11** (1987) 1–6.
141. P. Katerinis, Toughness of graphs and the existence of factors, *Discrete Math.* **80** (1990) 81–92.

142. P. Katerinis, Minimum degree of bipartite graphs and the existence of k-factors, *Graphs Combin.* **6** (1990) 253–258.

143. P. Katerinis, Regular factors in regular graphs, *Discrete Math.* **113** (1993) 269–297.

144. P. Katerinis, Regular factors in vertex-deleted subgraphs of regular graphs, *Discrete Math.* **131** (1994) 357–361.

145. P. Katerinis, N. Tsikopoulos, Minimum degree and F-factors in graphs, *New Zealand J. Math.* **29** (2000) 33–40.

146. P. Katerinis, N. Tsikopoulos, Independence number, connectivity and f-factors, *Utilitas Math.* **57** (2000) 81–95.

147. P. Katerinis and D.R. Woodall, Binding numbers of graphs and the existence of k-factors, *Quart. J. Math. Oxford (2)*, **38** (1987) 221–228.

148. K. Kawarabayashi, K_4^--factor in a graph, *J. Graph Theory* **39** (2002) 111–128.

149. K. Kawarabayashi, H. Matsuda, Y. Oda and K. Ota, Path factors in cubic graphs, *J. Graph Theory* **39** (2002) 188–193.

150. A. Kelmans, Optimal packing of induced stars in a graph, *RUTCOR Research Report 26-94, Rutgers University* (1994) 1–25.

151. A. Kelmans, Optimal packing of induced stars in a graph, *Discrete Math.* **173** (1997) 97–127.

152. A. Kelmans and D. Mubayi, Dhruv, How many disjoint 2-edge paths must a cubic graph have?, *J. Graph Theory* **45** (2004) 57–79.

153. D.G. Kirkpatrick and P. Hell, On the complexity of general graph factor problems, *SIAM J. Comput.* **12** (1983) 601–609.

154. S. Klinkenberg and L. Volkmann, On the order of close to regular graphs without a matching of given size, *Ars Combin.* **85** (2007) 99–106.

155. D. König, Über Graphen und ihre Anwendung auf Determinantentheorie und Mengenlehre, *Math. Ann.* **77** (1916) 453–465.

156. D. König, Graphen und Matrizen, *Math. Fiz. Lapok* **38** (1931) 116–119.

157. A. Kosowski and P. Żyliński, Packing three-vertex paths in 2-connected cubic graphs, *Ars Combin.* **89** (2008) 95–113.

158. K. Kotani, k-regular factors and semi-k-regular factors in graphs. *Discrete Math.* **186** (1998) 177–193.

159. K. Kotani, Factors and connected induced subgraphs, *Graphs Combin.* **17** (2001) 511–515.

160. M. Kouider and M. Mahéo, Connected [a, b]-factors in graphs, *Combinatorica* **22** (2002) 71–82.

161. K. Kouider and P. Vestergaard, Connected factors in graphs — A survey, *Graphs Combin.* **21** (2005) 1–26.

162. A. Kyaw, A sufficient condition for a graph to have a k-tree, *Graphs Combin.* **17** (2001) 113–121.

163. M. Las Vergnas, Sur une propriété des arbres maximaux dans un graphe, *C. R. Acad. Sci. Paris Sér. A-B*, **272** (1971) A1297–A1300.

164. M. Las Vergnas, A note on matchings in graphs, *Cahiers Centre Etudes Rech. Opér.* **17** (1975) 257–260.

165. M. Las Vergnas, An extension of Tutte's 1-factor theorem, *Discrete Math.* **23** (1978) 241–255.

166. Yanjun Li and Maocheng Cai, A degree condition for a graph to have $[a, b]$-factors, *J. Graph Theory* **27** (1998) 1–6.

167. C.H.C. Little, The parity of the number of 1-factors of a graph, *Discrete Math.*, **2** (1972) 179–181.

168. C.H.C. Little, D.D. Grant and D.A. Holton, On defect-d matchings in graphs, *Discrete Math.*, **13** (1975) 41–54.

169. Guizhen Liu, On (g, f)-covered graphs, *Acta Math. Sci.* **8** (1988) 181–184.

170. Guizhen Liu, (g, f)-factors and (g, f)-factorizations of graphs, *Acta. Mathematica Sinica* **37** (2) (1994) 230–237.

171. Guizhen Liu and Jianfang Wang, (a, b, k)-critical graphs, *Adv. Math.* **27** (1998) 536–540.

172. Guizhen Liu and Lanju Zhang, Characterizations of maximum fractional (g, f)-factors of graphs, *Discrete Appl. Math.* **156** (2008) 2293–2299.

173. Jiping Liu and Huishan Zhou, Maximum induced matchings in graphs, *Discrete Math.*, **170** (1997) 227–281.

174. L. Lovász, On graphs not containing independent circuits (in Hungarian), *Mat. Lapok* **16** (1965) 289–299.

175. L. Lovász, Subgraphs with prescribed valencies, *J. Combin. Theory.* **8** (1970) 391–416.

176. L. Lovász, The factorization of graphs II, *Acta Math. Acad. Sci. Hungar.* **23** (1972) 223–246.

177. L. Lovász, On the structure of factorizable graphs, *Acta Math. Acad. Sci. Hungar.*, **23** (1972) 179–195.

178. L. Lovász, On the structure of factorizable graphs II, *Acta Math. Acad. Sci. Hungar.*, **23** (1972) 465–478.

179. L. Lovász, Antifactors of graphs, *Period. Math. Hungar.* **4** (1973) 121–123.

180. L. Lovász, Three short proofs in graph theory, *J. Combin. Theory Ser. B* **19** (1975) 269–271.

181. L. Lovász, *Combinatorial Problems and Exercises*, North-Holland, Amsterdam (1979).

182. L. Lovász and M. Plummer, *Matching Theory*, Annals of Discrete math. **29** North-Holland, (1986).

183. W. Mader, 1-Faktoren von Graphen, *Math. Ann.* **201** (1973) 269–282.

184. E. Mahmoodian, On factors of a graph, *Canadian J. Math.* **29** (1977) 38–440.

185. H. Matsuda, A neighborhood condition for graphs to have $[a, b]$-factors, *Discrete Math.* **224** (2000) 289–292.

186. H. Matsuda, A neighborhood condition for graphs to have $[a, b]$-factors II, *Graphs Combin.* **18** (2002) 763–768.

187. N.S. Mendelsohn and A.L. Dulmage, Some generalizations of the problem of distinct representatives, *Canad. J. Math.* **10** (1958) 230–241.

188. E. Mendelsohn and A. Rosa, One-factorizations of the complete graph — a survey, *J. Graph Theory* **9** (1985) 43–65.

189. K. Menger, Zur allgemeinen Kurventheorie, *Fund. Math* **10** (1927) 95–115.

190. D. Miyata, *Connected factors of claw-free graphs*, Private communication.

191. Y. Nam, Ore-type condition for the existence of connected factors, *J. Graph Theory* **56** (2007) 241–248.

192. J. Nešetřil and J. Poljak, On the complexity of the subgraph problem, *Comment. Math. Univ. Carolin.* **26** (1985) 415–419.

193. V. Neumann-Lara and E. Rivera-Campo, Spanning trees with bounded degrees, *Combinatorica* **11** (1991) 55–61.

194. T. Niessen, Neighborhood unions and regular factors, *J. Graph Theory* **19** (1995) 45–64.

195. T. Niessen, Minimum degree, independence number and regular factors, *Graphs Combin.* **11** (1995) 367–378.

196. T. Niessen, A characterization of graphs having all (g, f)-factors, *J. Combin. Theory Ser. B* **72** (1998), 152–156.

197. T. Niessen and B. Randerath, Regular factors of simple regular graphs and factor-spectra. *Discrete Math.* **185** (1998) 89–103.

198. T. Niessen and L. Volkmann, Class 1 conditions depending on the minimum degree and the number of vertices of maximum degree. *J. Graph Theory* **14** (1990) 225–246.

199. T. Nishimura, Independence number, connectivity, and r-factors, *J. Graph Theory*, **13** (1989) 63–69.

200. T. Nishimura, Regular factors of line graphs II, *Math. Japon.* **36** (1991) 1033–1040.

201. T. Nishimura, A degree condition for the existence of k-factors, *J. Graph Theory* **16** (1992), 141–151.

202. T. Nishizeki, On the relation between the genus and the cardinality of the maximum matchings of a graph, *Discrete Math.* **25** (1979) 149–156.

203. T. Nishizeki and I. Baybars, Lower bounds on the cardinality of the maximum matchings of planar graphs, *Discrete Math.* **28** (1979) 255–267.

204. O. Ore, Graphs and matching theorems, *Duke Math. J.* **22** (1955) 625–639.

205. O. Ore, Graphs and subgraphs, *Trans. Amer. Math. Soc.* **84** (1957) 109–136.

206. O. Ore, Note on Hamilton circuits, *Amer. Math. Monthly* **67** (1960) 55.

207. O. Ore, Hamilton connected graph, *J. Math. Pures. Appl.* **42** (1963) 21–27.

208. K. Ozeki and T. Yamashita, Spanning trees : A survey, *Graphs Combinatorics* **27** (2011) 1–26.

209. J. Petersen, Die Theorie der regulären graphs, *Acta Math.* **15** (1891) 193–220.

210. J. Plesník, Connectivity of regular graphs and the existence of 1-factors, *Mat. časop.* **22** (1972) 310–318.

211. J. Plesník, Remarks on regular factors of regular graphs, *Czechoslovak Math. J.* **24** (1974) 292–300.

212. M. D. Plummer, Graph factors and factorization: 1985–2003: A survey, *Discrete Mathematics* **307** (2007) 791–821.

213. N. Robertson, D. Sanders, P. Seymour and R. Thomas, The four-colour theorem, *J. Combin. Theory Ser. B* **70** (1997) 2–44.

214. A. Saito, One-factors and k-factors, *Discrete Math.* **91** (1991) 323–326.

215. A. Saito, private communication.

216. A. Saito and M. Watanabe, Partitioning graphs into induced stars, *Ars Combinatoria*, **36** (1995) 3–6.

217. P. Seymour, On multi-colourings of cubic graphs, and conjectures of Fulkerson and Tutte, *Proc. London Math Soc.* **38** (1979) 423–460.

218. D.P. Sumner, Graphs with 1-factors, *Proc. Amer. Math.* **42** (1974) 8–12.

219. D.P. Sumner, 1-factors and antifactor sets, *J. London Math. Soc.* **22** (1976) 351–359.

220. J. Szabó, Graph packings and the degree proscribed subgraph problem, Eötvös University, Doctoral thesis (2006).

221. C. Thomassen, A remark on the factor theorems of Lovász and Tutte, *J. Graph Theory* **5** (1981) 441–442.

222. T. Tokuda, A degree condition for the existence of $[a, b]$-factors, in $K_{1,n}$-free graphs, *Tokyo J. Math.* **7** (1998) 377–380

223. T. Tokuda, Baoguang Xu and Jianfang Wang, Connected factors and spanning trees in graphs, *Graphs Combin.* **19** (2003) 259–262.

346   References

224. J. Topp and P.D. Vestergaard, Odd factors of a graph, *Graphs Combin.* **9** (1993) 371–381.
225. W.T. Tutte, The factorization of linear graphs, *J. London Math. Soc.* **22** (1947) 107–111.
226. W.T. Tutte, The factors of graphs, *Can. J. Math.* **4** (1952) 314–328.
227. W.T. Tutte, The 1-factors of oriented graphs, *Proc. Amer. Math. Soc.* **4** (1953) 922–931.
228. W.T. Tutte, A short proof of the factor theorem for finite graphs, *Can. J. Math.* **6** (1954) 347–352.
229. W.T. Tutte, A theorem on planar graphs, *Trans. Amer. Math. Soc.* **82** (1956) 99–116.
230. W.T. Tutte, The subgraph probelm, *Ann. Discrete Math.* **3** (1978) 289–295.
231. W.T. Tutte, Graph factors, *Combinatorica* **1** (1981) 79–97.
232. V.G. Vizing, On an estimate of the chromatic class of a p-graphs, *Diskret. Analiz.* **3** (1964) 25–30.
233. V.G. Vizing, Critical graphs with a given chromatic class, *Diskret. Analiz* **5** (1965) 9–17.
234. L. Volkmann, *Graphen und Digraphen*, Eine Einführung in die Graphentheorie (Springer Verlag) (1991).
235. L. Volkmann, Regular graphs, regular factors, and the impact of Petersen's Theorems, *Jber. d. Dt. Math.-Verein* **97** (1995) 19–42.
236. L. Volkmann, The maximum size of graphs with a unique k-factor, *Combinatorica* **24** (2004) 531–540
237. L. Volkmann, On the size of odd order graphs with no almost perfect matching, *Australas. J. Combin.* **29** (2004) 119-126.
238. L. Volkmann, A short proof of a theorem of Kano and Yu on factors in regular graphs, *Electron. J. Combin.* **14** (2007) N10.
239. M. Voorhoeve, A lower bound for the permanents of certain (0,1)-matrices, *Nederl. Akad. Wetensch. Indag. Math.* **41** (1979) 83–86.
240. W.D. Wallis, The smallest regular graphs without one-factors, *Ars Combin.* **11** (1981) 295–300.
241. Hong Wang, Path factors of bipartite graphs, *J. Graph Theory*, **18** (1994) 161–167.
242. W.D. Wallis, *One-factorizations*, Mathematics and its Applications, **390**. Kluwer Academic Publishers Group, (1997)
243. Bing Wei and Yongjin Zhu, Hamiltonian k-factors in graphs, *J. Graph Theory* **25** (1997) 217–227.
244. D. West, *Introduction to Graph Theory* (Prentice Hall) (1996).
245. H. Whitney, Congruent graphs and the connectivity of graphs, *Amer. J. Math.* **54** (1932) 150–168.
246. S. Win, Existenz von Gerüsten mit vorgeschriebenem Maximlgrad in Graphen, *Abh. Math. Seminar Univ. Humburg* **43** (1975) 263–267.
247. S. Win, On a conjecture of Las Vergnas concerning certain spanning trees in graphs, *Resultate Math.*, **2** (1979) 215–224.
248. S. Win, On a connection between the existence of k-trees and the toughness of a graph, *Graphs Combin.* **5** (1989) 201–205.
249. D.R. Woodall, The binding number of a graph and its Anderson number, *J. Combin. Theory Ser. B* **15** (1973) 225–255.
250. Bang Wu and Kun-Mao Chao, *Spanning trees and optimization problems*, Chapman & Hall/CRC, Boca Raton, FL, (2004).

251. T. Yamashita, Private communiction.
252. Guiying Yan, Some new results on (g, f)-factorizations of graphs, *J. Combin. Math. Combin. Comput.* **18** (1995) 177–185.
253. Qinglin Yu, On barrier sets of star-factors, *Graphs Combin.* **6** (1990) 71–76.
254. Qinglin Roger Yu and Guizhen Liu, *Graph Factors and Matching Extensions*, Springer (2009).
255. Cheng Zhao, The disjoint 1-factors of $(d, d + 1)$-graphs, *J. Combin. Math. Combin. Comput.* **9** (1991) 195-198

Glossary of functions

S and T denote disjoint vertex subsets of a graph G.

$$\delta(S,T) = \sum_{x\in S} f(x) + \sum_{x\in T}(\deg_G(x) - f(x)) - e_G(S,T) - q(S,T)$$

$$= \sum_{x\in S} f(x) + \sum_{x\in T}(\deg_{G-S}(x) - f(x)) - q(S,T),$$

where $q(S,T)$ denotes the number of components C of $G - (S\cup T)$ such that $\sum_{x\in V(C)} f(x) + e_G(C,T) \equiv 1 \pmod 2$.

$$\delta^*(S,T) = \sum_{x\in S} f(x) + \sum_{x\in T}(\deg_G(x) - f(x)) - e_G(S,T)$$

$$= \sum_{x\in S} f(x) + \sum_{x\in T}(\deg_{G-S}(x) - f(x))$$

$$\gamma(S,T) = \sum_{x\in S} f(x) + \sum_{x\in T}(\deg_G(x) - g(x)) - e_G(S,T) - q^*(S,T)$$

$$= \sum_{x\in S} f(x) + \sum_{x\in T}(\deg_{G-S}(x) - g(x)) - q^*(S,T),$$

where $q^*(S,T)$ denotes the number of components C of $G - (S\cup T)$ such that $g(x) = f(x)$ for all $x \in V(C)$ and $\sum_{x\in V(C)} f(x) + e_G(C,T) \equiv 1 \pmod 2$.

$$\gamma^*(S,T) = \sum_{x\in S} f(x) + \sum_{x\in T}(\deg_G(x) - g(x)) - e_G(S,T)$$

$$= \sum_{x\in S} f(x) + \sum_{x\in T}(\deg_{G-S}(x) - g(x))$$

$$\sum_{x\in V(C)} f(x) + e_G(C,T) \equiv 1 \pmod 2.$$

$$\eta(S,T) = \sum_{x\in S} f(x) + \sum_{x\in T}(\deg_G(x) - g(x)) - e_G(S,T) - q(S,T),$$

where $q(S,T)$ denotes the number of components C of $G - (S\cup T)$ such that $\sum_{x\in V(C)} f(x) + e_G(C,T) \equiv 1 \pmod 2$.

Glossary of notation

G denotes a graph.

$X \subseteq Y$	X is a subset of Y.
$X \subset Y$	X is a proper subset of Y.
$X - Y$	$X \setminus Y$, where $Y \subseteq X$.
$X - a$	$X - \{a\}$, where $a \in X$.
$X + Y$	$X \cup Y$, where $X \cap Y = \emptyset$.
$X \triangle Y$	$(X \cup Y) - (X \cap Y)$.
$\|X\|, \#X$	The cardinality of X.
$V(G)$	The set of vertices of G.
$E(G)$	The set of edges of G.
$\|G\|$	The order of $G = \|V(G)\|$.
$\|\|G\|\|$	The size of $G = \|E(G)\|$.
$\deg_G(v)$	The degree of a vertex v in G.
$n_j(G)$	The number of vertices of G with degree j.
$N_G(v)$	The neighborhood of v.
$N_G[v]$	The closed neighborhood of $v = N_G(v) \cup \{v\}$.
$N_G(S)$	$\bigcup_{x \in S} N_G(x)$.
$E_G(X, Y)$	The set of edges of G joining X to Y.
$e_G(X, Y)$	The number of edges of G joining X to Y.
$\Delta(G)$	The maximum degree of G.
$\delta(G)$	The minimum degree of G.
$\omega(G)$	The number of components of G.
$odd(G)$	The number of odd components of G.
$Odd(G)$	The set of odd components of G.
$iso(G)$	The number of isolated vertices of G.
$Iso(G)$	The set of isolated vertices of G.
$\kappa(G)$	The connectivity of G.
$\lambda(G)$	The edge connectivity of G.
$\alpha(G)$	The independence number of G.
$\alpha'(G)$	The edge independence number of G.
$\sigma_k(G)$	$\min\{\sum_{x \in I} \deg_G(x) : I$ are independent sets of size $k\}$.
$bind(G)$	The binding number of G.
$tough(G)$	The toughness of G.
$K_n = K(n)$	The complete graph of order n.
$K_{n,m} = K(n, m)$	The complete bipartite graph of order $n + m$.
P_n	The path of order n.
C_n	The cycle of order n.
$\mathbb{Z}$	The set of integers.
$\mathbb{Z}^+$	The set of non-negative integers $= \{0, 1, 2, \ldots\}$.

Index

LECTURE NOTES IN MATHEMATICS   🐎 Springer

Edited by J.-M. Morel, F. Takens, B. Teissier, P.K. Maini

Editorial Policy (for the publication of monographs)

1. Lecture Notes aim to report new developments in all areas of mathematics and their applications - quickly, informally and at a high level. Mathematical texts analysing new developments in modelling and numerical simulation are welcome.

   Monograph manuscripts should be reasonably self-contained and rounded off. Thus they may, and often will, present not only results of the author but also related work by other people. They may be based on specialised lecture courses. Furthermore, the manuscripts should provide sufficient motivation, examples and applications. This clearly distinguishes Lecture Notes from journal articles or technical reports which normally are very concise. Articles intended for a journal but too long to be accepted by most journals, usually do not have this "lecture notes" character. For similar reasons it is unusual for doctoral theses to be accepted for the Lecture Notes series, though habilitation theses may be appropriate.

2. Manuscripts should be submitted either online at www.editorialmanager.com/lnm to Springer's mathematics editorial in Heidelberg, or to one of the series editors. In general, manuscripts will be sent out to 2 external referees for evaluation. If a decision cannot yet be reached on the basis of the first 2 reports, further referees may be contacted: The author will be informed of this. A final decision to publish can be made only on the basis of the complete manuscript, however a refereeing process leading to a preliminary decision can be based on a pre-final or incomplete manuscript. The strict minimum amount of material that will be considered should include a detailed outline describing the planned contents of each chapter, a bibliography and several sample chapters.

   Authors should be aware that incomplete or insufficiently close to final manuscripts almost always result in longer refereeing times and nevertheless unclear referees' recommendations, making further refereeing of a final draft necessary.

   Authors should also be aware that parallel submission of their manuscript to another publisher while under consideration for LNM will in general lead to immediate rejection.

3. Manuscripts should in general be submitted in English. Final manuscripts should contain at least 100 pages of mathematical text and should always include
   - a table of contents;
   - an informative introduction, with adequate motivation and perhaps some historical remarks: it should be accessible to a reader not intimately familiar with the topic treated;
   - a subject index: as a rule this is genuinely helpful for the reader.

   For evaluation purposes, manuscripts may be submitted in print or electronic form (print form is still preferred by most referees), in the latter case preferably as pdf- or zipped ps-files. Lecture Notes volumes are, as a rule, printed digitally from the authors' files. To ensure best results, authors are asked to use the LaTeX2e style files available from Springer's web-server at:

   ftp://ftp.springer.de/pub/tex/latex/svmonot1/ (for monographs) and
   ftp://ftp.springer.de/pub/tex/latex/svmultt1/ (for summer schools/tutorials).
   Additional technical instructions, if necessary, are available on request from: lnm@springer.com.

4. Careful preparation of the manuscripts will help keep production time short besides ensuring satisfactory appearance of the finished book in print and online. After acceptance of the manuscript authors will be asked to prepare the final LaTeX source files and also the corresponding dvi-, pdf- or zipped ps-file. The LaTeX source files are essential for producing the full-text online version of the book (see
http://www.springerlink.com/openurl.asp?genre=journal&issn=0075-8434 for the existing online volumes of LNM).

The actual production of a Lecture Notes volume takes approximately 12 weeks.

5. Authors receive a total of 50 free copies of their volume, but no royalties. They are entitled to a discount of 33.3% on the price of Springer books purchased for their personal use, if ordering directly from Springer.

6. Commitment to publish is made by letter of intent rather than by signing a formal contract. Springer-Verlag secures the copyright for each volume. Authors are free to reuse material contained in their LNM volumes in later publications: a brief written (or e-mail) request for formal permission is sufficient.

Addresses:
Professor J.-M. Morel, CMLA,
École Normale Supérieure de Cachan,
61 Avenue du Président Wilson, 94235 Cachan Cedex, France
E-mail: Jean-Michel.Morel@cmla.ens-cachan.fr

Professor F. Takens, Mathematisch Instituut,
Rijksuniversiteit Groningen, Postbus 800,
9700 AV Groningen, The Netherlands
E-mail: F.Takens@rug.nl

Professor B. Teissier, Institut Mathématique de Jussieu,
UMR 7586 du CNRS, Équipe "Géométrie et Dynamique",
175 rue du Chevaleret,
75013 Paris, France
E-mail: teissier@math.jussieu.fr

For the "Mathematical Biosciences Subseries" of LNM:

Professor P.K. Maini, Center for Mathematical Biology,
Mathematical Institute, 24-29 St Giles,
Oxford OX1 3LP, UK
E-mail: maini@maths.ox.ac.uk

Springer, Mathematics Editorial, Tiergartenstr. 17,
69121 Heidelberg, Germany,
Tel.: +49 (6221) 487-259
Fax: +49 (6221) 4876-8259
E-mail: lnm@springer.com